Spectroscopy of the Excited State

NATO ADVANCED STUDY INSTITUTES SERIES

A series of edited volumes comprising multifaceted studies of contemporary scientific issues by some of the best scientific minds in the world, assembled in cooperation with NATO Scientific Affairs Division.

Series B: Physics

Volume 1 – Superconducting Machines and Devices
edited by S. Foner and B. B. Schwartz

Volume 2 – Elementary Excitations in Solids, Molecules, and Atoms (Parts A and B)
edited by J. Devreese, A. B. Kunz, and T. C. Collins

Volume 3 – Photon Correlation and Light Beating Spectroscopy
edited by H. Z. Cummins and E. R. Pike

Volume 4 – Particle Interactions at Very High Energies (Parts A and B)
edited by David Speiser, Francis Halzen, and Jacques Weyers

Volume 5 – Renormalization and Invariance in Quantum Field Theory
edited by Eduardo R. Caianiello

Volume 6 – Interaction between Ions and Molecules
edited by Pierre Ausloos

Volume 7 – Low-Dimensional Cooperative Phenomena
edited by H. J. Keller

Volume 8 – Optical Properties of Ions in Solids
edited by Baldassare Di Bartolo

Volume 9 – Electronic Structure of Polymers and Molecular Crystals
edited by Jean-Marie André and János Ladik

Volume 10 – Progress in Electro-Optics
edited by Ezio Camatini

Volume 11 – Fluctuations, Instabilities, and Phase Transitions
edited by Tormod Riste

Volume 12 – Spectroscopy of the Excited State
edited by Baldassare Di Bartolo

The series is published by an international board of publishers in conjunction with NATO Scientific Affairs Division

A	Life Sciences	Plenum Publishing Corporation
B	Physics	New York and London
C	Mathematical and Physical Sciences	D. Reidel Publishing Company Dordrecht and Boston
D	Behavioral and Social Sciences	Sijthoff International Publishing Company Leiden
E	Applied Sciences	Noordhoff International Publishing Leiden

Spectroscopy of the Excited State

Edited by

Baldassare Di Bartolo
Department of Physics
Boston College

Assistant Editors
Dennis Pacheco and Velda Goldberg
Department of Physics
Boston College

PLENUM PRESS • NEW YORK AND LONDON
Published in cooperation with NATO Scientific Affairs Division

Library of Congress Cataloging in Publication Data

Nato Advanced Study Institute on the Spectroscopy of the Excited State, Erice, Italy, 1975.
Spectroscopy of the excited state.

(Nato advanced study institutes series: Series B, Physics; v. 12)
Includes bibliographical references and index.
1. Excited-state chemistry–Congresses. 2. Spectrum analysis–Congresses. I. Di Bartolo, Baldassare. II. Title. III. Series.
QD461.5.N37 1975 541'.042 75-38526
ISBN-13: 978-1-4684-2795-0 e-ISBN-13: 978-1-4684-2793-6
DOI: 10.1007/978-1-4684-2793-6

Proceedings of the NATO Advanced Study Institute on the Spectroscopy of the Excited State held at Erice, Italy, June 9-24, 1975

Softcover reprint of the hardcover 1st edition 1976
A Division of Plenum Publishing Corporation
227 West 17th Street, New York, N.Y. 10011

United Kingdom edition published by Plenum Press, London
A Division of Plenum Publishing Company, Ltd.
Davis House (4th Floor), 8 Scrubs Lane, Harlesden, London, NW10 6SE, England

Preface

These proceedings report the lectures and seminars presented at the NATO Advanced Study Institute on "The Spectroscopy of the Excited State," held at Erice, Italy, June 9-24, 1975. This Institute was an activity of the International School of Atomic and Molecular Spectroscopy of the "Ettore Majorana" Centre for Scientific Culture.

The Institute consisted of a series of lectures on the spectroscopic properties of materials in excited electronic states, that, starting at a fundamental level, finally reached the current level of research. The sequence of lectures and the organization of the material taught were in keeping with a didactical presentation. In essence the course had the two-fold purpose of organizing what was known on the subject, and updating the knowledge in the field. The formal lectures were complemented by seminars whose abstracts are also included in these proceedings. The proceedings report also the contributions sent by Professors R.G.W. Norrish and S. Claesson who, unfortunately, were not able to come because of illness.

A total of 62 participants and 7 lecturers came from the following countries: Belgium, Canada, Czechoslovakia, France, Germany, Israel, Italy, Japan, Netherlands, Norway, Pakistan, Poland, Sweden, Switzerland, the United Kingdom, the United States and Venezuela. The secretaries of the course were: A. La Francesca for the administrative aspects of the meeting and P.Papagiannakopoulos for the scientific aspects of the meeting.

I would like to acknowledge the sponsorship of the Institute by the North Atlantic Treaty Organization, the National Science Foundation, the Italian Ministry of Public Education, the Italian Ministry of Scientific and Technological Research, the Regional Sicilian Government and the Department of Physics of Boston College.

I would like to thank for their help Prof. A. Zichichi, Director of the "Ettore Majorana" Centre for Scientific Culture, the members of the organizing committee (Professors D. A. Ramsay and M. Kasha), Professor R. L. Carovillano, Dr. A. Gabriele, Ms. M. Zaini, Ms. P. Savalli, Prof. V. Adragna, Dr. G. Denaro, Dr. C. La Rosa, and Dr. T. L. Porter of the National Science Foundation.

It was a real pleasure to direct this Institute and to meet so many fine colleagues in the friendly atmosphere provided by the town of Erice. It has also been a pleasure to edit these lectures.

August 1975
Chestnut Hill
Massachusetts

B. Di Bartolo
Editor, and Director
of the Institute

Contents

INTERACTIONS OF RADIATION WITH ATOMS AND MOLECULES

B. Di Bartolo

Department of Physics, Boston College

Chestnut Hill, Massachusetts 02167, USA

ABSTRACT

This series of four lectures presents in a fundamental and comprehensive way the interaction of radiation with atomic and molecular systems. The first lecture deals with the quantum theory of molecular systems; the adiabatic approximation is introduced and its implications for the use of the symmetry properties of the system are treated. The second lecture deals with the theory of the radiative field; it starts with the consideration of the classical radiative field and then treats the quantum theory of such a field. The third lecture deals with the interaction of a radiative field and a charged particle and considers the various radiative processes that such an interaction can produce. The fourth lecture treats the more general problem of the interaction between radiation and atomic and molecular systems; the Franck-Condon principle in its various formulations (classical, semiclassical and quantum mechanical) is introduced.

I. INTRODUCTION

The interaction of a physical system with a radiative field is the cause of the absorption and emission of radiation. In this treatment we present the theory of such interaction by following four steps: (i) we treat first the most general molecular system consisting of an ensemble of nuclei and electrons; (ii) we treat then the radiative field moving from the classical to the quantum mechanical description; (iii) we deal then with the basic problem of the interaction of a charged particle with a radiative field and (iv) we consider the interaction of the radiative field with a molecular system.

II. QUANTUM THEORY OF MOLECULAR SYSTEMS

II.A. The Hamiltonian of a Molecular System

Consider a molecular system consisting of N nuclei and n electrons; the Hamiltonian of such a system can be written as follows

$$H = \sum_{i=1}^{n} \frac{\vec{p}_i^{\,2}}{2m} + \sum_{s=1}^{N} \frac{\vec{P}_s^{\,2}}{2M_s} + V(\vec{r}_i, \vec{R}_s), \tag{1}$$

where m = mass of the electron

M_s = mass of the sth nucleus

$\vec{p}_i$ = linear momentum of the ith electron

$\vec{P}_s$ = linear momentum of the sth nucleus

and where the potential energy is given by

$$V(\vec{r}_i, \vec{R}_s) = \frac{1}{2} \sum_{i=1}^{m} \sum_{\substack{j=1 \\ i \neq j}}^{n} \frac{e^2}{|\vec{r}_i - \vec{r}_j|} + \frac{1}{2} \sum_{s=1}^{N} \sum_{\substack{t=1 \\ s \neq t}}^{N} \frac{e^2 Z_s Z_t}{|\vec{R}_s - \vec{R}_t|} - \sum_{i=1}^{n} \sum_{s=1}^{N} \frac{e^2 Z_s}{|\vec{R}_s - \vec{r}_i|} . \tag{2}$$

The Schroedinger equation of such a system can be written

$$H\Psi(\vec{r}_i, \vec{R}_s) = E\Psi(\vec{r}_i, \vec{R}_s). \tag{3}$$

or

$$- \sum_i \frac{\hbar^2}{2m} \nabla_i^2 \Psi - \sum_s \frac{\hbar^2}{2M_s} \nabla_s^2 \Psi + V\Psi = E\Psi. \tag{4}$$

We will seek solutions of the type

$$\Psi(\vec{r}_i, \vec{R}_s) = \phi(\vec{R}_s)\psi(\vec{r}_i, \vec{R}_s). \tag{5}$$

Using the expression (5) in (4) and dropping the subscripts we find

$$-\frac{\hbar^2}{2m}\phi(\vec{R})\sum_i \nabla_i^2 \psi(\vec{r},\vec{R}) - \frac{\hbar^2}{2}\sum_s \frac{\nabla_s^2}{M_s}\phi(\vec{R})\psi(\vec{r},\vec{R}) +$$

$$+ V(\vec{r},\vec{R})\phi(\vec{R})\psi(\vec{r},\vec{R}) = E\phi(\vec{R})\psi(\vec{r},\vec{R}). \tag{6}$$

But

$$\nabla_s^2 \phi\psi = \phi\nabla_s^2\psi + \psi\nabla_s^2 \phi + 2\vec{\nabla}_s\phi \cdot \vec{\nabla}_s\psi; \tag{7}$$

therefore (6) becomes:

$$\left\{- \sum_s \frac{\hbar^2}{M_s}\vec{\nabla}_s \phi(\vec{R})\cdot\vec{\nabla}_s\psi(\vec{r},\vec{R}) - \sum_s \frac{\hbar^2}{2M_s}\phi(\vec{R})\nabla_s^2\psi(\vec{r},\vec{R})\right\} -$$

$$- \psi(\vec{r},\vec{R})\sum_s \frac{\hbar^2}{2M_s}\nabla_s^2 \phi(\vec{R}) - \phi(\vec{R}) \sum_i \frac{\hbar^2}{2m}\nabla_i^2 \psi(\vec{r},\vec{R}) +$$

$$+ V(\vec{r},\vec{R})\phi(\vec{R})\psi(\vec{r},\vec{R}) = E\phi(\vec{R})\psi(\vec{r},\vec{R}). \tag{8}$$

We make at this point the assumption that the expression in the {} brackets in the left member of (8) is negligible with respect to the term that follows it, i.e.:

$$\left|- \sum_s \frac{\hbar^2}{M_s}[\vec{\nabla}_s\phi \cdot \vec{\nabla}_s\psi + \frac{1}{2}\phi\nabla_s^2 \psi]\right| << \left|- \psi \sum_s \frac{\hbar^2}{2M_s}\nabla_s^2\phi\right|. \tag{9}$$

We shall examine later the meaning of this approximation. As a consequence of (9),(8) becomes:

$$-\frac{\psi}{\phi}\sum_s \frac{\hbar^2}{2M_s}\nabla_s^2 \phi + [- \sum_i \frac{\hbar^2}{2m}\nabla_i^2 + V]\psi = E\psi. \tag{10}$$

The terms in square brackets in (10) represent the Hamiltonian of the system if the nuclei are kept fixed in space; we shall indicate this Hamiltonian by the symbol H_e:

$$H_e = - \sum_i \frac{\hbar^2}{2m}\nabla_i^2 + V(\vec{r},\vec{R}). \tag{11}$$

The Schroedinger equation of H_e is given by:

$$-\sum_i \frac{\hbar^2}{2m}\nabla_i^2\,\psi(\vec{r},\vec{R}) + V(\vec{r},\vec{R})\psi(\vec{r},\vec{R}) = \varepsilon(\vec{R})\psi(\vec{r},\vec{R}). \qquad (12)$$

Using this result in (10) we obtain

$$-\frac{\psi}{\phi}\sum_s \frac{\hbar^2}{2M_s}\nabla_s^2\phi + \varepsilon(\vec{R})\psi = E\psi, \qquad (13)$$

or, multiplying by ϕ/ψ:

$$-\sum_s \frac{\hbar^2}{2M_s}\nabla_s^2\,\phi + \varepsilon(\vec{R})\phi = E\phi. \qquad (14)$$

The solution of the Schrodinger equation (10) reduces then to the solution of the two following equations:

$$-\sum_i \frac{\hbar^2}{2m}\nabla_i^2\,\psi_k(\vec{r},\vec{R}) + V(\vec{r},\vec{R})\psi_k(\vec{r},\vec{R}) = \varepsilon_k(\vec{R})\psi_k(\vec{r},\vec{R}) \qquad (15)$$

$$-\sum_s \frac{\hbar^2}{2M_s}\nabla_s^2\phi_{kl}(\vec{R}) + \varepsilon_k(\vec{R})\phi_{kl}(\vec{R}) = E_{kl}\phi_{kl}(\vec{R}), \qquad (16)$$

where we have introduced the proper subscripts. A stationary state of the system is now represented by a function:

$$\Psi_a(\vec{r},\vec{R}) = \psi_k(\vec{r},\vec{R})\phi_{kl}(\vec{R}), \qquad (17)$$

where $\psi_k(\vec{r},\vec{R})$ and $\phi_{kl}(\vec{R})$are eigenfunctions of the Hamiltonians

$$H_e = -\sum_i \frac{\hbar^2}{2m}\nabla_i^2 + V(\vec{r},\vec{R}) \qquad (18)$$

and

$$H_v = -\sum_s \frac{\hbar^2}{2M_s}\nabla_s^2 + \varepsilon_k(\vec{R}), \qquad (19)$$

respectively.

Equation (15) above is an eigenvalue equation whose eigenfunctions represent the motion of the elctrons in the molecule when the nuclei are kept fixed in space; the energy eigenvalues of (15) depend parametrically on the coordinates of the nuclei, represented symbolically by $\vec{R}$.

Equation (16) is an eigenvalue equation, whose eigenfunctions represent the motion of the nuclei; in the corresponding Hamiltonian H_v, given by (19), the energy $\varepsilon_k(\vec{R})$ plays the role of a potential energy for the nuclear motion. This energy $\varepsilon_k(\vec{R})$ is an eigenvalue of equation (15) and, as such, depends on the quantum number k; however k does not play the role of a quantum number for $\phi(\vec{R})$ even if it is used as one of its subscripts. Therefore the functions $\phi_{kl}(\vec{R})$ and $\phi_{k'l}(\vec{R})$ with $k' \neq k$, are not mutually orthogonal.

II.B. The Adiabatic Approximation

It is proper at this time to go back to the approximation (9) which has allowed us to express the eigenstates of the system in the product form (17). The implication of the approximation is that the function $\psi(\vec{r},\vec{R})$, which represents the motion of the electrons, is a function which varies slowly with the nuclear coordinates, so that $|\vec{\nabla}_s \psi(\vec{r},\vec{R})|$ is much smaller than $|\vec{\nabla}_s \phi(\vec{R})|$. This is the case, since the electrons, having much smaller masses than the nuclei, go through their orbits many times before the nuclei have shifted from their equilibrium position by any considerable distance.

It is in the light of the above argument that the approximation (9) is called the adiabatic approximation. It is also sometimes called the Born-Oppenheimer approximation (1); it represents a particular aspect of the more general Ehrenfest principle (2) which states that if a system is perturbed slowly, it remains in a definite stationary state.

II.C. The Role of Symmetry

The quantum mechanical treatment of a physical system implies generally the solution of a Schroedinger equation. This solution gives the energy eigenvalues and the eigenfunctions of the Hamiltonian. In general the eigenfunctions are degenerate, namely several of them correspond to the same energy eigenvalue. The degeneracy and the transformation properties of the eigenfunctions are closely related to the symmetry properties of the Hamiltonian; indeed, both degeneracy and transformation properties can be derived from the knowledge of this symmetry.

Let us consider now the "exact" Hamiltonian H, given, according to (1) by

$$H = -\sum_i \frac{\hbar^2}{2m}\nabla_i^2 - \sum_s \frac{\hbar^2}{2M_s}\nabla_s^2 + \frac{1}{2}\sum_{\substack{i\ j \\ i\neq j}} \frac{e^2}{|\vec{r}_i-\vec{r}_j|} + \frac{1}{2}\sum_{\substack{s\ t \\ s\neq t}} \frac{e^2 Z_s Z_t}{|\vec{R}_s-\vec{R}_t|} -$$

$$- \sum_i \sum_s \frac{e^2 Z_s}{|\vec{R}_s - \vec{r}_i|} . \tag{20}$$

It can be seen immediately that such a Hamiltonian has translational invariance. If we transform all the (3n + 3N) coordinates by adding a vector $\vec{d}$

$$\begin{cases} \vec{R}'_s = \vec{R}_s + \vec{d} \\ \vec{r}'_i = \vec{r}_i + \vec{d}; \end{cases} \tag{21}$$

then H will have the same form in terms of the primed coordinates as it has in terms of the unprimed coordinates, i.e., is translationally invariant.

We may also operate another coordinate transformation by replacing the (3n + 3N) coordinates with the 3 coordinates of the center of mass and (3n + 3N - 3) coordinates depending only on the relative vectorial distances among the particles of the system. In the latter system of coordinates the Hmailtonian can be expressed as follows:

$$H = - \frac{\hbar^2}{2M} \nabla^2 + H' \tag{22}$$

where ∇^2 operates only on the coordinates of the center of mass and M is the total mass. The Hamiltonian H' depends only on the relative coordinates.

Since the rotational symmetry properties of H are the same as those of H', we can now restrict ourselves to the consideration of the rotational symmetries of H'. A typical "rotational" operation in real space produces a change of coordinates

$$\vec{x}' = \underset{\sim}{Q}\vec{x} \tag{23}$$

with

$$\begin{cases} x'_1 = Q_{11}x_1 + Q_{12}x_2 + Q_{13}x_3 \\ x'_2 = Q_{21}x_1 + Q_{22}x_2 + Q_{23}x_3 \\ x'_3 = Q_{31}x_1 + Q_{32}x_2 + Q_{33}x_3 . \end{cases} \tag{24}$$

$\underset{\sim}{Q}$ is a 3 x 3 real orthogonal matrix: if its determinant is +1, the rotation is called proper, if it is -1, the rotation is called improper.

Let us now consider the effect of a real orthogonal transformation Q on the Hamiltonian H'. This transformation does not change the distances between particles and therefore the terms in the Hamiltonian representing Coulomb interactions remain invariant. The Laplacian operator ∇^2 also remains invariant:

$$\nabla'^2 = \sum_{i=1}^{3} \frac{\partial^2}{\partial x_i'^2} = \sum_{i,jk} Q_{ij} Q_{ik} \frac{\partial}{\partial x_j} \frac{\partial}{\partial x_k} = \sum_{jk} \delta_{jk} \frac{\partial}{\partial x_j} \frac{\partial}{\partial x_k} = \nabla^2 . \tag{25}$$

Therefore the "exact" Hamiltonian H is invariant under all spatial transformations which include translations, rotations, (proper and improper) and combinations of the two. It is evident that under these non-restrictive conditions no relevant information on the eigenfunctions and eigenvalues of H can be gained by the consideration of the symmetry properties of H.

The situation is completely different when we consider the Hamiltonian

$$H_e = -\frac{\hbar^2}{2m} \sum_{i=1}^{n} \nabla_i^2 + V(\vec{r},\vec{R}) . \tag{26}$$

The Hamiltonian H_e represents the motion of the n electrons in the field of the nuclei which are assumed to be at fixed positions. Considering the symmetry properties of H_e, the nuclear coordinates are not to be operated upon, because they appear only as parameters in the Hamiltonian; we can only perform symmetry operations on the electronic coordinates. Let us perform a symmetry operation Q on the electronic coordinates. If we call

$$\vec{r}_i{}' = \underset{\sim}{Q}\, \vec{r}_i ,$$

the Hamiltonian H_e in the new coordinates is given by

$$-\frac{\hbar^2}{2m} \sum_{i=1}^{n} \nabla_i^2 + \frac{1}{2} \sum_{i=1}^{n} \sum_{\substack{j=1 \\ i \neq j}}^{n} \frac{e^2}{|\vec{r}_i{}' - \vec{r}_j{}'|} + \frac{1}{2} \sum_{s=1}^{N} \sum_{\substack{t=1 \\ s \neq t}}^{N} \frac{e^2 Z_\alpha Z_\beta}{|\vec{R}_s - \vec{R}_t|} -$$

$$-\sum_{i=1}^{n}\sum_{s=1}^{N}\frac{e^2Z_\alpha}{|\vec{R}_s-\vec{r}_i'|}\,. \tag{27}$$

The only term that has changed is the last sum. However, if the operation Q is such that,

$$|\vec{R}_s-\vec{r}_i'| = |\vec{R}_t-\vec{r}_i|, \tag{28}$$

where $\vec{R}_t$ indicates the position of a nucleus identical with the sth nucleus, then the last sum in (27) will go into itself. One such operation in the case of Fig. 1 is a clockwise rotation by 120° about an axis passing through 0 and perpendicular to the plane of the figure. The equality (28) above implies that

$$\underset{\sim}{Q}^{-1}\,\vec{R}_s = \vec{R}_t \tag{29}$$

or $$\underset{\sim}{Q}\,\vec{R}_t = \vec{R}_s \tag{30}$$

for all identical nuclei. Therefore the Hamiltonian H_e is invariant under all those operations on the electronic coordinates which, when used on the nuclear coordinate parameters, send identical nuclei into one another.

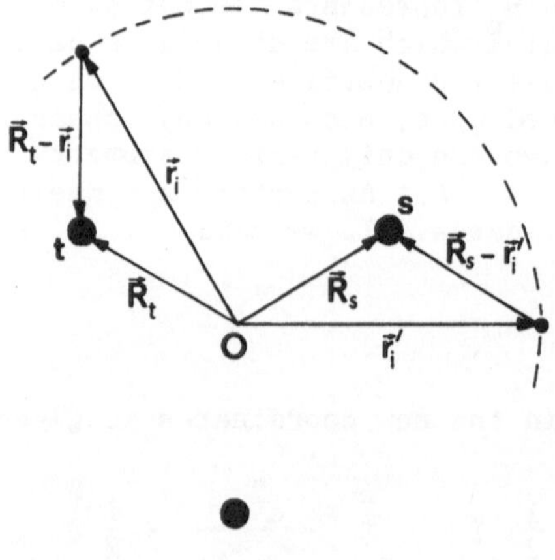

Fig. 1 Diagram illustrating the symmetry properties of a molecular system.

The Hamiltonian H_v represents the nuclear motion. The eigenvalues of H_e are used as the potential in which the nuclei experience their motion; the Hamiltonian H_V depends only on the nuclear coordinates. The symmetry operations of H_V are to be performed on the nuclear coordinates and it is clear that H_v is invariant under all those operations which send identical nuclei into one another.

We have gone through this discussion to show how symmetry considerations are relevant to the quantum mechanical treatment of molecular systems. We have found that in the adiabatic approximation the relevant symmetries are those related to the nuclear coordinates.

III. QUANTUM THEORY OF THE RADIATIVE FIELD

III.A. The Classical Radiative Field

The four Maxwell equations represent an electromagnetic field in the general case of the presence of currents of density $\vec{j}$ and charges of density ρ:

$$\vec{\nabla} \cdot \vec{B} = 0 \tag{31}$$

$$\vec{\nabla} \times \vec{E} + \frac{1}{c}\frac{\partial \vec{B}}{\partial t} = 0 \tag{32}$$

$$\vec{\nabla} \cdot \vec{E} = 4\pi\rho \tag{33}$$

$$\vec{\nabla} \times \vec{B} - \frac{1}{c}\frac{\partial \vec{E}}{\partial t} = \frac{4\pi}{c}\vec{j}. \tag{34}$$

Let us consider first the homogeneous equations (31) and (32) above. Since

$$\vec{\nabla} \cdot \vec{\nabla} \times \vec{u} = 0, \tag{35}$$

we can set

$$\vec{B} = \vec{\nabla} \times \vec{A}, \tag{36}$$

where $\vec{A}$ is defined as the vector potential. Replacing this expression of $\vec{B}$ in (32) we obtain

$$\vec{\nabla} \times \vec{E} + \frac{1}{c}\frac{\partial}{\partial t}(\vec{\nabla} \times \vec{A}) = 0. \tag{37}$$

Also, since

$$\vec{\nabla} \times \vec{\nabla}\varphi = 0, \tag{38}$$

we can set

$$\vec{E} = -\vec{\nabla}\varphi - \frac{1}{c}\frac{\partial\vec{A}}{\partial t} \tag{39}$$

where φ is defined as the <u>scalar potential</u>.

Let us now consider the inhomogeneous Maxwell equations (33) and (34). Using the expression (36) for $\vec{B}$ and (39) for $\vec{E}$ we obtain

$$\vec{\nabla}\cdot\left(-\vec{\nabla}\varphi - \frac{1}{c}\frac{\partial\vec{A}}{\partial t}\right) = 4\pi\rho \tag{40}$$

$$\vec{\nabla}\times(\vec{\nabla}\times\vec{A}) - \frac{1}{c}\frac{\partial}{\partial t}\left(-\vec{\nabla}\varphi - \frac{1}{c}\frac{\partial\vec{A}}{\partial t}\right) = \frac{4\pi}{c}\vec{j} \tag{41}$$

But

$$\vec{\nabla}\times(\vec{\nabla}\times\vec{A}) = \vec{\nabla}(\vec{\nabla}\cdot\vec{A}) - \nabla^2\vec{A}; \tag{42}$$

therefore (40) and (41) become

$$\nabla^2\varphi + \frac{1}{c}\frac{\partial}{\partial t}(\vec{\nabla}\cdot\vec{A}) = -4\pi\rho \tag{43}$$

$$\left(\nabla^2\vec{A} - \frac{1}{c^2}\frac{\partial^2\vec{A}}{\partial t^2}\right) - \vec{\nabla}\left(\vec{\nabla}\cdot\vec{A} + \frac{1}{c}\frac{\partial\varphi}{\partial t}\right) = -\frac{4\pi}{c}\vec{j}. \tag{44}$$

We note at this point that φ and $\vec{A}$ are not uniquely determined; in fact if we set

$$\varphi' = \varphi - \frac{1}{c}\frac{\partial f}{\partial t} \tag{45}$$

$$\vec{A}' = \vec{A} + \vec{\nabla}f, \tag{46}$$

f being the function of position and time, we find

$$\vec{E} = -\frac{1}{c}\frac{\partial\vec{A}}{\partial t} - \vec{\nabla}\varphi =$$

$$= -\frac{1}{c}\frac{\partial}{\partial t}(\vec{A}' - \vec{\nabla}f) - \vec{\nabla}\left(\varphi' + \frac{1}{c}\frac{\partial f}{\partial t}\right) =$$

$$= -\frac{1}{c}\frac{\partial\vec{A}'}{\partial t} + \frac{1}{c}\frac{\partial}{\partial t}(\vec{\nabla}f) - \vec{\nabla}\varphi' - \frac{1}{c}\frac{\partial}{\partial t}(\vec{\nabla}f) =$$

$$= -\frac{1}{c}\frac{\partial\vec{A}'}{\partial t} - \nabla\varphi' \tag{47}$$

and

$$\vec{B} = \vec{\nabla} \times \vec{A} = \vec{\nabla} \times (\vec{A}' - \vec{\nabla}f) = \vec{\nabla} \times \vec{A}'. \tag{48}$$

It is easy to show that (43) and (44) are also valid if we replace φ and $\vec{A}$ with φ' and $\vec{A}'$, respectively.

The indeterminacy of φ and $\vec{A}$ is removed by imposing an additional condition which we choose to be the so-called Coulomb gauge:

$$\vec{\nabla} \cdot \vec{A} = 0. \tag{49}$$

With this condition (43) and (44) become

$$\nabla^2\varphi = -4\pi\rho \tag{50}$$

$$\nabla^2\vec{A} - \frac{1}{c^2}\frac{\partial^2\vec{A}}{\partial t^2} = -\frac{4\pi}{c}\vec{j} + \frac{1}{c}\frac{\partial}{\partial t}(\vec{\nabla}\varphi). \tag{51}$$

Equation (50) above is called the Poisson's Equation; it can be integrated by using Green's theorem which is expressed by the equality

$$\int d\tau\, (G\nabla^2\varphi - \varphi\nabla^2 G) = \int dS\, G\left[\frac{\partial\varphi}{\partial n} - \varphi\frac{\partial G}{\partial n}\right], \tag{52}$$

where the integral on the left hand side is over a volume and the integral on the right hand side is over a surface enclosing the volume and $\frac{\partial\varphi}{\partial n}$ is the derivative of φ in the direction perpendicular to the surface. If we set

$$G(\vec{r},\vec{r}') = \frac{1}{|\vec{r}-\vec{r}'|} \tag{53}$$

we obtain

$$\nabla_r^2\, G(\vec{r},\vec{r}') = -4\pi\delta(\vec{r}-\vec{r}'). \tag{54}$$

Taking into account the fact that

$$\varphi(\infty) = G(\infty) = 0, \tag{55}$$

and the relation (54) we may write

$$\int d^3\vec{r}'\,(G\nabla^2\varphi - \varphi\nabla^2 G) = \int d^3\vec{r}'\left[\frac{-4\pi\rho(\vec{r}',t)}{|\vec{r} - \vec{r}'|} + \phi(\vec{r}',t)4\pi\delta(\vec{r}-\vec{r}')\right] = 0. \tag{56}$$

Then

$$\varphi(\vec{r},t) = \int \frac{\rho(\vec{r}',t)}{|\vec{r}-\vec{r}'|}\, d^3\vec{r}'. \tag{57}$$

φ is a function of the charge distribution; if $\rho = 0$, $\varphi = 0$.

On the other hand, if $\rho = \vec{j} = 0$ the equation (51) becomes

$$\nabla^2\vec{A}(\vec{r},t) - \frac{1}{c^2}\frac{\partial^2}{\partial t^2}\vec{A}(\vec{r},t) = 0, \tag{58}$$

called field equation which, together with

$$\vec{\nabla}\cdot\vec{A} = 0 \tag{59}$$

$$\vec{E} = -\frac{1}{c}\frac{\partial\vec{A}}{\partial t} \tag{60}$$

$$\vec{B} = \vec{\nabla}\times\vec{A}, \tag{61}$$

defines the radiative field.

III.B. Solutions of the Field Equation

The field equation is a second order differential equation in $\vec{r}$ and t; we shall perform a separation of variables by looking for solutions of the field equation which are of the form

$$\vec{A}(\vec{r},t) = q(t)\vec{\mathcal{A}}(\vec{r}). \tag{62}$$

The field equation becomes then

$$q\nabla^2\vec{\mathcal{A}} = \frac{1}{c^2}\vec{\mathcal{A}}\ddot{q}$$

or

$$\frac{\ddot{q}}{q} = \frac{c\nabla^2\vec{\mathcal{A}}}{\vec{\mathcal{A}}} = -\omega^2 \tag{63}$$

where ω^2 is a constant. From the last equation we obtain

$$\ddot{q} + \omega^2 q = 0 \tag{64}$$

$$\nabla^2\vec{\mathcal{A}} + \frac{\omega^2}{c^2}\vec{\mathcal{A}} = 0, \tag{65}$$

with the solutions

$$q(t) = |q|e^{-i\omega t} \tag{66}$$

$$\vec{\mathcal{A}}(\vec{r}) = \vec{\pi}\left(\frac{4\pi c^2}{V}\right)^{\frac{1}{2}} e^{i\vec{k}\cdot\vec{r}}, \tag{67}$$

where $\vec{\pi}$ = unit vector in the direction of polarization, V = volume of the space in which the field is confined and

$$|\vec{k}| = k = \frac{\omega}{c}. \qquad (68)$$

The solution (62) can now be written

$$\vec{A}(r,t) = |q|\left(\frac{4\pi c^2}{V}\right)^{\frac{1}{2}} \vec{\pi}\, e^{i(\vec{k}\cdot\vec{r} - \omega t)}, \qquad (69)$$

which represents a plane wave of wavelength $\lambda = \frac{2\pi}{k}$. We note that the condition $\vec{\nabla} \cdot \vec{A} = 0$, implies that

$$\pi_x k_x + \pi_y k_y + \pi_z k_z = \vec{\pi} \cdot \vec{k} = 0, \qquad (70)$$

namely the directions of propagation and polarization are perpendicular to each other.

We are now in the position of writing the most general expression for the vector potential $\vec{A}(\vec{r},t)$, which must include all the allowed components $\vec{A}_\alpha$ and all the allowed polarizations σ:

$$\vec{A}(\vec{r},t) = \sum_\alpha \sum_\sigma [q_\alpha^\sigma(t)\vec{A}_\alpha(\vec{r}) + q_\alpha^\sigma(t)^*\vec{A}_\alpha(\vec{r})^*]. \qquad (71)$$

In (71) above σ ranges over the two possible independent directions of polarization for each $\vec{k}$; the values of α have to be determined by the boundary conditions. We choose to use the periodic boundary Conditions.

III.C. Periodic Boundary Conditions and Density of States

We shall assume for simplicity that the volume V containing the radiative field consists of a cube with sides of lengths L_x, L_y and L_z:

$$V = L_x L_y L_z. \qquad (72)$$

The periodic boundary conditions require the vector potential to be such that

$$\begin{cases} \vec{A}(\vec{r} + L_x\vec{1}_x, t) = \vec{A}(\vec{r},t) \\ \vec{A}(\vec{r} + L_y\vec{1}_y t) = \vec{A}(\vec{r},t) \\ \vec{A}(\vec{r} + L_z\vec{1}_z, t) = \vec{A}(\vec{r},t). \end{cases} \qquad (73)$$

Here $\vec{1}_x$, $\vec{1}_y$, and $\vec{1}_z$ are the unit vectors in the x, y and z directions, respectively. Since the vector potential is the superposition of solutions of the type (69), it must be

$$\begin{cases} k_x = \dfrac{2\pi}{L_x} n_x \\ k_y = \dfrac{2\pi}{L_y} n_y \\ k_z = \dfrac{2\pi}{L_z} n_z \end{cases} \tag{74}$$

or

$$\begin{cases} n_x = \dfrac{L_x k_x}{2\pi} \\ n_y = \dfrac{L_y k_y}{2\pi} \\ n_z = \dfrac{L_z k_z}{2\pi} . \end{cases} \tag{75}$$

n_x, n_y, and n_z can take the values

$$0, \pm 1, \pm 2, \pm 3, \pm 4, \ldots . \tag{76}$$

Considering now the expression (71) for $\vec{A}(\vec{r},t)$ it is clear what α stands for:

$$\alpha \equiv (|n_x|, |n_y|, |n_z|). \tag{76}$$

The number of states with $\vec{k}$ in $(\vec{k}, \vec{k} + d\vec{k})$ is given by

$$dn_x\, dn_y\, dn_z = \frac{L_x dk_x}{2\pi} \frac{L_y dk_y}{2\pi} \frac{L_z dk_z}{2\pi} =$$

$$= \frac{L_x L_y L_z}{8\pi^3} dk_x\, dk_y\, dk_z = \frac{V}{8\pi^3} d^3\vec{k} =$$

$$= \frac{V}{8\pi^3} k^2 dk \sin\theta d\theta d\varphi = \frac{V\omega^2}{8\pi^3 c^3} d\omega d\Omega, \tag{77}$$

where $d\Omega = \sin\theta d\theta d\varphi$ and $\omega = k/c$. The number of states per unit frequency range at $\omega = \omega_\alpha$ is then given by

$$g(\omega_\alpha) = \frac{V\omega_\alpha^2}{8\pi^3 c^3}\, d\Omega_\alpha . \tag{78}$$

In the formula (78) above only one polarization is counted.

III.D. The Hamiltonian of the Radiative Field

Having specified the boundary conditions and found the possible values of $\vec{k}$, we can now rewrite the expression (71) in the following manner.

$$\vec{A}(\vec{r},t) = \sum_\alpha \sum_\sigma \vec{\pi}_\alpha^\sigma \left(\frac{4\pi c^2}{V}\right)^{\frac{1}{2}} |q_\alpha^\sigma| \left\{ \exp\left[i\left(\frac{2\pi}{L_x} n_{x\alpha} x + \frac{2\pi}{L_y} n_{y\alpha} y + \frac{2\pi}{L_z} n_{z\alpha} z - \omega_\alpha t\right)\right] + \text{complex conjugate} \right\}. \tag{79}$$

At this point we shall drop for simplicity of notation the subscript σ; we shall resume it later. We note that

$$\vec{k}_{-\alpha} = -\vec{k}_\alpha$$
$$\vec{A}_{-\alpha} = \vec{A}_\alpha^* \tag{80}$$
$$\vec{\pi}_{-\alpha} = \vec{\pi}_\alpha ,$$

and that, since

$$\vec{A}_\alpha(\vec{r}) = \vec{\pi}_\alpha \left(\frac{4\pi c^2}{V}\right)^{\frac{1}{2}} e^{i\vec{k}_\alpha \cdot \vec{r}} , \tag{81}$$

it is

$$\int \vec{A}_\alpha \cdot \vec{A}_{\alpha'} d^3\vec{r} = 4\pi c^2\, \delta_{\alpha,-\alpha'} \tag{82}$$

$$\int \vec{A}_\alpha \cdot \vec{A}_\alpha\, d^3\vec{r} = \int \vec{A}_\alpha^* \cdot \vec{A}_\alpha^*\, d^3\vec{r} = 0 \tag{83}$$

$$\int \vec{A}_\alpha \cdot \vec{A}_\alpha\, d^3\vec{r} = 4\pi c^2 . \tag{84}$$

The αth component of the vector field is given by

$$\vec{A}_\alpha(\vec{r},t) = q_\alpha(t)\vec{A}_\alpha(\vec{r}) + q_\alpha(t)^* \vec{A}_\alpha(\vec{r})^* ; \tag{85}$$

therefore the αth component of the electric field can be expressed as follows

$$\vec{E}_\alpha = -\frac{1}{c}\frac{\partial \vec{A}_\alpha}{\partial t} = \frac{i\omega_\alpha}{c}(q_\alpha \vec{A}_\alpha - q_\alpha \vec{A}^*_\alpha). \tag{86}$$

The "total" electric field is then given by

$$\vec{E} = \sum_\alpha \frac{i\omega_\alpha}{c}(q_\alpha \vec{A}_\alpha - q^*_\alpha \vec{A}^*_\alpha). \tag{87}$$

Therefore

$$\begin{aligned}(\vec{E})^2 &= \sum_\alpha \sum_{\alpha'} \left[\frac{i\omega_\alpha}{c}(q_\alpha \vec{A}_\alpha - q^*_\alpha \vec{A}^*_\alpha)\right]\cdot\left[\frac{i\omega_{\alpha'}}{c}(q_{\alpha'}\vec{A}_{\alpha'} - q^*_{\alpha'}\vec{A}^*_{\alpha'})\right] = \\ &= \sum_\alpha \sum_{\alpha'}\left(-\frac{\omega_\alpha\omega_{\alpha'}}{c^2}\right)\Big[q_\alpha q_{\alpha'}\vec{A}_\alpha\cdot\vec{A}_{\alpha'} + q^*_\alpha q^*_{\alpha'}\vec{A}^*_\alpha\cdot\vec{A}^*_{\alpha'} - \\ &\qquad - q^*_\alpha q_{\alpha'}\vec{A}_\alpha\cdot\vec{A}_{\alpha'} - q_\alpha q^*_{\alpha'}\vec{A}_\alpha\cdot\vec{A}^*_{\alpha'}\Big].\end{aligned} \tag{88}$$

By integrating over space

$$\begin{aligned}\int(\vec{E})^2 d^3\vec{r} &= \sum_\alpha\sum_{\alpha'}\left(\frac{\omega_\alpha\omega_{\alpha'}}{c^2}\right)[q_\alpha q_{\alpha'}4\pi c^2\delta_{\alpha,-\alpha'} + q^*_\alpha q^*_{\alpha'}4\pi c^2\delta_{\alpha,-\alpha'} - \\ &\qquad - q^*_\alpha q_{\alpha'}4\pi c^2\delta_{\alpha\alpha'} - q_\alpha q^*_{\alpha'}4\pi c^2\delta_{\alpha\alpha'} = \\ &= 4\pi\sum_\alpha(-\omega^2_\alpha)\,[q_\alpha q_{-\alpha} + q^*_\alpha q^*_{-\alpha} - q^*_\alpha q_\alpha - q_\alpha q^*_\alpha] = \\ &= 4\pi\sum_\alpha[\omega^2_\alpha(q_\alpha q^*_\alpha + q^*_\alpha q_\alpha) - \omega^2_\alpha(q_\alpha q_{-\alpha} + q^*_\alpha q^*_{-\alpha})].\end{aligned} \tag{89}$$

On the other hand

$$\begin{aligned}\vec{B} &= \vec{\nabla}\times\vec{A} = \vec{\nabla}\times\left[\sum_\alpha(q_\alpha\vec{A}_\alpha + q^*_\alpha\vec{A}_\alpha)\right] = \\ &= \sum_\alpha\left[q_\alpha(\vec{\nabla}\times\vec{A}_\alpha) + q^*_\alpha(\vec{\nabla}\times A^*_\alpha)\right],\end{aligned} \tag{90}$$

and

$$(\vec{B})^2 = (\vec{\nabla} \times \vec{A}) \cdot (\vec{\nabla} \times \vec{A}) =$$

$$= \sum_{\alpha} \sum_{\alpha'} [q_\alpha(\vec{\nabla}\times\vec{A}_\alpha) + q^*_\alpha(\vec{\nabla}\times\vec{A}^*_\alpha)] \cdot [q_{\alpha'}(\vec{\nabla}\times\vec{A}_{\alpha'}) + q^*_{\alpha'}(\vec{\nabla}\times\vec{A}^*_{\alpha'})] =$$

$$= \sum_{\alpha} \sum_{\alpha'} [q_\alpha q_{\alpha'}(\vec{\nabla}\times\vec{A}_\alpha)\cdot(\vec{\nabla}\times\vec{A}_{\alpha'}) + q^*_\alpha q^*_{\alpha'}(\vec{\nabla}\times\vec{A}^*_\alpha)\cdot(\vec{\nabla}\times\vec{A}^*_{\alpha'}) +$$

$$+ q^*_\alpha q_{\alpha'}(\vec{\nabla}\times\vec{A}^*_\alpha)\cdot(\vec{\nabla}\times\vec{A}_{\alpha'}) + q_\alpha q^*_{\alpha'}(\vec{\nabla}\times\vec{A}_\alpha)\cdot(\vec{\nabla}\times\vec{A}^*_{\alpha'})]. \quad (91)$$

Given two vectors $\vec{C}$ and $\vec{D}$ we know from vector analysis that

$$\vec{D} \cdot (\vec{\nabla} \times \vec{C}) = \vec{\nabla} \cdot (\vec{C} \times \vec{D}) + \vec{C} \cdot (\vec{\nabla} \times \vec{D}). \quad (92)$$

It is then

$$(\vec{\nabla}\times\vec{A}_\alpha)\cdot(\vec{\nabla}\times\vec{A}_{\alpha'}) = \vec{\nabla}\cdot[\vec{A}_\alpha\times (\vec{\nabla}\times\vec{A}_{\alpha'})] + \vec{A}_\alpha\cdot [\vec{\nabla}\times (\vec{\nabla}\times\vec{A}_{\alpha'})] =$$

$$= \vec{\nabla}\cdot[\vec{A}_\alpha\times (\vec{\nabla}\times\vec{A}_{\alpha'})] + \vec{A}_\alpha\cdot[\vec{\nabla} (\vec{\nabla}\cdot\vec{A}_{\alpha'}) - \nabla^2\vec{A}_{\alpha'}] =$$

$$= \vec{\nabla}\cdot[\vec{A}_\alpha\times (\vec{\nabla}\times\vec{A}_{\alpha'})] - \vec{A}_\alpha\cdot\nabla^2\vec{A}_{\alpha'} =$$

$$= \vec{\nabla}\cdot[\vec{A}_\alpha\times (\vec{\nabla}\times\vec{A}_{\alpha'})] + k^2_{\alpha'} \vec{A}_\alpha\cdot\vec{A}_{\alpha'}, \quad (93)$$

where we have taken into account the fact that $\vec{\nabla} \cdot \vec{A}_{\alpha'} = 0$. Then

$$\int(\vec{\nabla}\times\vec{A}_\alpha)\cdot(\vec{\nabla}\times\vec{A}_{\alpha'})\, d^3\vec{r} = \int dS[\vec{A}_\alpha\times(\vec{\nabla}\times\vec{A}_{\alpha'})]_n + \frac{\omega^2_{\alpha'}}{c^2}\int \vec{A}_\alpha\cdot\vec{A}_{\alpha'}d^3\vec{r}$$

$$= \frac{\omega^2_{\alpha'}}{c^2}\, 4\pi c^2\, \delta_{\alpha,-\alpha'} = \omega^2_{\alpha'}\, 4\pi\delta_{\alpha,-\alpha'}, \quad (94)$$

where the subscript n above indicates the component of the vector $\vec{A}_\alpha \times (\vec{\nabla} \times \vec{A}_{\alpha'})$ normal to the boundary surface. The integral over the boundary surface vanishes because of the periodic boundary conditions. We can now write

$$\int(\vec{B})^2 d^3\vec{r} = \sum_{\alpha} \sum_{\alpha'} [q_\alpha q_{\alpha'}\omega^2_{\alpha'}\, 4\pi\delta_{\alpha,-\alpha'} + q^*_\alpha q^*_{\alpha'}\omega^2_{\alpha'}\, 4\pi\delta_{\alpha,-\alpha'} +$$

$$+ q^*_\alpha q_{\alpha'}\omega^2_{\alpha'}\, 4\pi\delta_{\alpha\alpha'} + q_\alpha q^*_{\alpha'}\omega^2_{\alpha'}\, 4\pi\delta_{\alpha\alpha'}] =$$

$$= 4\pi\sum_{\alpha} [\omega^2_\alpha(q_\alpha q_{-\alpha} + q^*_\alpha q^*_{-\alpha}) + \omega^2_\alpha(q_\alpha q^*_\alpha + q^*_\alpha q_\alpha)]. \quad (95)$$

The energy of the radiative field is then given by

$$\frac{1}{8\pi}\int(\vec{E})^2 d^3\vec{r} + \frac{1}{8\pi}\int(\vec{B})^2 d^3\vec{r} = \sum_\alpha \omega_\alpha^2 (q_\alpha q_\alpha^* + q_\alpha^* q_\alpha). \quad (96)$$

The Hamiltonian of the radiative field can then be expressed reintroducing for a moment the polarization index σ

$$H = \sum_\alpha \sum_\sigma \omega_\alpha^2 (q_\alpha^\sigma q_\alpha^{\sigma *} + q_\alpha^{\sigma *} q_\alpha^\sigma) = \sum_\alpha H_\alpha^\sigma, \quad (97)$$

where

$$H_\alpha^\sigma = \omega_\alpha^2 (q_\alpha^\sigma q_\alpha^{\sigma *} + q_\alpha^{\sigma *} q_\alpha^\sigma). \quad (98)$$

We note here that we have been able to express the Hamiltonian of the radiative field as the sum of independent terms; therefore the coordinates q_α^σ represent "normal mode" coordinates of the radiative field.

Dropping again the superscript σ (98) reduces to

$$H_\alpha = \omega_\alpha^2 (q_\alpha q_\alpha^* + q_\alpha^* q_\alpha). \quad (99)$$

Since we are dealing with a classical treatment H_α could be expressed as $2\omega_\alpha^2(q_\alpha q_\alpha^*)$; we prefer the form (99) in preparation for the move into quantum mechanics (In quantum mechanics q_α and q_α^* are two non-commuting operators.)

Let us introduce two new variables for each α

$$\begin{cases} Q_\alpha = q_\alpha + q_\alpha^* \\ P_\alpha = -i\omega_\alpha (q_\alpha - q_\alpha^*) = \dot{Q}_\alpha . \end{cases} \quad (100)$$

It is also

$$\begin{cases} q_\alpha = \frac{1}{2} (Q_\alpha - \frac{1}{i\omega_\alpha} P_\alpha) \\ q_\alpha^* = \frac{1}{2} (Q_\alpha + \frac{1}{i\omega_\alpha} P_\alpha). \end{cases} \quad (101)$$

In terms of the new coordinates

$$H_\alpha = \frac{1}{2}\omega_\alpha^2 Q_\alpha^2 + \frac{1}{2} P_\alpha^2. \tag{102}$$

We make at this point the following claim: Q_α and P_α are (real) variables that satisfy Hamilton's equations. The proof follows. From (102)

$$\frac{\partial H_\alpha}{\partial P_\alpha} = P_\alpha \tag{103}$$

$$\frac{\partial H_\alpha}{\partial Q_\alpha} = \omega_\alpha^2 Q_\alpha. \tag{104}$$

For Q_α and P_α to satisfy Hamilton's equation it must be

$$\frac{\partial H_\alpha}{\partial P_\alpha} = P_\alpha = \dot{Q}_\alpha \tag{105}$$

$$\frac{\partial H_\alpha}{\partial Q_\alpha} = \omega_\alpha^2 Q_\alpha = - \dot{P}_\alpha . \tag{106}$$

Equation (105) has already been verified in the second of equations (100) above. For (106) to be true it must be

$$\dot{P}_\alpha + \omega_\alpha^2 Q_\alpha = 0 \tag{107}$$

or

$$\ddot{Q}_\alpha + \omega_\alpha^2 Q_\alpha = 0. \tag{108}$$

This implies that

$$\ddot{q}_\alpha + \omega_\alpha^2 q_\alpha = 0 \tag{109}$$

which is indeed true, considering the time dependence of the coordinates q given by (66).

A consequence of the above considerations is that now it is possible to write the following Poisson brackets

$$\left\{Q_\alpha, P_\alpha\right\} = \sum_i \left(\frac{\partial Q_\alpha}{\partial Q_i} \frac{\partial P_\alpha}{\partial P_i} - \frac{\partial Q_\alpha}{\partial P_i} \frac{\partial P_\alpha}{\partial Q_i} \right) = 1 \tag{110}$$

$$\left\{Q_\alpha, P_{\alpha'}\right\} = \delta_{\alpha\alpha'} \tag{111}$$

$$\left\{Q_\alpha, Q_{\alpha'}\right\} = \left\{P_\alpha, P_{\alpha'}\right\} = 0. \tag{112}$$

III.E. The Quantum Radiative Field

The entire treatment of the radiative field up to this point has been classical. It is easy to go over to Quantum Mechanics by simply considering that Q_α and P_α become quantum mechanical (Hermitian) operators and that the Poisson brackets have to be replaced by commutators as follows

$$\left\{Q_\alpha, P_{\alpha'}\right\} \xrightarrow[\text{Q.M.}]{} \frac{1}{i\hbar}\left[Q_\alpha, P_{\alpha'}\right]. \tag{113}$$

From (110), (111) and (112) we then derive

$$\left[Q_\alpha, P_{\alpha'}\right] = i\hbar\delta_{\alpha\alpha'} \tag{114}$$

$$\left[Q_\alpha, Q_{\alpha'}\right] = \left[P_\alpha, P_{\alpha'}\right] = 0 \tag{115}$$

These last two relations together with (101) give us

$$\left[q_\alpha, q^+_{\alpha'}\right] = \frac{\hbar}{2\omega_\alpha}\,\delta_{\alpha\alpha'}. \tag{116}$$

q_α and q^+_α are now two operators, replacing q_α and q^*_α. These two operators can be expressed in terms of the dimensionless operators a_α and a^+_α, called <u>annihilation</u> and <u>creation</u> operators, respectively, and given by

$$a_\alpha = \sqrt{\frac{2\omega_\alpha}{\hbar}}\; q_\alpha \tag{117}$$

and

$$a^+_\alpha = \sqrt{\frac{2\omega_\alpha}{\hbar}}\; q^+_\alpha, \tag{118}$$

It is also

$$\left[a_\alpha, a^+_{\alpha'}\right] = \delta_{\alpha\alpha'} \tag{119}$$

$$\left[a_\alpha, a_{\alpha'}\right] = \left[a^+_\alpha, a^+_{\alpha'}\right] = 0. \tag{120}$$

The Hamiltonian of the radiative field now takes the form

$$H = \sum_\alpha \omega_\alpha^2 \frac{\hbar}{2\omega_\alpha}\left(a_\alpha a_\alpha^+ + a_\alpha^+ a_\alpha\right) =$$

$$= \sum_\alpha \frac{\hbar\omega_\alpha}{2}\left(a_\alpha a_\alpha^+ + a_\alpha^+ a_\alpha\right) = \sum_\alpha \hbar\omega_\alpha\left(a_\alpha^+ a_\alpha + \frac{1}{2}\right)$$

or, reintroducing the superscript σ,

$$H = \sum_\alpha \sum_\sigma \hbar\omega_\alpha \left(a_\alpha^{\sigma+} a_\alpha^\sigma + \frac{1}{2}\right). \tag{121}$$

III.F. Energy Levels and Eigenfunctions of a Radiative Field

Let us consider now the Hamiltonian of a single mode of the radiative field

$$H = \frac{\hbar\omega}{2}(aa^+ + a^+a) = \hbar\omega(a^+a + \tfrac{1}{2}). \tag{122}$$

The commutators of a and a^+ with H are given by

$$[H,a] = \hbar\omega\,[a^+a,a] =$$

$$= \hbar\omega\left\{[a^+,a]\,a + a^+[a,a]\right\} = -\hbar\omega a \tag{123}$$

$$[H,a^+] = \hbar\omega\,[a^+a,a^+] =$$

$$= \hbar\omega\left\{[a^+,a^+]\,a + a^+[a,a^+]\right\} = \hbar\omega a^+. \tag{124}$$

We have also the following relation:

$$\langle m|[H,a]|n\rangle = \langle m|Ha - aH|n\rangle =$$

$$= \langle m|Ha|n\rangle - \langle m|aH|n\rangle = (E_m - E_n)\langle m|a|n\rangle = -\hbar\omega\langle m|a|n\rangle,$$

or

$$(E_m - E_n + \hbar\omega)\langle m|a|n\rangle = 0. \tag{125}$$

Using the same procedure for $[H,a^+]$ we get:

$$(E_m - E_n - \hbar\omega)\langle m|a^+|n\rangle = 0. \tag{126}$$

From (122)

$$a^+a = \frac{H}{\hbar\omega} - \frac{1}{2} = \frac{1}{\hbar\omega}(H - \frac{\hbar\omega}{2}) \tag{127}$$

and

$$< n|a^+a|n > = \frac{1}{\hbar\omega} (E_n - \frac{\hbar\omega}{2}) =$$

$$= \sum_t < n|a^+|t > < t|a|n > = \sum_t | < n|a^+|t >|^2 \geqslant 0. \quad (128)$$

Let us now show that the energy E_o of the lowest level is $\hbar\omega/2$. Because of (128), $E_o \geqslant \hbar\omega/2$. If we assume that $E_o > \hbar\omega/2$,

$$\sum_t |a^+_{ot}|^2 = \frac{1}{\hbar\omega} (E_o - \frac{\hbar\omega}{2}) > 0, \quad (129)$$

which implies that there is at least one t for which a_{ot} is $\neq 0$. On the other hand, because of (126)

$$(E_o - E_t - \hbar\omega)\, a^+_{ot} = 0 \quad (130)$$

and

$$E_o = E_t + \hbar\omega \quad (131)$$

which is contrary to our assumption that E_o is the energy of the lowest state. With this we have then proved that the energy of the lowest state is $\hbar\omega/2$.

Let us call now E_1 the energy of the first excited state. From (128)

$$\sum_t |a^+_{1t}|^2 = \frac{1}{\hbar\omega} (E_1 - \frac{\hbar\omega}{2}) > 0 \quad (132)$$

which implies that there is at least one t for which $a^+_{1t} \neq 0$. But from (126)

$$(E_1 - E_t - \hbar\omega)\, a^+_{1t} = 0; \quad (133)$$

then

$$E_1 - E_t - \hbar\omega = 0 \quad (134)$$

and

$$E_t = E_1 - \hbar\omega. \quad (135)$$

E_t can only be $\hbar\omega/2$. Therefore

$$E_1 = (1 + \tfrac{1}{2})\hbar\omega. \tag{136}$$

The energy eigenvalues of the Hamiltonian (122) are then given

$$E_n = (n + \tfrac{1}{2})\hbar\omega, \tag{137}$$

which correspond to the energy eigenvalues of a harmonic oscillator with angular frequency ω. The eigenfunctions of the Hamiltonian (122) are simply given by the ket $|n\rangle$ which indicates the degree of excitation of the particular normal mode of frequency ω.

The entire radiative field is represented by a Hamiltonian which is the sum of infinite terms (each term corresponding to a particular frequency and a particular polarization):

$$H = \sum_\alpha \sum_\sigma H_\alpha^\sigma = \sum_\alpha \sum_\sigma \hbar\omega_\alpha (a_\alpha^{\sigma+} a_\alpha^\sigma + \tfrac{1}{2}). \tag{138}$$

The eigenvalues and the eigenfunctions of this Hamiltonian are given by

$$E_{\{n_1^{\sigma_1}\, n_1^{\sigma_2}\, n_2^{\sigma_1}\, \ldots\}} = \sum_\alpha \sum_\sigma \hbar\omega_\alpha (n_\alpha^\sigma + \tfrac{1}{2}) \tag{139}$$

and

$$\Psi_{\{n_1^{\sigma_1}\, n_1^{\sigma_2} n_2^{\sigma_1}\, \ldots\}} = \prod_{\alpha\sigma} |n_\alpha^\sigma\rangle , \tag{140}$$

respectively.

III.G. The Operator Vector Potential

The vector potential can now be expressed in terms of the operators a_α^σ and $a_\alpha^{\sigma+}$ as follows

$$\vec{A} = \sum_\alpha \sum_\sigma [\vec{A}_\alpha^\sigma(\vec{r})\, q_\alpha^\sigma + \vec{A}_\alpha^\sigma\, (\vec{r})^*\, q_\alpha^{+}] =$$

$$= \sum_\alpha \sum_\sigma \sqrt{\frac{4\pi c^2}{V}} \sqrt{\frac{\hbar}{2\omega_\alpha}}\; \vec{\pi}_\alpha^\sigma\, (e^{i\vec{k}_\alpha \cdot \vec{r}}\, a_\alpha^\sigma + e^{-i\vec{k}_\alpha \cdot \vec{r}}\, a_\alpha^{\sigma+}) =$$

$$= \sum_\alpha \sum_\sigma \sqrt{\frac{hc^2}{\omega_\alpha V}}\; \vec{\pi}_\alpha^\sigma\, (e^{i\vec{k}_\alpha \cdot \vec{r}}\, a_\alpha^\sigma + e^{-i\vec{k}_\alpha \cdot \vec{r}}\, a_\alpha^{\sigma+}). \tag{141}$$

It is useful to see how the operator vector potential operates on the eigenfunctions of the radiative field. Since all the normal modes of the field are independent it is enough to see how the operator a and a^+ operate within each mode. We shall then consider again the Hamiltonian (122) of the previous section; from (125) we derive the fact that $<m|a|n> \neq 0$ only when $m = n - 1$, and from (126) the fact that $<m|a^+|n> \neq 0$ only when $m = n + 1$. On the other hand

$$\begin{aligned} E_n &= <n|H|n> = (n + \tfrac{1}{2})\hbar\omega = \\ &= \hbar\omega <n|a^+a|n> + \tfrac{1}{2}\hbar\omega = \hbar\omega \sum_m (<n|a^+|m><m|a|n>) + \tfrac{1}{2}\hbar\omega = \\ &= \hbar\omega(<n|a^+|n-1><n-1|a|n>) + \tfrac{1}{2}\hbar\omega = \hbar\omega|<n-1|a|n>|^2 + \tfrac{1}{2}\hbar\omega, \end{aligned} \tag{142}$$

and

$$\begin{aligned} E_{n+1} &= <n+1|H|n+1> = (n+1+\tfrac{1}{2})\hbar\omega \\ &= \hbar\omega <n+1|a^+a|n+1> + \tfrac{1}{2}\hbar\omega = \hbar\omega \sum_m (<n+1|a^+|m><m|a|n+1>) + \tfrac{1}{2}\hbar\omega = \\ &= \hbar\omega(<n+1|a^+|n> <n|a|n+1>) + \tfrac{1}{2}\hbar\omega|<n+1|a^+|n>|^2 + \tfrac{1}{2}\hbar\omega. \end{aligned} \tag{143}$$

Then we can set

$$\begin{aligned} <n-1|a|n> &= \sqrt{n} \\ <n+1|a^+|n> &= \sqrt{n+1} \end{aligned} \tag{144}$$

and

$$\begin{aligned} a|n> &= \sqrt{n}\ |n-1> \\ a^+|n> &= \sqrt{n+1}\ |n+1>. \end{aligned} \tag{145}$$

Also

$$<n|a^+a|n> = n. \tag{146}$$

IV. INTERACTION OF A RADIATIVE FIELD WITH A CHARGED PARTICLE

IV.A. The Hamiltonian of a Charged Particle in an Electromagnetic Field

Let us consider a particle with a charge q and mass m in an electromagnetic field defined by a scalar potential $\varphi(\vec{r},t)$ and a vector potential $\vec{A}(\vec{r},t)$. The Lagrangian of this particle is given by

$$L = T - q\varphi + \frac{q}{c}\vec{v}\cdot\vec{A} = \sum_{i=1}^{3}\frac{m\dot{x}_i^2}{2} - q\varphi + \frac{q}{c}\sum_{i=1}^{3}\dot{x}_iA_i, \tag{143}$$

where T = kinetic energy of the particle and

$\vec{v}$ = velocity of the particle.

In fact, considering for example the x component of the particle's position vector we have

$$\frac{d}{dt}\frac{\partial L}{\partial \dot{x}} = m\ddot{x} + \frac{q}{c}\frac{dA_x}{dt} \tag{144}$$

$$\frac{\partial L}{\partial x} = -q\frac{\partial\varphi}{\partial x} + \frac{\partial}{\partial x}\left(\frac{q}{c}\vec{v}\cdot\vec{A}\right). \tag{145}$$

The equation of motion in the x direction is given by

$$m\ddot{x} + \frac{q}{c}\frac{dA_x}{dt} = -q\frac{\partial\varphi}{\partial x} + \frac{\partial}{\partial x}\left(\frac{q}{c}\vec{v}\cdot\vec{A}\right) \tag{146}$$

or

$$m\ddot{x} = q\left[-\frac{\partial\varphi}{\partial x} + \frac{1}{c}\frac{\partial}{\partial x}(\vec{v}\cdot\vec{A}) - \frac{1}{c}\frac{dA_x}{dt}\right]. \tag{147}$$

Then

$$m\ddot{\vec{r}} = q\left[-\vec{\nabla}\varphi + \frac{1}{c}\vec{\nabla}(\vec{v}\cdot\vec{A}) - \frac{1}{c}\frac{d\vec{A}}{dt}\right]. \tag{148}$$

But

$$\vec{v}\times(\vec{\nabla}\times\vec{A}) = \vec{\nabla}(\vec{v}\cdot\vec{A}) - (\vec{v}\cdot\vec{\nabla})\vec{A} \tag{149}$$

and

$$\frac{d\vec{A}}{dt} = \frac{\partial\vec{A}}{\partial t} + (\vec{v}\cdot\vec{\nabla})\vec{A}. \tag{150}$$

Therefore

$$m\ddot{\vec{r}} = q\Big\{-\vec{\nabla}\varphi + \frac{1}{c}[\vec{v}\times(\vec{\nabla}\times\vec{A}) + (\vec{v}\cdot\vec{\nabla})\vec{A}] -$$

$$- \frac{1}{c}\frac{\partial\vec{A}}{\partial t} - \frac{1}{c}(\vec{v}\cdot\vec{\nabla})\vec{A}\Big\} =$$

$$= q\left\{-\vec{\nabla}\varphi - \frac{1}{c}\frac{\partial\vec{A}}{\partial t} + \frac{1}{c}\vec{v}\times(\vec{\nabla}\times\vec{A})\right\} =$$

$$= q\left\{\vec{E} + \frac{1}{c}\ \vec{v}\times\vec{B})\right\}, \tag{151}$$

where we have used the relations

$$\vec{E} = -\vec{\nabla}\varphi - \frac{1}{c}\frac{\partial\vec{A}}{\partial t} \tag{152}$$

$$\vec{B} = \vec{\nabla}\times\vec{A}. \tag{153}$$

Since the right side of (151) expresses the Lorentz force on the charge, L in (143) is the correct Lagrangian.

Having found the Lagrangian, it is possible to find the Hamiltonian, by first deriving the generalized momenta

$$p_i = \frac{\partial L}{\partial \dot{x}_i} = \frac{\partial T}{\partial \dot{x}_i} + \frac{q}{c}\frac{\partial}{\partial \dot{x}_i}(\vec{v}\cdot\vec{A}) =$$

$$= \frac{\partial T}{\partial \dot{x}_i} + \frac{q}{c}A_i = m\dot{x}_i + \frac{q}{c}A_i. \tag{154}$$

The Hamiltonian is then given by

$$H = \sum_i p_i\dot{x}_i - L =$$

$$= \sum_i (m\dot{x}_i + \frac{q}{c}A_i)\dot{x}_i - (T - q\varphi + \frac{q}{c}\vec{v}\cdot\vec{A}) =$$

$$= \sum_i (m\dot{x}_i^2 + \frac{q}{c}A_i\dot{x}_i) - T + q\varphi - \frac{q}{c}\sum_i \dot{x}_i A_i =$$

$$= \sum_i m\dot{x}_i^2 - T + q\varphi = T + q\varphi . \tag{155}$$

From (154)

$$m\dot{x}_i = p_i - \frac{q}{c}A_i. \tag{156}$$

Then

$$T = \sum_i \frac{m\dot{x}_i^2}{2} = \frac{1}{2m} \sum_i (m\dot{x}_i)^2 =$$

$$= \frac{1}{2m} \sum_i (p_i - \frac{q}{c} A_i)^2 = \frac{1}{2m} (\vec{p} - \frac{q}{c} \vec{A})^2 \tag{157}$$

and

$$H = \frac{(\vec{p} - \frac{q}{c} \vec{A})^2}{2m} + q\varphi. \tag{158}$$

IV.B. The Interaction of a Charged Particle with a Radiative Field

Let us assume that a particle of mass m and charge q is in some bound state due to a scalar potential $\varphi(\vec{r})$ and that this particle is imbedded in a radiative field represented by a vector potential $\vec{A}(\vec{r},t)$. We shall consider the "total" system which consists of the bound particle and of the radiative field; the Hamiltonian of this system is given by (3,4):

$$H = \frac{1}{2m} (\vec{p} - \frac{q}{c} \vec{A})^2 + q\varphi + \frac{1}{8\pi} \int (E^2 + B^2)\, d^3\vec{r}, \tag{159}$$

where $\vec{p}$ = momentum of the particle. We can write the Hamiltonian as follows

$$H = \frac{\vec{p}^2}{2m} + q\varphi + \frac{1}{8\pi} \int (E^2 + B^2)\, A^3\vec{r} - \frac{q}{2mc} (\vec{p} \cdot \vec{A} + \vec{A} \cdot \vec{p}) +$$

$$+ \frac{q^2}{2mc^2} (\vec{A})^2. \tag{160}$$

It is

$$\vec{p} \cdot \vec{A} = \vec{A} \cdot \vec{p} + [\vec{p},\vec{A}]. \tag{161}$$

The x component of the operator $[\vec{p},\vec{A}]$ is given by

$$[\vec{p},\vec{A}]_x = p_x A_x - A_x p_x =$$

$$= - i\hbar \frac{\partial}{\partial x} A_x + A_x i\hbar \frac{\partial}{\partial x}. \tag{162}$$

Using this operator on some function $\psi(x)$,

$$[p_x,A_x]\psi = - i\hbar \frac{\partial}{\partial x} (A_x\psi) + A_x i\hbar \frac{\partial \psi}{\partial x} =$$

$$= -i\hbar \frac{\partial A_x}{\partial x}\psi - i\hbar A_x \frac{\partial \psi}{\partial x} + A_x i\hbar \frac{\partial \psi}{\partial x} =$$

$$= -i\hbar \frac{\partial A_x}{\partial x}\psi, \tag{163}$$

or

$$[\vec{p},\vec{A}] = -i\hbar(\vec{\nabla}\cdot\vec{A}) = 0, \tag{164}$$

because of the Coulomb gauge $\vec{\nabla}\cdot\vec{A} = 0$. We can write

$$\vec{p}\cdot\vec{A} = \vec{A}\cdot\vec{p} \tag{165}$$

and

$$\vec{p}\cdot\vec{A} + \vec{A}\cdot\vec{p} = 2\vec{A}\cdot\vec{p}. \tag{166}$$

The Hamiltonian (160) can now be written

$$H = H_o + H_1 + H_2, \tag{167}$$

where

$$H_o = \left(\frac{p^2}{2m} + q\varphi\right) + \left(\frac{1}{8\pi}\int[(\vec{E})^2 + (\vec{B})^2]\, d^3\vec{r}\right) =$$

$$= -\frac{\hbar^2}{2m}\nabla^2 + q\varphi + \sum_\alpha\sum_\sigma h\omega_\alpha(a_\alpha^{\sigma+}a_\alpha^\sigma + \tfrac{1}{2}) \tag{168}$$

$$H_1 = -\frac{q}{mc}\vec{A}\cdot\vec{p} =$$

$$= -\frac{q}{m}\sum_\alpha\sum_\sigma\left(\frac{h}{\omega_\alpha V}\right)^{\frac{1}{2}}(a_\alpha^\sigma e^{i\vec{k}_\alpha\cdot\vec{r}} + a_\alpha^{\sigma+}e^{-i\vec{k}_\alpha\cdot\vec{r}})\vec{\pi}_\alpha^\sigma\cdot\vec{p} \tag{169}$$

$$H_2 = \frac{q^2}{2mc^2}(\vec{A})^2 =$$

$$= \frac{q^2}{2m}\frac{h}{V}\sum_\alpha\sum_\sigma\sum_{\alpha'}\sum_{\sigma'}\frac{1}{(\omega_\alpha\omega_{\alpha'})^{\frac{1}{2}}}(\vec{\pi}_\alpha^\sigma\cdot\vec{\pi}_{\alpha'}^{\sigma'})\times$$

$$\times(a_\alpha^\sigma e^{i\vec{k}_\alpha\cdot\vec{r}} + a_\alpha^{\sigma+}e^{-i\vec{k}_\alpha\cdot\vec{r}})(a_{\alpha'}^{\sigma'}e^{i\vec{k}_{\alpha'}\cdot\vec{r}} + a_{\alpha'}^{\sigma'+}e^{-i\vec{k}_{\alpha'}\cdot\vec{r}}). \tag{170}$$

The terms H_1 and H_2 of the Hamiltonian represent the interaction between the charged particle and the radiative field. We notice that this interaction is time dependent. The eigenfunctions of the system, neglecting the interaction between particle and radiative field, namely the eigenfunctions of H_o are given by

$$\Psi_{e;n_1^1,n_1^2,\ldots} = |\psi^e\rangle \prod_{\alpha\sigma} |n_\alpha^\sigma\rangle =$$

$$= |\psi^e\rangle\, |n_1^1\rangle |n_1^2\rangle |n_2^1\rangle |n_2^2\rangle \ldots \quad (171)$$

where

$|\psi^e\rangle$ = eigenfunction of the particle, and

$\prod_{\alpha\sigma} |n_\alpha^\sigma\rangle$ = eigenfunction of the radiative field.

The energies of such eigenstates are given by

$$E_{e;n_1^1,n_1^2,\ldots} = E^e + \sum_\alpha \sum_\sigma \hbar\omega_\alpha(n_\alpha^\sigma + \tfrac{1}{2}) \quad (172)$$

where E^e = energy of the particle.

IV.C. Radiative Processes

The interaction Hamiltonian given by $H_1 + H_2$ can produce transitions between states described by the eigenfunctions Ψ in (171) provided energy is conserved in the process. Such transitions or processes can be classified as follows:

1) First Order Processes. They are due to H_1 in first order.
   - 1a. Annihilation of a photon of frequency ω_α, called absorption,
   - 1b. creation of a photon of frequency ω_α called emission.
2) Second Order Processes. They are due to H_1 in second order and/or to H_2 in first order.
   - 2a. Photon scattering, consisting in the simultaneous absorption of a photon of frequency ω_α and emission of a photon of frequency $\omega_{\alpha'}$ ($\omega_\alpha \gtrless \omega_{\alpha'}$),
   - 2b. two photon absorption,
   - 2c. two photon emission.

Higher order processes are possible but are much less probable.

IV.D. First Order Processes

First order processes come about by the use of the Hamiltonian H_1

given in (169) to first order. We can expand the exponentials appearing in H_1 as follows

$$e^{\pm i\vec{k}_\alpha \cdot \vec{r}} \approx 1^{\pm i\vec{k}_\alpha \cdot \vec{r}} + \ldots \tag{173}$$

Then

$$\begin{aligned} H_1 = &- \frac{q}{m}\sum_\alpha \sum_\sigma \left(\frac{h}{\omega_\alpha V}\right)^{\frac{1}{2}} (\vec{\pi}_\alpha^\sigma \cdot \vec{p})\, a_\alpha^\sigma - \\ &- \frac{q}{m}\sum_\alpha \sum_\sigma \left(\frac{h}{\omega_\alpha V}\right)^{\frac{1}{2}} (\vec{\pi}_\alpha^\sigma \cdot \vec{p})\, a_\alpha^{\sigma+} - \\ &- i\, \frac{q}{m}\sum_\alpha \sum_\sigma \left(\frac{h}{\omega_\alpha V}\right)^{\frac{1}{2}} (\vec{\pi}_\alpha^\sigma \cdot \vec{p})(\vec{k}_\alpha \cdot \vec{r})\, a_\alpha^\sigma + \\ &+ i\, \frac{q}{m}\sum_\alpha \sum_\sigma \left(\frac{h}{\omega_\alpha V}\right)^{\frac{1}{2}} (\vec{\pi}_\alpha^\sigma \cdot \vec{p})(\vec{k}_\alpha \cdot \vec{r})\, a_\alpha^{\sigma+} + \ldots \end{aligned} \tag{174}$$

Given a function $F(x_i, p_i, t)$, we have classically

$$\begin{aligned} \frac{dF(x_i, p_i, t)}{dt} &= \frac{\partial F}{\partial t} + \sum_i \left(\frac{\partial F}{\partial x_i} \frac{\partial x_i}{\partial t} + \frac{\partial F}{\partial p_i} \frac{\partial p_i}{\partial t} \right) = \\ &= \frac{\partial F}{\partial t} + \sum_i \left(\frac{\partial F}{\partial x_i} \frac{\partial H}{\partial p_i} - \frac{\partial F}{\partial p_i} \frac{\partial H}{\partial x_i} \right) = \\ &= \frac{\partial F}{\partial f} + \{F, H\} \end{aligned} \tag{175}$$

where the {} brackets indicate a Poisson bracket. If we want to express the above relation quantum mechanically we have to replace the Poisson bracket by a commutator as follows

$$\{F, H\} \xrightarrow[\text{Q.M.}]{} \frac{1}{i\hbar} [F, H] = \frac{i}{\hbar} [H, F]. \tag{176}$$

When this is done (175) becomes

$$\frac{dF}{dt} = \frac{i}{\hbar} [H, F] + \frac{\partial F}{\partial t}. \tag{177}$$

If we set $F = x$,

$$\dot{x} = \frac{i}{\hbar} [H, x] \tag{178}$$

and

$$p_x = m\dot{x} = i\frac{m}{\hbar}[H,x]. \tag{179}$$

Taking a matrix element of [H,x] between two eigenfunctions of the particle we find

$$\begin{aligned} <\psi_f^e\,|[H,x]|\,\psi_i^e> &= <\psi_f^e\,|Hx - xH|\,\psi_i^e> = \\ &= (E_f^e - E_i^e) <\psi_f^e\,|x|\,\psi_i^e> \end{aligned} \tag{180}$$

and

$$\begin{aligned} <\psi_f^e\,|\vec{p}|\,\psi_i^e> &= \frac{im}{\hbar}(E_f^e - E_i^e) <\psi_f^e\,|\vec{r}|\,\psi_i^e> \\ &= im\omega_{fi} <\psi_f^e\,|\vec{r}|\,\psi_i^e>, \end{aligned} \tag{181}$$

where

$$\omega_{fi} = \frac{E_f^e - E_i^e}{\hbar}. \tag{182}$$

On the basis of this result the first two terms in the expression of H_1 (174) can be written

$$\begin{aligned} &-\frac{q}{m}\sum_\alpha\sum_\sigma\left(\frac{h}{\omega_\alpha V}\right)^{\frac{1}{2}} a_\alpha^\sigma\,\vec{\pi}_\alpha^\sigma\cdot\vec{p} - \frac{q}{m}\sum_\alpha\sum_\sigma\left(\frac{h}{\omega_\alpha V}\right)^{\frac{1}{2}} a_\alpha^{\sigma+}\,\vec{\pi}_\alpha^\sigma\cdot\vec{p} = \\ &= -i\sum_\alpha\sum_\sigma\left[\frac{h\omega_\alpha}{V}\right]^{\frac{1}{2}} a_\alpha^\sigma\vec{\pi}_\alpha^\sigma\cdot(q\vec{r}) + i\sum_\alpha\sum_\sigma\left[\frac{h\omega_\alpha}{V}\right]^{\frac{1}{2}} a_\alpha^{\sigma+}\,\vec{\pi}_\alpha^\sigma\cdot(q\vec{r}). \end{aligned} \tag{183}$$

Consider now the quantity $(\vec{k}\cdot\vec{r})\vec{p}$. We can write

$$(\vec{k}\cdot\vec{r})\vec{p} = \tfrac{1}{2}\{(\vec{k}\cdot\vec{r})\vec{p} - (\vec{k}\cdot\vec{p})\vec{r}\} + \tfrac{1}{2}\{(\vec{k}\cdot\vec{r})\vec{p} + (\vec{k}\cdot\vec{p})\vec{r}\}. \tag{184}$$

Also

$$\begin{aligned} \tfrac{1}{2}\{(\vec{k}\cdot\vec{r})\vec{p} - (\vec{k}\cdot\vec{p})\vec{r}\} &= -\tfrac{1}{2}\{\vec{k}\times(\vec{r}\times\vec{p})\} = \\ &= -\frac{\omega}{2c}\{\vec{1}_{\vec{k}}\times\vec{L}\}, \end{aligned} \tag{185}$$

where $\vec{L}$ = angular momentum of the particle, and $\vec{1}_{\vec{k}}$ = unit vector in the $\vec{k}$ direction. The second term in (184) can be expressed as follows

$$\tfrac{1}{2}\{(\vec{k}\cdot\vec{r})\vec{p} + (\vec{k}\cdot\vec{p})\vec{r}\} = \frac{m}{2}\{(\vec{k}\cdot\vec{r})\dot{\vec{r}} + (\vec{k}\cdot\dot{\vec{r}})\vec{r}\} =$$

$$= \frac{m}{2}\frac{d}{dt}[(\vec{k}\cdot\vec{r})\vec{r}] = \frac{im\omega}{2}[(\vec{k}\cdot\vec{r})\vec{r}]. \quad (186)$$

On the basis of the above result we can write the third and fourth terms in (174) as follows

$$- i\frac{q}{m}\sum_\alpha\sum_\sigma\left(\frac{h}{\omega_\alpha V}\right)^{\frac{1}{2}} a_\alpha^\sigma(\vec{k}_\alpha\cdot\vec{r})\vec{p}\cdot\vec{\pi}_\alpha^\sigma + i\frac{q}{m}\sum_\alpha\sum_\sigma\left(\frac{h}{\omega_\alpha V}\right)^{\frac{1}{2}} a_\alpha^{\sigma+}(\vec{k}_\alpha\cdot\vec{r})\vec{p}\cdot\vec{\pi}_\alpha^\sigma =$$

$$= i\sum_\alpha\sum_\sigma\left[\frac{h\omega_\alpha}{V}\right]^{\frac{1}{2}}\left[\left(\vec{1}_{\vec{k}_\alpha}\times\frac{q}{2mc}\vec{L}\right).\vec{\pi}_\alpha^\sigma\right]a_\alpha^{\sigma+} +$$

$$+ i\sum_\alpha\sum_\sigma\left[\frac{h\omega_\alpha}{V}\right]^{\frac{1}{2}}\left[\left(\vec{1}_{\vec{k}_\alpha}\times\frac{q}{2mc}\vec{L}\right)\cdot\vec{\pi}_\alpha^\sigma\right]a_\alpha^{\sigma+} +$$

$$+ \sum_\alpha\sum_\sigma\left[\frac{h\omega_\alpha}{V}\right]^{\frac{1}{2}}\tfrac{1}{2}(\vec{k}_\alpha\cdot\vec{r})(q\vec{r}\cdot\vec{\pi}_\alpha^\sigma)\,a_\alpha^\sigma +$$

$$+ \sum_\alpha\sum_\sigma\left[\frac{h\omega_\alpha}{V}\right]^{\frac{1}{2}}\tfrac{1}{2}(\vec{k}_\alpha\cdot\vec{r})(q\vec{r}\cdot\vec{\pi}_\alpha^\sigma)\,a_\alpha^{\sigma+}. \quad (187)$$

The interaction Hamiltonian H_1 can then be written

$$H_1 = H_1(E1) + H_1(M1) + H_1(E2) \quad (188)$$

where

$H_1(E1)$ = <u>electric dipole</u> interaction =

$$= - i\sum_\alpha\sum_\sigma\left[\frac{h\omega_\alpha}{V}\right]^{\frac{1}{2}}(q\vec{r}\cdot\vec{\pi}_\alpha^\sigma)a_\alpha^\sigma + i\sum_\alpha\sum_\sigma\left[\frac{h\omega_\alpha}{V}\right]^{\frac{1}{2}}(q\vec{r}\cdot\vec{\pi}_\alpha^\sigma)\,a_\alpha^{\sigma+} \quad (189)$$

$H_1(M1)$ = <u>magnetic dipole</u> interaction =

$$= i\sum_{\alpha}\sum_{\sigma}\left[\frac{h\omega_\alpha}{V}\right]^{\frac{1}{2}}\left[\left(\vec{1}_{\vec{k}_\alpha} \times \frac{q}{2mc}\vec{L}\right)\cdot \vec{\pi}_\alpha^{\sigma}\right] a_\alpha^{\sigma} +$$

$$+ i\sum_{\alpha}\sum_{\sigma}\left[\frac{h\omega_\alpha}{V}\right]^{\frac{1}{2}}\left[\left(\vec{1}_{\vec{k}_\alpha} \times \frac{q}{2mc}\vec{L}\right)\cdot \vec{\pi}_\alpha^{\sigma}\right] a_\alpha^{\sigma+} \qquad (190)$$

H(E2) = <u>electric quadrupole</u> interaction =

$$= \sum_{\alpha}\sum_{\sigma}\left[\frac{h\omega_\alpha}{V}\right]^{\frac{1}{2}} \tfrac{1}{2}(\vec{k}_\alpha \cdot \vec{r})(q\vec{r} \cdot \vec{\pi}_\alpha^{\sigma})\, a_\alpha^{\sigma} +$$

$$+ \sum_{\alpha}\sum_{\sigma}\left[\frac{h\omega_\alpha}{V}\right]^{\frac{1}{2}} \tfrac{1}{2}(\vec{k}_\alpha \cdot \vec{r})(q\vec{r} \cdot \vec{\pi}_\alpha^{\sigma})\, a_\alpha^{\sigma+}. \qquad (191)$$

These three terms produce electric dipole, magnetic dipole and electric quadrupole transitions, respectively.

IV.E. Electric Dipole Processes

The electric dipole approximation is based on the assumption that

$$e^{i\vec{k}\cdot\vec{r}} \approx 1; \qquad (192)$$

this means that if d is the dimension of the physical system under consideration (atom, molecule, etc.), for the electric dipole transitions to be predominant it must be

$$|kd| << 1, \qquad (193)$$

or $d << \lambda$ (= wavelength of emitted or absorbed light). This is generally the case for atoms and molecules so that the electric dipole transitions are predominant.

In the case of an atom, the dimension d is given by

$$d = \frac{a_o}{Z_{eff}} \qquad (194)$$

where a_o = Bohr radius and Z_{eff} = effective charge (= 1 for transitions by valence electrons). For an atom the transition energy is given by

$$\hbar\omega = \hbar kc \lesssim Z_{eff}^2 \frac{e^2}{a_o} = Z_{eff}\frac{e^2}{d}. \qquad (195)$$

Then

$$kd \lesssim Z_{eff} \frac{e^2}{\hbar c} = \frac{Z_{eff}}{137}. \tag{196}$$

A consequence of this argument is that in atoms electric dipole transitions are most intense and that electric quadrupole and magnetic dipole transitions are weaker by $\sim(Z_{eff}/137)^2$.

V. ABSORPTION AND EMISSION OF RADIATION

V.A. Transition Probabilities for Absorption and Emission

In the model set up in the previous section the "unperturbed" system consists of the particle and the radiative field and is related to the Hamiltonian H_o of (168). The interaction Hamiltonian H_1 of (169) is considered to be a small perturbation which may cause transitions of the unperturbed system from an initial to a final state.

In the case of one photon absorption the initial and the final states are given by

$$\Psi_i = |\psi_i^e >|n_\alpha^\sigma >|n_{\alpha'}^{\sigma'} > \ . \ . \ . \tag{197}$$

$$\Psi_f = |\psi_f^e >|n_\alpha^\sigma - 1>|n_{\alpha'}^{\sigma'} > \ . \ . \ . \ . \ , \tag{198}$$

respectively. It is

$$<n_\alpha^\sigma - 1|a_\alpha^\sigma|n_\alpha^\sigma > = \sqrt{n_\alpha^\sigma} \ ; \tag{199}$$

therefore the relevant matrix element for the process of absorption of one photon is

$$< \psi_f^e;n_\alpha^\sigma - 1|H_1|\psi_i^e;n_\alpha^\sigma > =$$

$$= - \frac{q}{m}\left(\frac{h}{\omega_\alpha V}\right)^{1/2} < \psi_f^e|e^{i\vec{k}_\alpha \cdot \vec{r}} \vec{\pi}_\alpha^\sigma \cdot \vec{p}|\psi_i^e > \sqrt{n_\alpha^\sigma} \ . \tag{200}$$

In the case of one photon emission, the initial and the final states are given by

$$\Psi_i = |\psi_i^e >|n_\alpha^\sigma >|n_{\alpha'}^{\sigma'}> \ . \ . \ . \tag{201}$$

$$\Psi_f = |\psi_f^e>|n_\alpha^\sigma + 1>|n_{\alpha'}^{\sigma'}> \ldots , \tag{202}$$

respectively. It is

$$<n_\alpha^\sigma + 1|a_\alpha^{\sigma+}|n_\alpha^\sigma> = \sqrt{n_\alpha^\sigma + 1}; \tag{203}$$

therefore the relevant matrix element for the process of emission of one photon is

$$< \psi_f^e;n_\alpha^\sigma + 1|H_1|\psi_i^e;n_\alpha^\sigma> =$$

$$= -\frac{q}{n}\left(\frac{h}{\omega_\alpha V}\right)^{\frac{1}{2}} < \psi_f^e|e^{-i\vec{k}_\alpha \cdot \vec{r}}\,\vec{\pi}_\alpha^\sigma \cdot \vec{p}|\psi_i^e> \sqrt{n_\alpha^\sigma + 1} . \tag{204}$$

Since the radiative field has a continuous density of states (see for this section III.C.), both absorption and emission processes are associated with a probability per unit time. By applying the Fermi Golden Rule we derive that the probability per unit time of finding the system (particle + radiative field) with one less or one more photon of energy $\hbar\omega_\alpha$, polarization $\vec{\pi}_\alpha^\sigma$ in the solid angle $(\Omega_\alpha,\Omega_\alpha + d\Omega_\alpha)$ is given by

$$P_\alpha^\sigma\, d\Omega_\alpha = \frac{2\pi}{\hbar^2} |M_\alpha^\sigma|^2 g(\omega_\alpha) \tag{205}$$

where

$$g(\omega_\alpha) = \frac{V\omega_\alpha^2}{8\pi^3 c^3}\, d\Omega_\alpha \tag{206}$$

and

$$|M_\alpha^\sigma|^2 = \begin{cases} \dfrac{q^2}{m^2}\dfrac{h}{\omega_\alpha V} |<\psi_f^e|e^{i\vec{k}_\alpha \cdot \vec{r}}\,\vec{\pi}_\alpha^\sigma \cdot \vec{p}|\psi_i^e>|^2\, n_\alpha^\sigma \\ \\ \dfrac{q^2}{m^2}\dfrac{h}{\omega_\alpha V} |<\psi_f^e|e^{-i\vec{k}_\alpha \cdot \vec{r}}\,\vec{\pi}_\alpha^\sigma \cdot \vec{p}|\psi_i^e>|^2\, (n_\alpha^\sigma + 1), \end{cases} \tag{207}$$

where the upper (lower) row corresponds to the process of absorption (emission) of one photon. Replacing (207) in (205) we find:

$$P^{\sigma}_{\alpha}\, d\Omega_{\alpha} =$$

$$= \frac{2\pi}{\hbar^2}\,\frac{q^2}{m^2}\,\frac{h}{\omega_\alpha V}\left\{\begin{array}{l} |\langle\psi^e_f|e^{i\vec{k}_\alpha\cdot\vec{r}}\,\vec{\pi}^{\sigma}_{\alpha}\cdot\vec{p}|\psi^e_i\rangle|^2\, n^{\sigma}_{\alpha} \\ \\ |\langle\psi^e_f|e^{-i\vec{k}_\alpha\cdot\vec{r}}\,\vec{\pi}^{\sigma}_{\alpha}\cdot\vec{p}|\psi^e_i\rangle|^2 (n^{\sigma}_{\alpha}+1) \end{array}\right\}\frac{V\omega^2_\alpha}{8\pi^3 c^3}\, d\Omega_\alpha =$$

$$= \frac{\omega_\alpha q^2}{hc^3 m^2}\,\left|\langle\psi^e_f\left|\begin{array}{l} e^{i\vec{k}_\alpha\cdot\vec{r}}\,\vec{\pi}^{\sigma}_{\alpha}\cdot\vec{p} \\ \\ e^{-i\vec{k}_\alpha\cdot\vec{r}}\,\vec{\pi}^{\sigma}_{\alpha}\cdot\vec{p}\end{array}\right|\psi^e_i\rangle\right|^2 \left\{\begin{array}{l}(n^{\sigma}_{\alpha}) \\ \\ (n^{\sigma}_{\alpha}+1)\end{array}\right\} d\Omega_\alpha\,. \qquad (208)$$

V.B. "Upward" and "Downward" Induced Transitions

Let us consider two quantum states of the particle ψ^e_i and ψ^e_f with $E_i > E_f$. The process of emission of one photon is related to the squared matrix element

$$|\langle\psi^e_f|e^{-i\vec{k}\cdot\vec{r}}\,\vec{\pi}\cdot\vec{p}|\psi^e_i\rangle|^2. \qquad (209)$$

Since $\vec{\pi}\cdot\vec{k} = 0$ we can interchange in (209) $\vec{\pi}\cdot\vec{p}$ with $e^{-i\vec{k}\cdot\vec{r}}$:

$$|\langle\psi^e_f|e^{-i\vec{k}\cdot\vec{r}}(\vec{\pi}\cdot\vec{p})|\psi^e_i\rangle|^2 = |\langle\psi^e_f|(\vec{\pi}\cdot\vec{p})e^{-i\vec{k}\cdot\vec{r}}|\psi^e_i\rangle|^2 =$$

$$= |\langle(\vec{\pi}\cdot\vec{p})\psi^e_f|e^{-i\vec{k}\cdot\vec{r}}|\psi^e_i\rangle|^2 = |\langle\psi^e_i|e^{i\vec{k}\cdot\vec{r}}(\vec{\pi}\cdot\vec{p})|\psi^e_f\rangle|^2. \qquad (210)$$

The very last squared matrix element is the one that would enter the transition probability for a f → i (absorption) process.

On the basis of the above result, (208) becomes

$$P^{\sigma}_{\alpha} d\Omega_{\alpha} = \frac{\omega_\alpha q^2}{hc^3 m^2}\,|\langle\psi^e_f|e^{i\vec{k}\cdot\vec{r}}\,\vec{\pi}_{\alpha}\cdot\vec{p}|\psi^e_i\rangle|^2 \begin{pmatrix} n^{\sigma}_{\alpha} \\ n^{\sigma}_{\alpha}+1\end{pmatrix} d\Omega_\alpha\,. \qquad (211)$$

If two or more charged particles are present

$$P_\alpha^\sigma d\Omega_\alpha = \frac{\omega_\alpha}{hc^3} \left| < \sum_\ell \frac{q_\ell}{m_\ell} (\vec{\pi}_\alpha^\sigma \cdot \vec{p}_\ell) e^{i\vec{k}_\alpha \cdot \vec{r}_\ell} >_{fi} \right|^2 \begin{pmatrix} n_\alpha^\sigma \\ n_\alpha^\sigma + 1 \end{pmatrix} d\Omega_\alpha . \quad (212)$$

The probability per unit time of spontaneous emission of a photon of frequency ω_α and polarization $\vec{\pi}_\alpha^\sigma$ in the solid angle $d\Omega_\alpha$ is given by

$$P_\alpha^\sigma(sp) d\Omega_\alpha = \frac{\omega_\alpha}{hc^3} \left| < \sum_\ell \frac{q_\ell}{m_\ell} (\vec{\pi}_\alpha^\sigma \cdot \vec{p}_\ell) e^{i\vec{k}_\alpha \cdot \vec{r}_\ell} >_{fi} \right|^2 d\Omega_\alpha . \quad (213)$$

The probability per unit time of absorption or induced emission of a photon of frequency ω_α and polarization $\vec{\pi}_\alpha^\sigma$ in the solid angle $d\Omega_\alpha$ is given by

$$P_\alpha^\sigma(\text{abs; emi})\, d\Omega_\alpha = [P_\alpha^\sigma(sp)\, d\Omega_\alpha]\, n_\alpha^\sigma . \quad (214)$$

V.C. Einstein's A and B Coefficients

The probability per unit time of spontaneous emission of a photon of frequency ω_α is given by

$$\sum_\alpha \int P_\alpha^\sigma(sp) d\Omega_\alpha =$$

$$= \frac{\omega_\alpha}{hc^3} \sum_\alpha \int d\Omega_\alpha \left| < \sum_\ell \frac{q_\ell}{m_\ell} (\vec{\pi}_\alpha^\sigma \cdot \vec{p}_\ell) e^{i\vec{k}_\alpha \cdot \vec{r}_\ell} >_{fi} \right|^2 . \quad (215)$$

The probability per unit time of induced absorption or emission of a photon of frequency ω_α is given by

$$\frac{\omega_\alpha}{hc^3} \sum_\sigma \int d\Omega_\alpha \left| < \sum_\ell \frac{q_\ell}{m_\ell} (\vec{\pi}_\alpha^\sigma \cdot \vec{p}_\ell) \right| e^{i\vec{k}_\alpha \cdot \vec{r}_\ell} >_{fi} |^2\, n_\alpha^\sigma . \quad (216)$$

We consider now the case in which

a) the relevant matrix element is not polarization or direction dependent, a case always occuring in free atoms, and
b) the intensity of the field at a certain frequency is equal in the two polarizations.

In this case the probability per unit time of spontaneous emission is given by

$$A = \frac{2\omega_\alpha}{3\hbar c^3} 4\pi \left| < \sum_\ell \frac{q_\ell}{m_\ell} \vec{p}_\ell \, e^{i\vec{k}_\alpha \cdot \vec{r}} >_{fi} \right|^2 \tag{217}$$

and the probability for induced emission or absorption by

$$B\rho(\nu_\alpha) = An_\alpha, \tag{218}$$

where A and B are the so-called Einstein's coefficients and $\rho(\nu_\alpha)$ is the energy density per unit frequency range. It is

$$\rho(\nu_\alpha) = 2\pi \, \rho(\omega_\alpha)$$

with

$$\rho(\omega_\alpha) = 2 \frac{4\pi}{V} \frac{V\omega_\alpha^2}{8\pi^3 c^3} \hbar\omega_\alpha \, n_\alpha = \frac{\omega_\alpha^3 \hbar}{\pi^2 c^3} n_\alpha. \tag{219}$$

Then

$$\rho(\nu_\alpha) = \frac{\omega_\alpha^3 h}{\pi^2 c^3} n_\alpha = \frac{8\pi h \nu_\alpha^3}{c^3} \, n_\alpha \tag{220}$$

and

$$\frac{A}{B} = \frac{8\pi h \nu_\alpha^3}{c^3} \quad . \tag{221}$$

V.D. Absorption and Emission in the Electric Dipole Approximation

In the electric dipole approximation the expression (211) becomes

$$P_\alpha^\sigma \, d\Omega_\alpha = \frac{\omega_\alpha}{\hbar c^3} \left| < \sum_\ell \frac{q_\ell}{m_\ell} (\vec{\pi}_\alpha^\sigma \cdot \vec{p}_\ell) >_{fi} \right|^2 \begin{pmatrix} n_\alpha^\sigma \\ n_\alpha^\sigma + 1 \end{pmatrix}$$

$$= \frac{\omega_\alpha^3}{\hbar c^3} \left| < \vec{\pi}_\alpha^\sigma \cdot \vec{M} >_{fi} \right|^2 \begin{pmatrix} n_\alpha^\sigma \\ n_\alpha^\sigma + 1 \end{pmatrix} d\Omega_\alpha, \tag{222}$$

where

$$\vec{M} = \sum_\ell q_\ell \vec{r}_\ell = \text{electric dipole operator.} \tag{223}$$

The radiation density per unit frequency range at $\omega = \omega_\alpha$, unit solid angle and polarization $\vec{\pi}_\alpha^\sigma$ is given by

$$\frac{g(\omega_\alpha)\hbar\omega_\alpha n_\alpha^\sigma}{Vd\Omega_\alpha} = \frac{\omega_\alpha^2}{8\pi^3c^3}\hbar\omega_\alpha n_\alpha^\sigma = \frac{\hbar\omega_\alpha^3}{8\pi^3c^3} n_\alpha^\sigma , \tag{224}$$

and the radiation intensity per unit frequency range at $\omega = \omega_\alpha$, unit solid angle and polarization $\vec{\pi}_\alpha^\sigma$ is given by

$$I(\omega_\alpha,\sigma) = \frac{\hbar\omega_\alpha^3}{8\pi^3c^2} n_\alpha^\sigma. \tag{225}$$

It is

$$\frac{4\pi^2 I}{\hbar^2 c} = \frac{4\pi^2}{\hbar^2 c}\frac{\hbar\omega_\alpha^3}{8\pi^3c^2} n_\alpha^\sigma = \frac{\omega_\alpha^3}{hc^3} n_\alpha^\sigma. \tag{226}$$

Therefore we can write the formula (222) as follows

$$p_\alpha^\sigma = \begin{cases} \dfrac{4\pi^2 I(\omega_\alpha;\sigma)}{\hbar^2 c}|M_{fi}|^2 \\[2ex] \left\{\dfrac{4\pi^2 I(\omega_\alpha;\sigma)}{\hbar^2 c} + \dfrac{\omega_\alpha^3}{hc^3}\right\}|M_{fi}|^2 , \end{cases} \tag{227}$$

where the upper (lower) row corresponds to absorption (emission) and where

$$|M_{fi}|^2 = |< \vec{\pi}_\alpha^\sigma \cdot \vec{M} >_{fi}|^2. \tag{228}$$

V.E. Interaction of Radiation with Molecular Systems The Franck - Condon Principle

The process of absorption or emission of ultraviolet light by molecules is accompanied by a change of their electronic state (and structure). Stated very simply the Franck - Condon principle recognizes the fact that the change in the electronic structure occurring during an electronic transition is much more rapid than the possible changes in the internuclear distances occurring during the same transition.

By using purely classical arguments the Franck - Condon principle is stated as follows:

"During an electronic transition the electronic state changes so fast that 1) the nuclei do not move and 2) the nuclei do not change their momenta."

This is best illustrated considering a diatomic molecule. Potential curve diagrams for such a system are given in Fig. 2; in it R is the distance between the two nuclei.

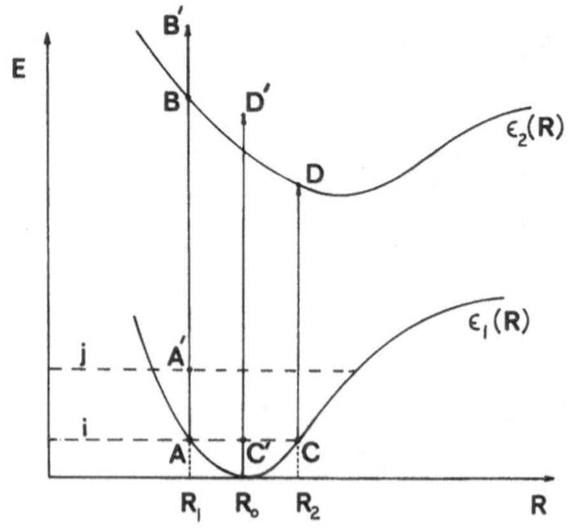

Fig. 2. Diagram illustrating the Franck - Condon principle.

The condition 1) above means that during an electronic transition the internuclear distance R must remain constant; the condition 2) implies that the kinetic energy must also remain constant and this means that if the molecule's initial state is given by the point A' the molecule after the electronic transition will be found at B' with AA' = BB'. A third condition may be added considering that the harmonic oscillator which represents the vibrational motion of the molecule spends most of its time at its turning points at which the kinetic energy is zero. This condition 3) states that at the instant at which the electronic transition takes place the molecule is found at these turning points. Therefore according to this condition the following transitions are allowed

AB , CD

and the following transitions are forbidden

A'B' , C'D'.

In its semiclassical formulation the Franck - Condon principle consists of three conditions:

1) R = const, 2) p = const and 3) an electronic transition can take place for any value of R with a probability W(R), as given by quantum mechanics.

If the system is in a particular vibrational quantum state, say the ith vibrational state, then

$$W(R) = |\phi_i(R)|^2 . \tag{229}$$

If the system is in thermal equalibrium at a temperature T

$$W_T(R) = \frac{\sum_i e^{-E_i/kT} |\phi_i(R)|^2}{\sum_i e^{-E_i/kT}} . \tag{230}$$

For most diatomic molecules at ordinary temperature hω is so large (of the order 0.1 eV) that

$$\hbar\omega >> kT \tag{231}$$

and the vibrational states other than the ground state are practically unoccupied; under these conditions the relevant probability W(R) may well be given by (229) with i = 0.

The above considerations can be generalized by treating a molecule with many vibrational degrees of freedom. In these more complex cases the relevant probability is given by

$$W_T = \prod_K W_{KT}(q_K) , \tag{232}$$

where W_{KT} is of the type (230) with $\omega = \omega_K$ and $R = q_K$ (vibrational coordinate); in the limit of low temperatures or high ω, W_{KT} is simply of the type (229).

Considering now again the diagram in Fig. 2 we note that the semiclassical Franck - Condon principle allows the transitions AB, CD, A'B' and C'D' which have probabilities

$$|\varphi_i(R_1)|^2, |\varphi_i(R_2)|^2, |\varphi_j(R_1)|^2 \text{ and } |\varphi_i(R_o)|^2,$$

respectively; in particular, if i = 0, the probability will have its maximum in correspondence to R_o and transitions such as C'D' have the greatest importance.

Finally the quantum mechanical formulation of the Franck - Condon principle is perhaps the most illuminating. The transition probability for a radiative transition in the dipole approximation, is proportional to the square of the matrix element

$$M_{fi} = \langle \vec{\pi}_\alpha^\sigma \cdot \vec{M} \rangle_{fi} = \langle \vec{\pi}_\alpha^\sigma \cdot \sum_\ell q_\ell \vec{r}_\ell \rangle_{fi} \tag{233}$$

where q_ℓ and $\vec{r}_\ell$ are the charge and the position of the ℓth charged particle, respectively. The initial state of the system is given in the adiabatic approximation by

$$\Psi_i(\vec{r},\vec{R}) = \psi_k(\vec{r},\vec{R})\phi_{k\ell}(\vec{R}). \tag{234}$$

and the final state by

$$\Psi_f(\vec{r},\vec{R}) = \psi_m(\vec{r},\vec{R})\phi_{mn}(\vec{R}). \tag{235}$$

The electric dipole operator can be written as follows:

$$\vec{M} = -e\sum_i \vec{r}_i + e\sum_s Z_s\vec{R}_s , \tag{236}$$

where the sums over i and s extend to the electrons and the nuclei, respectively. We can then write

$$\vec{\pi}\cdot\vec{M} = -e\sum_i \vec{r}_i\cdot\vec{\pi} + e\sum_s Z_s\vec{R}_s\cdot\vec{\pi} =$$

$$= D_e(\vec{r}) + D_n(\vec{R}). \tag{237}$$

The relevant matrix element is now given by

$$M_{fi} = M_{mn;k\ell} =$$

$$= \iint d^3\vec{r}\,d^3\vec{R}\;\Psi_f(\vec{r},\vec{R})^*\,[D_e(\vec{r}) + D_n(\vec{R})]\Psi_i(\vec{r},\vec{R}) =$$

$$= \iint d^3\vec{r}\,d^3\vec{R}\;\psi_m(\vec{r},\vec{R})^*\phi_{mn}(\vec{R})^*[D_e(\vec{r}) + D_n(\vec{R})]\psi_k(\vec{r},\vec{R})\phi_{k\ell}(\vec{R}) =$$

$$= \int d^3\vec{R}\;\phi_{mn}(\vec{R})^*\left[\int d^3\vec{r}\;\psi_m(\vec{r},\vec{R})^*D_e(\vec{r})\psi_k(\vec{r},\vec{R})\right]\phi_{k\ell}(\vec{R}) +$$

$$+ \int d^3\vec{R}\, \phi_{mn}(\vec{R})^* D_n(\vec{R}) \phi_{k\ell}(\vec{R}) \left[\int d^3\vec{r} \psi_m(\vec{r},\vec{R})^* \psi_k(\vec{r},\vec{R}) \right] =$$

$$= \int d^3\vec{R}\, \phi_{mn}(\vec{R})^* D_{mk}(\vec{R}) \phi_{k\ell}(\vec{R}) + \int d^3\vec{R} \phi_{mn}(\vec{R})^* D_n(\vec{R}) \phi_{k\ell}(\vec{R}) \delta_{mk} \tag{238}$$

where

$$D_{mk}(\vec{R}) = \int d^3\vec{r} \psi_m(\vec{r},\vec{R})^* D_e(\vec{r}) \psi_k(\vec{r},\vec{R}). \tag{239}$$

The following observations can be made:

1) The matrix element M_{fi} consists of two terms. The second term contributes to the matrix element only when $m = k$, namely when the electronic state does not change in the transition and gives rise to infrared absorption by the vibrations.
2) The first term contains the dipole moment of the electrons and corresponds to transitions between different electronic states.
3) If $m = k$ the first term becomes

$$\int d^3\vec{R} \phi_{mn}(\vec{R})^* D_{mm}(\vec{R}) \phi_{m\ell}(\vec{R}) \tag{240}$$

where

$$D_{mm}(\vec{R}) = \int d^3\vec{r} |\psi_m(\vec{r},\vec{R})|^2 D_e(\vec{r}) \tag{241}$$

Since $D_e(\vec{r})$ is an odd operator, $D_{mm}(\vec{R})$ is different from zero only if the molecule lacks inversion symmetry. If this is the case the first term in M_{fi} also contributes to the infrared transition.

For transitions between different electronic states the relevant matrix element is given by

$$M_{fi} = M_{mn;k\ell} =$$

$$= \int \phi_{mn}(\vec{R})^* D_{mk}(\vec{R}) \phi_{k\ell}(\vec{R}) R^2 \sin\theta dR d\theta d\varphi \tag{242}$$

where $D_{mk}(\vec{R})$ is given by (239).

We shall consider now for simplicity the case of a diatomic molecule; for such a system the eigenfunctions appearing in the integral (242) are of the type (7)

$$\phi(R,\theta,\varphi) = \mathcal{R}_{NK}(R)\Theta_{KM}(\theta)\phi_M(\varphi). \tag{243}$$

The equation for $\mathcal{R}(R)$ is

$$\frac{1}{R^2}\frac{d}{dR}\left(R^2\frac{d\mathcal{R}}{dR}\right) + \left\{-\frac{K(K+1)}{R^2} + \frac{2\mu}{\hbar^2}[E - \varepsilon(R)]\right\}\mathcal{R} = 0, \tag{244}$$

where μ = reduced mass of the molecule. This equation may be simplified by setting

$$\mathcal{R}(R) = \frac{S(R)}{R}; \tag{245}$$

with this (244) reduces to

$$\frac{d^2S}{dR^2} + \left\{-\frac{K(K+1)}{R^2} + \frac{8\pi^2\mu}{h^2}[E - \varepsilon(R)]\right\}S = 0. \tag{246}$$

In order to find S(R) and $\mathcal{R}(R)$ we need to know $\varepsilon(R)$, the form of the adiabatic potential. If it is assumed that the force between the atoms is proportional to the displacement of the internuclear distance from its equilibrium value, then the functions S(R) take the form of harmonic oscillator eigenfunctions (7). In any case, regardless of the form of the adiabatic potential $\varepsilon(R)$, the matrix element (242) will be proportional to the radial integral

$$\int S_{mn}(R)^* D_{mk}(R) S_{k\ell}(R)dR. \tag{247}$$

If we assume that $D_{mk}(R)$ does not depend strongly on the internuclear distance, we may expand it in terms of the displacement of this distance from its equilibrium value. According to the Franck - Condon principle we retain only the first (constant) term in this expansion:

$$D_{mk}(R) \simeq D^o_{mk}. \tag{248}$$

Then for electronic transitions

$$|M_{fi}|^2 \sim |D^o_{mk}|^2 \left|\int S_{mn}(R)^* S_{k\ell}(R)dR\right|^2 \tag{249}$$

The transition probability is then proportional to the square of the overlap integral of the vibrational eigenfunctions of the initial and final states. It is this overlap that controls (apart from D^o_{mk}) the strength of the transition probabilities. It is to be noted here, that $S_{mn}(R)$ and $S_{k\ell}(R)$ are solutions of different Schroedinger equations; they actually belong to different sets of orthonormal eigenfunctions and the subscripts m and k do not play the role of good quantum numbers.

Therefore if $m \neq k$ the overlap integral in general may not go to zero even if $n \neq \ell$.

The diagram in Fig. 3 illustrates the role played by the overlap integral in determining the strength of a transition.

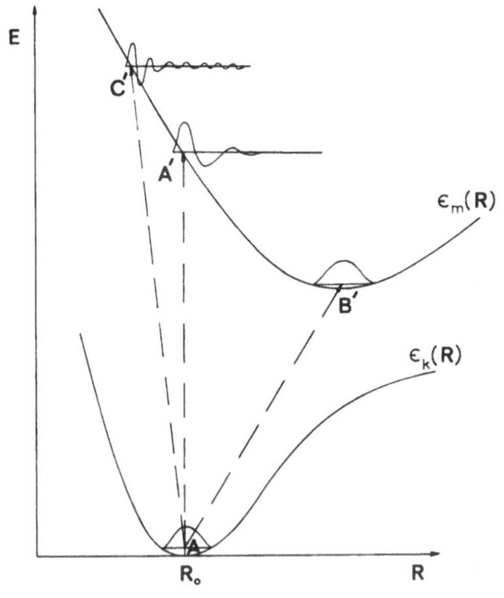

Fig. 3. Diagram illustrating the quantum mechanical Franck - Condon principle.

It is evident from this diagram that the "vertical" transition AA' is the one of highest probability and that the "sloped" transitions AB' and AC' have much lower probabilities; in particular in the former case (AB') there is practically no overlap, in the latter case (AC') the overlap is small because of the oscillating nature of the upper vibrational eigenfunction.

The following considerations can be made:

1) The quantum mechanical Franck - Condon principle replaces the requirements R = const., p = const. of its semiclassical formulation with the notion that transitions for which such conditions are not fulfilled are very unlikely.
2) The Franck - Condon principle in its quantum mechanical

formulation is not a selection rule in the conventional sense. It does not say that certain transtitions cannot occur but, rather, that they are highly improbable. Selection rules are generally derived from the matrix element of an operator taken between two eigenfunctions of the same orthonormal set, contrary to the present situation where the relevant entity is an overlap integral of two wavefunctions belonging to different orthonormal sets.

3) If the potential curves of the two electronic states have the same shape, i.e., if

$$\varepsilon_k(R) = \varepsilon_m(R) + \text{const.}, \tag{250}$$

then the two functions S_{mn} and $S_{k\ell}$ belong to the same set and

$$\int S_{mn}(R)^* S_{k\ell}(R)dR = \delta_{n\ell}. \tag{251}$$

In this case the Franck - Condon principle can be expressed as the selection rule: "If the adiabatic potentials of two electronic states are the same, no transition can take place in which the molecule changes its vibrational state." This selection rule is independent from the nature (electric dipole, magnetic dipole, or electric quadrupole) of the transition.

REFERENCES

1. M. Born and J. R. Oppenheimer, Ann. Phys. 84, 457 (1927).
2. M. Born and V. Fock, Zeit. f. Phys. 51, 165 (1928); also T. Kato, J. Phys. Soc. Jap. 5, 435 (1950).
3. W. Heitler, Quantum Theory of Radiation, Oxford University Press, London and New York (1944).
4. G. N. Fowler, in Quantum Theory, Vol. III (D. R. Bates, ed.), Academic Press, New York and London, p. 47 (1962).
5. J. Franck, Trans. Faraday Soc. 21, 536 (1925).
6. E. U. Condon, Phys. Rev. 32, 858 (1928).
7. L. Pauling and E. B. Wilson, Introduction to Quantum Mechanics, McGraw-Hill, New York and London, p. 265 (1935).

INTRODUCTION TO MOLECULAR SPECTROSCOPY

D. A. Ramsay

National Research Council of Canada

Ottawa, Ontario, Canada

ABSTRACT

After a brief historical introduction, the theory of electronic spectra is discussed. The classification of electronic states and electronic selection rules is given. Energy formulae and selection rules for the vibrational structure and rotational fine structure accompanying an electronic transition are also given. The molecular orbitals for AH_2, HAB and AB_2 molecules are discussed with the aid of Walsh diagrams. The spectra which have been observed for these molecules are summarized together with the molecular geometries which have been derived from the analyses of the optical spectra.

I. HISTORICAL INTRODUCTION

The first example of the rotational analysis of the electronic spectrum of a polyatomic molecule was that carried out by Dieke and Kistiakowsky (1934) on the near ultraviolet bands of formaldehyde. This work was followed by extensive investigations at Chicago on the electronic spectra of CO_2^+, CS_2, SO_2, and ClO_2. Although it was suspected that CS_2 was nonlinear in its first excited state, the first molecule for which it was definitely established that electronic excitation is accompanied by a change in molecular symmetry was acetylene. Ingold and King (1953) first showed, and Innes (1954) later independently confirmed, that the molecule in the upper state of the near ultraviolet bands has a nonlinear trans-configuration. Several examples have now been found of molecules for which large changes in molecular geometry accompany electronic excitation. Thus, HCN and CS_2 are linear in their ground states and nonlinear in their first excited states; the free radicals NH_2 and HCO are nonlinear in their ground states and linear in their first excited states. Altogether geometries are known for more than 50 molecules in excited electronic states (Herzberg 1966).

II. THEORY OF ELECTRONIC SPECTRA

II.A. Electronic States

1. Classification. The electronic states of a polyatomic molecule may be classified according to (1) the resultant electron spin S, and (2) the symmetry species of the molecule. The quantity 2S+1 is referred to as the multiplicity of the state, and is indicated in the upper left hand corner of the symbol representing the state, e.g. $^1\Sigma$ (singlet sigma), $^2\Pi$ (doublet pi). Molecules with even numbers of electrons give rise to singlet, triplet, ... electronic states, while molecules with odd numbers of electrons give rise to doublet, quartet, ... states. To a first approximation the selection rule $\Delta S=0$ is valid, i.e., only states of the same multiplicity combine with each other. This rule holds fairly rigorously for light molecules but is less rigorous for molecules containing heavier atoms.

The various point groups describing the possible types of symmetry in polyatomic molecules have been discussed by many authors and will not be repeated here. Instead, we shall illustrate the principles involved by means of an example. Let us consider the water molecule (Fig. 1) which belongs to the point group C_{2v}. The symmetry elements are the twofold axis of symmetry lying in the z-axis, and the xz- and yz-planes of symmetry. The symmetry operations of the group are (i) the identity operation, I, included for the sake of completeness, (ii) rotation by 180° about the z-axis denoted by $C_2(z)$, (iii) reflection across the xz-plane, $\sigma_v(xz)$, and (iv) reflection across the yz-plane, $\sigma_v(yz)$. The electronic eigenfunction can either remain unchanged (symmetric) or change sign (antisymmetric) when subjected to each of these operations. The four possible symmetry species forming the C_{2v} group, and their symmetry properties are given in Table I, where "+" stands for symmetric and "-" for antisymmetric.

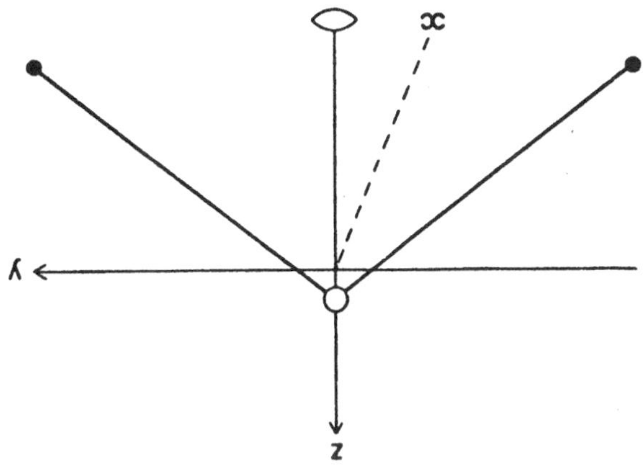

Fig. 1. Principal axes for the water molecule (•: hydrogen; O: oxygen). The x-axis lies perpendicular to the plane of the paper.

2. Electronic Selection Rules. The electronic transition probability between two states depends on the square of the matrix elements of the dipole moment $|\int \psi_e'^* M \psi_e'' d\tau_e|^2$. The quantities ψ_e' and ψ_e'' are the electronic wave functions for the upper and lower states, respectively, and the asterisk ($\psi_e'^*$) denotes the complex conjugate value. M is the dipole moment of the system of nuclei and electrons, and the integration is carried out for all configurations of the electrons while the

TABLE I

Species Table for Symmetry C_{2v}

C_{2v}	I	$C_2(z)$	$\sigma_v(xz)$	$\sigma_v(yz)$
A_1	+1	+1	+1	+1
A_2	+1	+1	-1	-1
B_1	+1	-1	+1	-1
B_2	+1	-1	-1	+1

nuclei are kept fixed. Herzberg and Teller (1933) showed that the integral differs from zero only if the direct product of the symmetry species of the two combining states contains a term which transforms like one of the components of the dipole moment M_x, M_y, or M_z (see Table II). For example, for a molecule with C_{2v} symmetry,

$$B_1(+-+-) \times B_1(+-+-) = A_1(++++)$$

and

$$B_1(+-+-) \times B_2(+--+) = A_2(++--)$$

From Table II we see that a transition $B_1 \leftrightarrow B_1$ is allowed and has a transition moment parallel to the z-axis. On the other hand, a transition between a B_1- and a B_2-state is forbidden since the direct product, A_2, does not have a component of the dipole moment in any of the x-, y-, or z-directions. Frequently, a gross change of molecular geometry accompanies an electronic transition, and in

TABLE II

Symmetry Species of the Components of the Dipole Moment For Some of the More Important Point Groups

	C_s	C_{2h}	C_{2v}	C_{3v}	$C_{\infty v}$	D_{2h}	D_{3h}	D_{6h}	$D_{\infty h}$
M_x	A"	B_u	B_1	E	Π	B_{3u}	E'	E_{1u}	Π_u
M_y	A'	B_u	B_2	E	Π	B_{2u}	E'	E_{1u}	Π_u
M_z	A'	A_u	A_1	A_1	Σ^+	B_{1u}	A_2''	A_{2u}	Σ_u^+

some cases, the two combining states belong to different point groups; in such cases, the electronic selection rules are determined by considering the point group containing the common elements of symmetry. The magnitude of the electronic transition probability depends on the above integral, and is usually found to be large only when the electron configurations for the two combining states differ in the configuration of a single electron. Transitions which are allowed by the electronic selection rules, but which involve double (or more) electron jumps, usually have very low transition probabilities and behave like "forbidden" transitions.

II.B. Vibrational Structure

1. Vibrational Energy Formulae. Due to the vibrational and rotational motions in polyatomic molecules, an electronic transition consists, in general, of a number of bands, in contrast to an atomic transition which gives rise to a single line (neglecting multiplet structure). The wave-numbers of the band origin are given by the equation

$$\nu = \nu_e + G'(v_1, v_2, \ldots) - G''(v_1, v_2, \ldots) \qquad (1)$$

and are commonly expressed in cm^{-1}. The quantity ν_e is known as the origin of the electronic transition and represents the frequency of the line which would be obtained in the absence of vibrational and rotational motions. The quantities $G'(v_1, v_2 \ldots)$ and $G''(v_1, v_2 \ldots)$ are, respectively, the vibrational term values for the upper and lower electronic states. These term values may be expressed to a first approximation by the equation

$$G(v_1, v_2, \ldots) = \sum_i \nu_i \left(v_i + \frac{d_i}{2} \right) \qquad (2)$$

where ν_i is the vibration frequency in cm^{-1}, v_i the vibrational quantum number, and d_i the degeneracy of the ith mode of vibration ($d_i = 1$ for a nondegenerate vibration). The summation extends over the (3N-6) normal modes of vibration (N = number of atoms in the molecule) for a nonlinear polyatomic molecule or the (3N-5) normal modes of vibration for a linear polyatomic molecule. A more complete expression for the vibrational term values, including the effects of the anharmonicities of the vibrations is

$$G(v_1, v_2, \ldots) = \sum_i \omega_i \left(v_i + \frac{d_i}{2} \right) + \sum_i \sum_{k \geq i} x_{ik} \left(v_i + \frac{d_i}{2} \right) \left(v_k + \frac{d_k}{2} \right) + \sum_i \sum_{k \geq i} g_{ik} \ell_i \ell_k \ldots \qquad (3)$$

where the ω_i's are the zero-order frequencies, and the x_{ik}'s are the anharmonicity constants. The last term in (3) is a small correction term which is applied when one or more degenerate vibrations are excited.

As an example of the vibrational structure of an electronic transition, let us consider the ultraviolet absorption spectrum of HCN (Fig. 2). The principal bands in the $\tilde{A}$-$\tilde{X}$ system arise from the lowest vibrational level (0,0,0) of the ground state. Transitions have been observed to vibrational levels in the excited state involving the CN stretching frequency ν_3, the bending frequency ν_2, and combinations of these frequencies. No transitions have yet been observed to levels involving the CH stretching frequency ν_1. The origins of the principal bands are given by the formula (Herzberg and Innes, 1957)

$$\nu = 52{,}277.7_0 + 1505._9 v_3' - 10._0 v_3'^2 + 948.53 v_2' - 7.39 v_2'^2 - 0.52 v_2'^3 - 6.27 v_2' v_3'. \qquad (4)$$

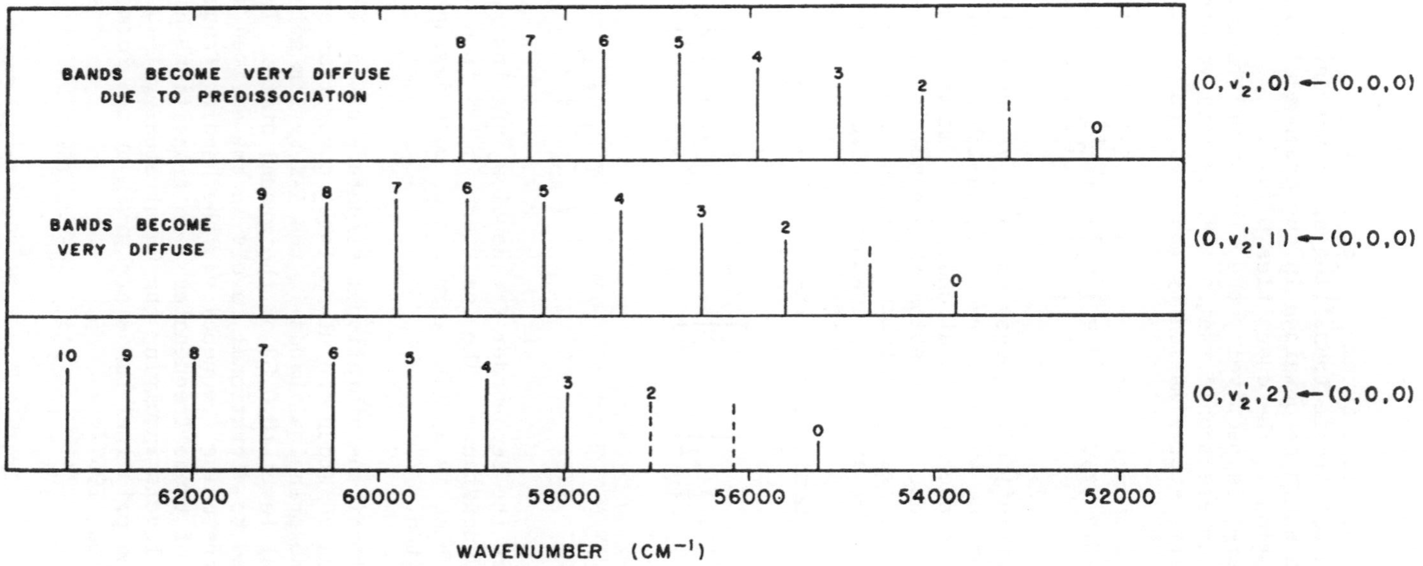

Fig. 2. Schematic diagram of the principal progressions in the ultraviolet absorption system of HCN.

For DCN, the corresponding formula is

$$\nu = 52{,}404.7_0 + 1505.8v_3' - 14.2v_3'^2 + 735.0_4v_2' - 4.20_5v_2'^2 - 0.29v_2'^3 - 6.94v_2'v_3' + 0.6v_2'v_3'^2. \quad (5)$$

It may be noted that the CN stretching frequency ν_3' does not change appreciably ($1505._9$ $cm^{-1} \rightarrow 1505.8$ cm^{-1}) when H is replaced by D, whereas the H—C—N bending frequency ν_2' decreases roughly in the ratio $\sqrt{2}:1$ (948.53 $cm^{-1} \rightarrow 735.0_4$ cm^{-1}). This example illustrates the usefulness of isotopic data in assigning vibration frequencies.

Some of the weaker bands originate in the $v_2'' = 1$, and $v_2'' = 2$ levels of the bending vibration in the ground state and thus give information concerning these lower state term values. Bands arising from vibrationally excited levels of the ground state, usually referred to as "hot" bands, may be readily identified by studying the absorption spectrum at various temperatures. The intensities of the "hot" bands are much more sensitive to variation of temperature than are the intensities of the principal bands, since the numbers of molecules in the various vibrationally excited levels of the ground state depend on the appropriate Boltzmann factors, $\exp[-hcG''(v_1, v_2 \ldots)/kT]$.

In general, we see that most of the vibrational data derived from an absorption spectrum pertain to the upper state, though some of the vibrational data may refer to the lower state. From an emission spectrum, it is generally possible to obtain vibrational data concerning both electronic states, depending on the mechanism of excitation employed. It should be pointed out, however, that it is frequently difficult to excite the emission spectra of polyatomic molecules, due to predissociations in many of the excited states.

2. Franck-Condon Principle. **The intensity distribution of the vibrational bands accompanying an electronic transition is governed by the Franck-Condon principle. According to the simple classical picture of Franck (1925) the nuclear positions and velocities do not change appreciably during an electronic rearrangement. Hence, if the molecule is considered to be at rest in its initial state, the equilibrium configuration of the nuclei in this state will correspond to the turning point of the vibration excited in the final state. For example, if an electronic transition is accompanied by a large change in the length of one bond in the molecule, bands will be excited which involve the stretching vibration of this bond. Similarly, if an electronic transition involves a large change in a particular bond angle in the two combining states,the bands which will be excited will be those involving the appropriate bending vibration.**

On the basis of the Franck model, one would expect to observe only a few bands in an emission or an absorption spectrum, even when the transition is accompanied by a large change in molecular geometry. Experimentally, however, it is found that under the latter circumstances, long progressions of bands are observed, i.e., long series of bands involving the excitation of successive quanta of a particular vibration. The quantum number of the strongest band in a given progression is governed approximately by the Franck principle, but the intensity distribution in the progression can only be explained satisfactorily on a wave-mechanical basis, as was first shown by Condon (1928).

The intensities of the various vibrational bands of an electronic transition are proportional to the squares of the matrix elements

$$\left|\int\int \psi_e'^* \psi_v'^* M \psi_e'' \psi_v'' d\tau_e d\tau_v\right|^2$$

where $\psi_v'^*$ and ψ_v'' refer to vibrational wavefunctions and $d\tau_v$ denotes integration with respect to the coordinates of the nuclei. If the electronic transition probability does not vary appreciably during the vibrations of the molecule, the above integral may be rewritten as a product

$$\left|\int \psi_e'^* M \psi_e'' d\tau_e\right|^2 \cdot \left|\int \psi_v'^* \psi_v'' d\tau_v\right|^2 .$$

The relative intensities of the vibrational bands are then proportional to

$$\left|\int \psi_v'^* \psi_v'' d\tau_v\right|^2 .$$

When this quantity is summed over all the vibrational levels the result is unity, i.e., the intensity of an electronic transition summed over all the vibrational bands depends only on the electronic transition probability. It is well known that the overlap integrals

$$\left|\int \psi_v'^* \psi_v'' d\tau_v\right|^2$$

give an adequate interpretation of the vibrational intensity distributions observed in the spectra of diatomic molecules, but for polyatomic molecules few comparisons have been made. For progressions involving totally symmetric stretching vibrations the results are expected to be essentially similar to those for diatomic molecules, but for progressions involving the non-totally symmetric stretching vibrations or the bending vibrations of the molecule, the electronic transition probability may vary appreciably with molecular vibration. Under these circumstances the factorization carried out above would no longer be valid and the overlap integrals would not give the correct intensity distributions.

As an example of the intensity distribution in an electronic spectrum we shall again consider the ultraviolet bands of HCN. A schematic diagram showing the relative intensities of the bands in the principal progressions $(0,v_2',0) \leftarrow (0,0,0)$, $(0,v_2',1) \leftarrow (0,0,0)$, and $(0,v_2',2) \leftarrow (0,0,0)$ is given in Fig. 2. First, we note that long progressions of the bending vibration are excited in the upper state, and that the maximum intensity in any progression is near $v_2' = 7$. On the basis of the Franck-Condon principle we should expect a large change in the equilibrium values of the H—C—N angle in the two states, and this expectation is indeed confirmed by the results of rotational analysis ($\angle$ H—C—N = 125° for the upper state and 180° for the lower state). Second, we note that the bands of the $(0,v_2',1) \leftarrow (0,0,0)$ progression are more intense than the corresponding bands of the $(0,v_2',0) \leftarrow (0,0,0)$ and $(0,v_2',2) \leftarrow (0,0,0)$ progressions. This observation suggests that the electronic transition is accompanied by an appreciable change in the CN bond length. The results of rotational analysis show that the CN bond length is 1.156 Å in the ground state and 1.297 Å in the excited state.

3. Vibrational Selection Rules. From the integral $\int \psi_v'^* \psi_v'' d\tau_v$ certain vibrational selection rules may be derived. These rules have been discussed by Herzberg and Teller (1933) and may be briefly summarized as follows:

(1) Vibrations which are symmetric with respect to all the symmetry elements in a molecule, may change by any number of quanta.
(2) Vibrations which are antisymmetric with respect to some symmetry element may change only by $\Delta v = 0,2,4, \ldots$

In the latter case the transition with $\Delta v = 0$ is very much stronger than the transitions with $\Delta v \neq 0$. If ν' and ν'' are the frequencies of a particular non-totally symmetric vibration in the two electronic states, then for simple harmonic vibrations

$$\frac{\text{Intensity of 0-0 band}}{\text{Intensity of all the bands v-0}} = \frac{(\nu'\nu'')^{1/2}}{1/2(\nu' + \nu'')} \tag{6}$$

where $v = 0,2,4, \ldots$ If $\nu' = \nu''$ then all the intensity is concentrated in the (0,0) band and none in the (v,0) bands. Even in an extreme case when $\nu' = 1/2\nu''$, the intensity of the (0,0) band accounts for 94.3% of the total intensity of the progression, leaving only 5.7% for the (2,0), (4,0), . . . bands.

II.C. Rotational Structure

1. Rotational Energy Levels. The rotational energy levels of a polyatomic molecule may be expressed in terms of the moments of inertia I_A, I_B, and I_C of the molecule about the three principal axes of inertia. It is always assumed that $I_C \geq I_B \geq I_A$. The principal axes pass through the center of gravity of the molecule and are mutually perpendicular. Several special cases arise. If the three moments of inertia are unequal, the molecule is referred to as an *asymmetric top*. If two of the moments of inertia are equal ($I_A = I_B \neq I_C$, or $I_A \neq I_B = I_C$) the molecule is called a *symmetric top*. If all three moments of inertia are equal ($I_A = I_B = I_C$) the molecule is referred to as a *spherical top*. We shall not, however, consider spherical top molecules any further, since no electronic spectra of spherical top molecules have yet been analyzed. *Linear molecules* may be regarded as limiting cases of symmetric top molecules for which $I_A = 0$ and $I_B = I_C$.

For linear molecules the rotational term values F(J) in cm^{-1}, neglecting the effects of electron and nuclear spin, are given by

$$F(J) = B_v[J(J + 1) - K^2] - D_v[J(J + 1) - K^2]^2 + \ldots \quad (7)$$

where J is the quantum number of the total angular momentum, and K is the quantum number of the resultant angular momentum about the axis. The rotational constant B is given by

$$B = \frac{h}{8\pi^2 c I_B} = \frac{27.9932 \times 10^{-40}}{I_B} \quad (8)$$

where I_B is in gm.cm^2. If I_B is expressed in atomic mass units $\times$ Å^2, the numerical constant in Eq. (8) is 16.8576. The term

$$D_v[J(J + 1) - K^2]^2$$

in eq. (7) is a small correction term due to the centrifugal distortion of the molecule when it rotates. For a diatomic molecule, assuming the harmonic oscillator approximation, D is given by the Kratzer relation,

$$D = \frac{4B^3}{\omega^2} \quad (9)$$

where ω is the vibrational frequency. For linear polyatomic molecules similar though slightly more complicated expressions apply.

For symmetric top molecules the rotational term values F(J,K) are given by

$$F(J,K) = B_v^{\perp}[J(J+1) - K^2] + B_v^{\parallel}[K^2 - 2K\sum_i(\pm\zeta_i \ell_i)]$$
$$- D_v^{J}J^2(J+1)^2 - D_v^{JK}J(J+1)K^2 - D_v^{K}K^4 + \ldots \qquad (10)$$

where J is the quantum number of the total angular momentum, and K is the quantum number of the resultant angular momentum about the symmetry axis. $B^{\perp}$ and $B^{\parallel}$ are the rotational constants corresponding to the moments of inertia perpendicular to and parallel to the symmetry axis. The term

$$\sum_i (\pm\zeta_i \ell_i)$$

represents the contributions from the vibrational angular momenta of the degenerate vibrations.

For asymmetric top molecules the rotational term values $F(J_{K_aK_c})$ may be written

$$F(J_{K_aK_c}) = \frac{1}{2}(A + C)J(J+1) + \frac{1}{2}(A - C)E^{J}_{K_aK_c}(\kappa) + \ldots \qquad (11)$$

where

$$\kappa = \frac{2B - A - C}{A - C}\,.$$

The quantities $E^{J}_{K_aK_c}(\kappa)$ have been tabulated by King, Hainer, and Cross (1943) up to $J = 12$ in intervals of 0.1 for κ. Nowadays these quantities are usually obtained by computer diagonalisation of asymmetric rotor matrices.

2. <u>Symmetry Properties</u>. The rotational levels of a linear polyatomic molecule are called <u>positive</u> or <u>negative</u> depending upon whether the total eigenfunction ψ remains unaltered or changes its sign when all the particles (electrons and nuclei) are reflected at the center of mass, i.e., all coordinates x_i, y_i, and z_i are replaced by $-x_i$, $-y_i$, and $-z_i$. In addition, for linear molecules with a center of symmetry, the sign of the total eigenfunction, ψ, may or may not change sign when all the nuclei on one side of the center of symmetry are simultaneously exchanged with the identical nuclei on the other side. The corresponding rotational levels are called <u>symmetric</u> (s), or <u>antisymmetric</u> (a). The symmetry properties of the rotational levels of a linear polyatomic molecule are formally identical with those for a corresponding diatomic molecule, provided that the electronic symmetry species of the latter is replaced by the vibronic symmetry species of the former. The symmetry properties for the more important types of singlet state are given in Fig. 3.

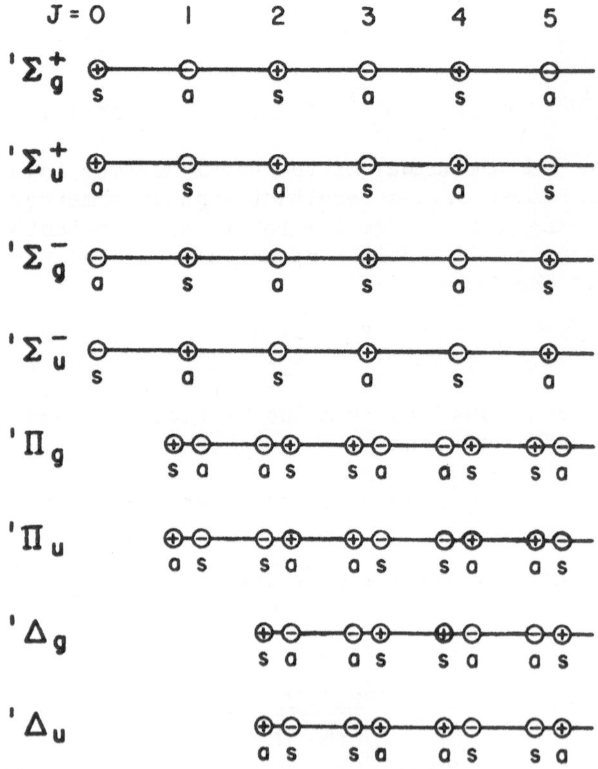

Fig. 3. Symmetry properties for the rotational levels of linear polyatomic molecules. The group symbols refer to the vibronic species of the molecule. (+) denotes positive and (-) negative rotational levels; s stands for symmetric, and a for antisymmetric rotational levels. For molecules which do not have a center of symmetry, the g and u and the s and a symbols may be deleted.

For symmetric top molecules the rotational levels may be classified as positive or negative in the same manner as for a linear polyatomic molecule. The appropriate classifications have been given by Herzberg (1966) for planar and nonplanar molecules. In the latter case an additional twofold degeneracy is present since the molecule can exist in two modifications which can be converted into each other only by inversion through a potential barrier. Each level is split into a positive and a negative component, a process known as "inversion doubling". In most molecules, however, the barrier to inversion is high, and the inversion doublets are not resolved.

For asymmetric top molecules the rotational energy levels may be classified in the same manner as for linear and symmetric top molecules, i.e., by means of the symmetry properties of the total eigenfunction. It is more usual, however, to classify the levels according to the symmetry properties of the rotational eigenfunction ψ_r. Since the moment of inertia of the molecule about any axis through the center of mass can be represented by an ellipsoid, the rotational eigenfunction ψ_r must either remain unchanged or change sign when the molecule is rotated by 180° about one of the axes of the ellipsoid. If a, b, and c are the axes of least, intermediate, and greatest moment of inertia respectively, then the rotational levels of an asymmetric top may be distinguished by their behavior (+ or -) with respect to the three operations C_2^a, C_2^b, and C_2^c. It is sufficient to specify the behavior with respect to two of these operations, usually C_2^c and C_2^a, since the behavior with respect to the third is then automatically determined. In addition for asymmetric top molecules with a twofold axis of symmetry, the sign of the rotational eigenfunction may either remain unchanged (symmetric), or it may change (antisymmetric) when pairs of identical nuclei are exchanged by a rotation of 180° about the twofold axis.

3. Rotational Selection Rules. The following selection rules rigorously restrict the rotational transitions permitted in an allowed electric dipole transition:

$$\Delta J = 0, \pm 1, \text{ with the restriction } J = 0 \not\leftrightarrow J = 0. \qquad (12)$$

$$+ \leftrightarrow -, \qquad + \not\leftrightarrow +, \qquad - \not\leftrightarrow -. \qquad (13)$$

$$s \leftrightarrow s, \qquad a \leftrightarrow a, \qquad s \not\leftrightarrow a. \qquad (14)$$

Other selection rules which may further limit the number of rotational transitions permitted for special types of transitions are:

a.) Linear molecules and symmetric top molecules. If the transition moment is parallel to the axis for a linear molecule, or to the top axis for a symmetric top molecule, (|| band), we have

$$\Delta K = 0 \ , \ \Delta J = \pm 1 \qquad \text{if } K = 0 \qquad (15)$$

$$\Delta K = 0 \ , \ \Delta J = 0 \ , \ \pm 1 \qquad \text{if } K \neq 0 \ . \qquad (16)$$

If the transition moment is perpendicular to the above axes (⊥ band), we have

$$\Delta K = \pm 1 \ , \ \Delta J = 0 \ , \ \pm 1 \ . \qquad (17)$$

b.) Asymmetric top molecules. The rotational selection rules are determined by the direction of the transition moment with respect to the a-, b-, and c-axes of the molecule.

If the transition moment lies in the a-axis, we get a Type A band for which the only transitions permitted are

$$++ \leftrightarrow -+ \text{ and } +- \leftrightarrow -- \ . \tag{18}$$

If the transition moment lies in the b-axis, the appropriate selection rules (Type B band) are

$$++ \leftrightarrow -- \text{ and } +- \leftrightarrow -+ \ . \tag{19}$$

Finally, if the transition moment lies in the c-axis the selection rules (Type C band) are

$$++ \leftrightarrow +- \text{ and } -+ \leftrightarrow -- \ . \tag{20}$$

In some molecules, the transition moment may not be confined to the a-, b-, or c-axis in which case a hybrid band results.

III. MOLECULAR ORBITALS AND SPECTRA OF TRIATOMIC MOLECULES

According to simple theory the electrons in a molecule may be assigned to various orbitals each of which is characterized by a binding energy and a set of symmetry properties. The total energy of the molecule is equal to the sum of the binding energies of the individual electrons. Such a model is clearly an oversimplification since it neglects the interactions between the electrons and, in particular, the effects of exchange forces. Provided these limitations are recognized, however, the simple theory can be very useful for predicting the energies, symmetry properties, and shapes of molecules in their various electronic states. The theory has been developed to varying degrees by several workers, notably by Mulliken and his school and by Walsh (1953). We shall consider here the approach used by the latter author, since the treatment is basically simple and, moreover, it has been very successful both in explaining and predicting the characteristics of the spectra of polyatomic molecules.

III.A. AH_2 Molecules

The lowest energy molecular orbitals for AH_2 molecules may be formed by overlapping the s- and p-atomic orbitals in the valence shell of the A atom with the 1s-orbitals of the H atoms. Initially, we shall consider two cases: (1) a linear AH_2 molecule, and (2) a bent AH_2 molecule with an angle of 90°. The appropriate axes are given in Fig. 4. It is convenient to consider the orbitals of the H atoms as a group, the two possible group orbitals being formed by the in-phase addition (1s+1s) and out-of-phase addition (1s-1s) of the atomic orbitals. In addition, for the bent (90°) molecule, it is convenient to replace the p-orbitals of A which initially may be considered to lie along the AH bond directions, by two equivalent orbitals p_y and p_z lying parallel to the y- and z-directions, respectively. The components from which the molecular orbitals can be constructed and their symmetry properties are given in the tabulation.

Linear molecule		Bent molecule	
A atom	H_2 group	A atom	H_2 group
s_A , σ_g	(1s+1s) , σ_g	s_A , a_1	(1s+1s) , a_1
p_y , σ_u	(1s-1s) , σ_u	p_z , a_1	(1s-1s) , b_2
p_x,p_z , π_u		p_y , b_2	
		p_x , b_1	

Molecular orbitals may be formed by mixing components with the same symmetry. For the linear molecule, the procedure is straightforward, but for the bent molecule, Walsh introduces the assumption that in the 90° configuration the geometry is determined by the directional properties of the p-valencies on the A atom and that the s_A-orbital does not mix with the other orbitals of the same symmetry. The molecular orbitals for the linear and 90° configurations may be written:

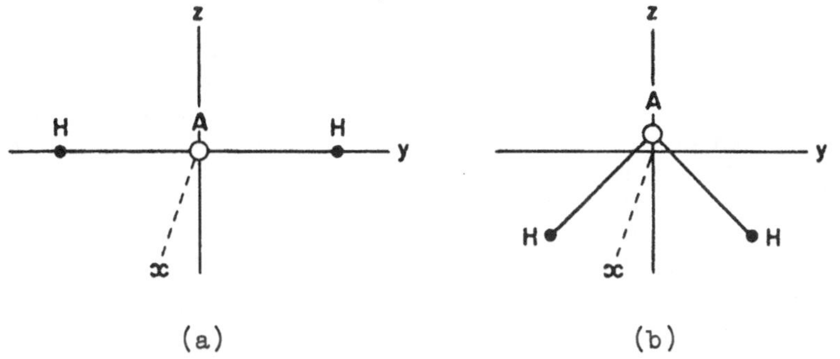

Fig. 4. Coordinate axes for (a) linear and (b) bent (90°) AH_2 molecules. The x-axes are perpendicular to the plane of the paper.

Linear molecule		Bent Molecule (90°)	
$s_A + (1s+1s)$,	σ_g →	$p_z + (1s+1s)$	, a_1
$p_y + (1s-1s)$,	σ_u →	$p_y + (1s-1s)$	, b_2
	↗	s_A	, a_1
p_x , p_z ,	π_u		
	↘	p_x	, b_1
$s_A - (1s+1s)$,	$\bar{\sigma}_g$ →	$p_z - (1s+1s)$	, $\bar{a}_1$
$p_y - (1s-1s)$,	$\bar{\sigma}_u$ →	$p_y - (1s-1s)$	, $\bar{b}_2$

The correlations between the orbitals of the linear and bent configurations may be determined by consideration of the symmetry properties and are shown above. It now remains to decide the order of the energies of the various orbitals and the variation in the energy of a particular orbital with valence angle. Estimates of the relative binding energies are based on the principles: (i) s-electrons are more tightly bound than p-electrons, and (ii) the binding energy increases with increasing overlap of the component orbitals. On this basis it is seen that the order given above will approximate the order of decreasing binding energy downwards. The last two orbitals in both cases are antibonding orbitals and have been distinguished from the other orbitals by including a bar over the symbol for the orbital.

Estimates of the variation of binding energy with angle are based on two principles enunciated by Walsh: (i) "whether or not an orbital becomes more tightly bound with change of angle is determined primarily by whether or not it changes from being built from a p-orbital of A to being built from an s-orbital of A," and (ii) "if the orbital is antibonding between the end atoms it is most tightly bound when the latter are as far apart as possible (i.e., in the linear molecule)" and vice versa. The second effect is considered by Walsh to be subsidiary to the first effect. On this basis the predictions given in the tabulation below can be made concerning the changes in binding energy of the individual orbitals in going from the linear to the 90° configuration.

		Variation in binding energy with decrease of angle	
Linear	Bent	First effect	Second effect
σ_g	$\rightarrow a_1$	Decrease	Increase
σ_u	$\rightarrow b_2$	No change	Decrease
π_u	$\nearrow a_1$	Increase	No change
	$\searrow b_1$	No change	No change
$\bar{\sigma}_g$	$\rightarrow \bar{a}_1$	Decrease	Increase
$\bar{\sigma}_u$	$\rightarrow \bar{b}_2$	No change	Decrease

The orbital diagram constructed by Walsh on the basis of these principles is given in Fig. 5. The antibonding orbitals are not included since they correlate with Rydberg orbitals and lie at fairly high energies.

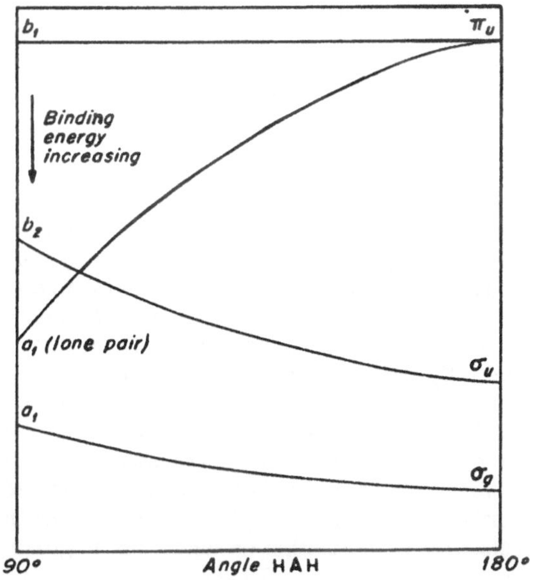

Fig. 5. Orbital diagram for AH_2 molecules (from A.D. Walsh, 1953).

We note that electrons in the a_1-σ_g and b_2-σ_u orbitals tend to stabilize the molecule in the linear configuration while electrons in the a_1-π_u orbital strongly favor a bent configuration. Electrons in the b_1-π_u orbital have little influence on the valence angle. Experimentally, it is found that molecules with no electrons in the a_1-π_u orbital are linear. With one electron in the a_1-π_u orbital, AH_2 molecules have angles between 130° and 145° for first row atoms, and between 123° and 127° for second row atoms. With two electrons in the a_1-π_u orbital the corresponding angles are 102-110° and 91-93° respectively (Table III). We will now consider individual molecules.

1. Five Electrons: BH_2, AlH_2. Both molecules have been identified in flash photolysis studies and rotational analyses of the BH_2 bands have been carried out. The transition is of the type

$$(a_1)^2(b_2)^2(b_1),\ \tilde{A}^2B_1 - (a_1)^2(b_2)^2(a_1),\ \tilde{X}^2A_1$$

and involves the ground state in which the molecule is bent with an angle of 131°, and an excited state with a linear configuration. Both electronic states are derived from a common $^2\Pi$ state.

2. Six Electrons: CH_2, SiH_2. Two transitions of the CH_2 radical have been observed in absorption. The first lies in the region 6000-9000 Å and is of the type

$$(a_1)^2(b_2)^2(a_1)(b_1),\ \tilde{b}^1B_1 - (a_1)^2(b_2)^2(a_1)^2,\ \tilde{a}^1A_1$$

while the second lies in the vacuum ultraviolet region. The latter system involves a different lower state and is believed to be triplet-triplet in character although no triplet splittings have been resolved. The system is extensively predissociated and was originally interpreted as arising from a linear↔linear transition. However, esr studies (Wasserman, Yager and Kuck 1970) indicated that the CH_2 radical is bent in its lowest triplet state. The optical spectrum was therefore reinterpreted on the basis of a linear↔bent transition and the angle in the lower state shown to be 136° (Herzberg and Johns 1971). This angle is consistent with predictions from the simple Walsh diagram. It is generally assumed that the lowest triplet state ($\tilde{X}^3B_1$) is the ground state of the molecule but it should be remembered that no rigorous proof of this has yet been given. It should also be noted that the two states involved in the visible system both arise from a common $^1\Delta_g$ state in the linear configuration. The correlation between the states of linear and bent CH_2 is:

Linear		Bent	
$(\sigma_g)^2(\sigma_u)^2(\pi_u)^2$,	$^1\Sigma_g^+$	$(a_1)^2(b_2)^2(b_1)^2$	, $^1A_1$
	$^1\Delta_g$	$(a_1)^2(b_2)^2(a_1)(b_1)$	, $^1B_1$, $^3B_1$
	$^3\Sigma_g^-$	$(a_1)^2(b_2)^2(a_1)^2$	, $^1A_1$

The three low-lying observed states are underlined. The $\tilde{b}^1B_1$state has been assumed to have a linear configuration but it seems more likely that it is quasilinear with an angle in the range 130-145°.

One band system of SiH_2 is known and is the analogue of the visible bands of CH_2. Since it is the only system that has so far been seen in absorption, it is tacitly assumed that the $^1A_1$ lower state is the ground state for SiH_2.

3. Seven Electrons: NH_2, PH_2, H_2O^+, H_2S^+. All the molecules in this group have a transition of the type

$$(a_1)^2(b_2)^2(a_1)(b_1)^2,\ \tilde{A}^2A_1 - (a_1)^2(b_2)^2(a_1)^2(b_1),\ \tilde{X}^2B_1 .$$

The two combining states correlate with a common $^2\Pi$ state in the linear configuration. The molecules are strongly bent in their ground states but have much larger angles in their excited states. Indeed in the higher bending levels of the excited state the molecules behave like linear molecules. The rotational structures of the bands are rather complex since the molecules are very asymmetric tops in their ground states. Furthermore in the excited states there is a strong interaction between the vibrational angular momenta associated with the bending vibrations and the electronic angular momentum associated with the fact that the excited state is derived from a Π state. The consequences of this type of interaction, which were discussed by Renner (1934) and first found experimentally in the spectrum of NH_2 by Dressler and Ramsay (1959), are known as the Renner effect or Renner-Teller interaction.

4. Eight Electrons: H_2O, H_2S. In the ground states of these molecules all the low-lying valence orbitals are filled. The first transitions in absorption lie in the vacuum ultraviolet region and involve the excitation of a valence electron to a Rydberg-type orbital.

It is interesting to compare the angles in the ground states of these molecules with the angles in states of related molecules such that the differences in the electron configurations involve

TABLE III

MOLECULAR ORBITALS AND GEOMETRIES FOR AH_2 MOLECULES[a]

Molecule	State	$a_1-\sigma_g$	$b_2-\sigma_u$	$sa_1-\pi_u$	$b_1-\pi_u$	$\angle$ HAH	r_o(AH)
BeH_2 (4)	$\tilde{X}\ ^1\Sigma_g^+$	2	2	0	0	(180°)[b]	-
BH_2 (5)	$\tilde{X}\ ^2A_1$	2	2	1	0	131°	1.18
	$\tilde{A}\ ^2B_1(\Pi)$	2	2	0	1	180°	1.17
CH_2 (6)	$\tilde{X}\ ^3B_1$	2	2	1	1	136°	1.07_8[c]
	$\tilde{a}\ ^1A_1$	2	2	2	0	102.4°	1.11
	$\tilde{b}\ ^1B_1$	2	2	1	1	∿180°?	1.05_6
NH_2 (7)	$\tilde{X}\ ^2B_1$	2	2	2	1	103.4°	1.024
	$\tilde{A}\ ^2A_1(\Pi)$	2	2	1	2	144°	1.00_4
H_2O^+ (7)	$\tilde{X}\ ^2B_1$	2	2	2	1	110.5°	0.999[d]
H_2O (8)	$\tilde{X}\ ^1A_1$	2	2	2	2	105.2°	0.956
SiH_2 (6)	$\tilde{X}\ ^1A_1$	2	2	2	0	92.1°	1.516[e]
	$\tilde{A}\ ^1B_1(\Pi)$	2	2	1	1	123°	1.48_7[e]
PH_2 (7)	$\tilde{X}\ ^2B_1$	2	2	2	1	91.7°	1.418[f]
	$\tilde{A}\ ^2A_1(\Pi)$	2	2	1	2	123.2°	1.389[f]
H_2S^+ (7)	$\tilde{X}\ ^2B_1$	2	2	2	1	92.9	1.35_8[g]
	$\tilde{A}\ ^2A_1(\Pi)$	2	2	1	2	∿127°	1.36_9[g]
H_2S (8)	$\tilde{X}\ ^1A_1$	2	2	2	2	92.2°	1.328

a. From Herzberg (1966) except where otherwise stated.
b. Predicted but not yet confirmed experimentally.
c. Herzberg and Johns (1971). d. Lew and Heiber (1973)
e. Dubois (1968). f. Berthou, Pascat, Guenebaut and Ramsay (1972).
g. Duxbury, Horani and Rostas (1972).

only differences in the number of electrons in the non-bonding b_1-π_u orbital:

H_2O,	$\tilde{X}^1A_1$	105.2°	H_2S,	$\tilde{X}^1A_1$	92.2°
H_2O^+,	$\tilde{X}^2B_1$	110.5°	H_2S^+,	$\tilde{X}^2B_1$	92.9°
NH_2,	$\tilde{X}^2B_1$	103.4°	PH_2,	$\tilde{X}^2B_1$	91.7°
CH_2,	$\tilde{a}^1A_1$	102.4°	SiH_2,	$\tilde{X}^1A_1$	92.1°

It is seen that the regularities in these series are quite striking. The only molecule which shows a small discrepancy is H_2O^+; however it appears that the angle in the ground state of this ion is well determined experimentally.

III.B. HAB Molecules

Nine molecular orbitals for an HAB molecule may be formed from the 1s-atomic orbital of the H atom and the s- and p-atomic orbitals in the valence shells of the A and B atoms. As before, we shall first consider two special cases, viz., (1) a linear HAB molecule (Fig. 6a), and (2) a bent HAB molecule with an angle of 90° (Fig. 6b).

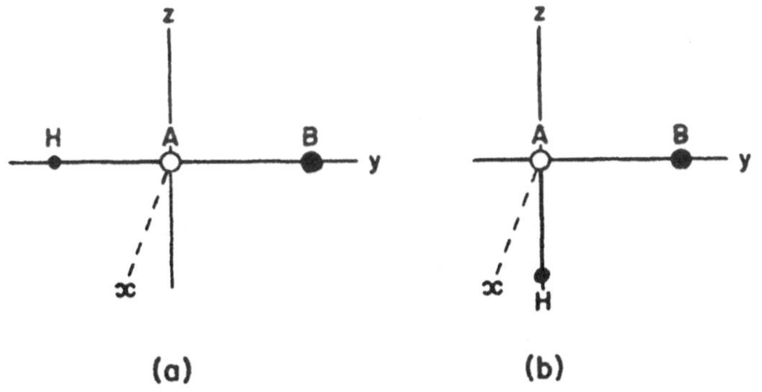

Fig. 6. Coordinate axes for (a) linear and (b) bent (90°) HAB molecules. The x-axes are perpendicular to the plane of the paper.

For the linear molecule the approximate molecular orbitals in order of decreasing binding energy are:

(1) An s-orbital on B. It is assumed that this plays no direct part in the bonding of A to B and remains virtually

unchanged whatever the HAB angle.

(2) A σ-orbital bonding the H and the A atoms. Since the xz-plane is not a plane of symmetry, the s- and p_y-atomic orbitals of A can mix giving sp_y-hybrid orbitals. The σ-orbital between H and A is formed by overlapping an sp_y-hybrid orbital of A with the H 1s-orbital.

(3) A σ-orbital bonding the A and B atoms, formed by overlapping an sp_y-hybrid orbital of A with a p_y-atomic orbital of B.

(4), (5) A π-orbital bonding the A and B atoms, formed by in-phase overlap of the pπ-atomic orbitals on A and B. Since in many HAB molecules, B is of greater electronegativity than A, this orbital is usually more localized on B than on A.

(6), (7) A $\bar{\pi}$-orbital which is antibonding between A and B, and is formed by out-of-phase overlap of the pπ-atomic orbitals on A and B. This orbital is usually more localized on A than on B.

(8) A $\bar{\sigma}$-orbital which is antibonding between H and A and is formed by the out-of-phase overlap of the 1s atomic H orbital and an sp_y-hybrid orbital of the A atom.

(9) A $\bar{\sigma}$-orbital which is antibonding between A and B and is formed by the out-of-phase overlap of an sp_y-hybrid orbital of A and a p_y-orbital of B.

For the nonlinear molecule the order of the binding energies of the orbitals is less well-defined. The approximate order may be established, however, by correlating the orbitals of the linear and bent molecules. If the HAB angle is 90°, the orbitals may be described approximately as follows:

(1) An a'-orbital due to the s-lone-pair on B.

(2) An a'-orbital bonding the H and A atoms, formed from the H 1s-orbital and a p_z-orbital of A.

(3) An a'-orbital bonding A and B, built from A p_y and B p_y.

(4) An a"-orbital bonding A and B, formed from the in-phase overlap of the p_x-orbitals on A and B.

(7) An $\bar{a}$"-orbital which is antibonding between A and B and is built by the out-of-phase overlap of the p_x-orbitals on A and B.

(8) An $\bar{a}$'-orbital which is antibonding between H and A and is formed by the out-of-phase overlap of an H 1s-orbital and a p_z-orbital of A.

(9) An $\bar{a}$'-orbital which is antibonding between A and B and is built by the out-of-phase overlap of p_y-atomic orbitals on A and B.

In addition, there are two further a'-orbitals corresponding to the lone-pair s-orbital on A, and the lone-pair p_z-orbital on B. These may be correlated with orbitals (6) and (5), respectively, in the linear molecule since usually (6) is more localized on A and (5) is more localized on B. The molecular orbitals for the linear and 90° configurations are summarized in Table IV, together with the correlations between the two sets of orbitals.

The changes in the binding energies of the various orbitals with change of angle may be estimated using the principles introduced by Walsh and discussed for AH_2 molecules. Since there is no change in the component atomic orbitals used in constructing molecular orbitals (1), (4), and (7) for the linear and bent configurations, no change in binding energy with change of angle is to be expected. The molecular orbitals (2), (3), (8), and (9) are built from sp-hybrid orbitals of A in the linear configuration, and from p-orbitals of A in the 90° configuration. These orbitals will therefore be more tightly bound in the linear configuration. Orbital (5) is AB bonding in the linear configuration and AB nonbonding in the 90° configuration; it will be more tightly binding, therefore, in the linear configuration. Orbital (6) is AB antibonding in the linear configuration and nonbonding in the 90°

TABLE IV

MOLECULAR ORBITALS FOR LINEAR AND BENT (90°) HAB MOLECULES[a]

	Linear H	Linear A	Linear B		Bent (90°) H	Bent (90°) A	Bent (90°) B		Variation in binding energy with decrease of angle
1.	0	+ 0	+ s,	σ →	0	+ 0	+ s,	a'	No change
2.	s	+ sp_y	+ 0,	σ →	s	+ p_z	+ 0,	a'	Decrease
3.	0	+ sp_y	+ p_y,	σ →	0	+ p_y	+ p_y,	a'	Decrease
4.	0	+ p_x	+ p_x }	→	0	+ p_x	+ p_x,	a"	No change
5.	0	+ p_z	+ p_z },	π →	0	+ 0	+ p_z,	a'	Decrease
6.	0	+ p_z	− p_z }	→	0	+ s	+ 0,	a'	Large increase
7.	0	+ p_x	− p_x },	$\bar{\pi}$ →	0	+ p_x	− p_x,	$\bar{a}$"	No change
8.	s	− sp_y	+ 0,	$\bar{\sigma}$ →	s	− p_z	+ 0,	$\bar{a}$'	Decrease
9.	0	+ sp_y	− p_y,	$\bar{\sigma}$ →	0	+ p_y	− p_y,	$\bar{a}$'	Decrease

[a]The symbol + denotes in-phase overlap and - denotes out-of-phase overlap.

configuration. In addition it is formed from a p-atomic orbital of A in the linear configuration and from an s-atomic orbital of A in the 90° configuration. The binding energy, therefore, will be much greater in the 90° than in the linear configuration. These conclusions are incorporated in the orbital diagram given by Walsh and reproduced in Fig. 7. [Walsh does not include the highest antibonding orbital (9)]. According to this diagram the only

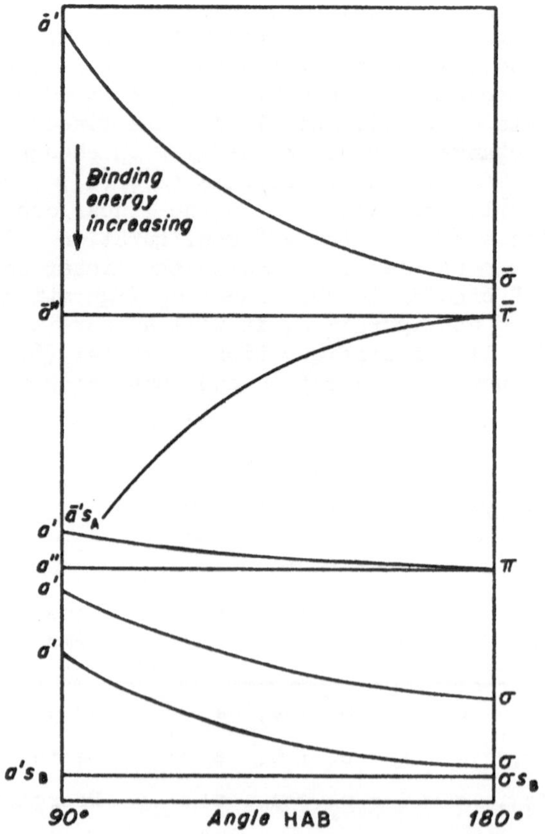

Fig. 7. Orbital diagram for HAB molecules (from A.D. Walsh, 1953).

orbital with a tendency to produce a bent molecule is the sixth orbital ($a's_A-\bar{\pi}$). As we shall see the value of the HAB angle is a sensitive function of the number of electrons in this orbital (Table V).

1. <u>Ten Electrons:</u> <u>HCN, HCP</u> . These molecules are linear in their ground states but bent in their first singlet excited states. The transitions may be written

$$\cdots\ (a'')(a')^2(a's_A),\ \tilde{A}^1A'' \leftarrow \cdots\ (\pi)^4,\ \tilde{X}^1\Sigma_g^+ \ .$$

In addition, transitions are expected in which an electron is promoted from the $a'-\pi$ orbital to the $a's_A-\bar{\pi}$ orbital, but such excited states have not yet been identified experimentally.

2. Eleven Electrons: HCO. The ground state of HCO should show a similar structure to the above excited state of HCN, since the two electronic configurations differ only in the presence of the number of electrons in the a"-π orbital, which according to Fig. 7 should have little effect on the bond angle. Indeed the correspondence is quite remarkable; the recently revised value for the ground state angle of HCO is 124.95° [(±0.25°, Brown and Ramsay, 1975)] whereas the angle in the excited state of HCN is 125.0° (Herzberg and Innes, 1957). In the first absorption system of HCO the unpaired electron in the $a's_A-\bar{\pi}$ orbital is promoted to the $\bar{a}''-\bar{\pi}$ orbital, producing an excited state with a linear configuration.

3. Twelve Electrons: HNO, HPO, HCF, HCCl, HSiCl. For all these molecules the first absorption system is of the type

$$\cdots\cdots (a's_A)(\bar{a}''),\ \tilde{A}^1A'' \leftarrow \cdots\cdots (a's_A)^2,\ \tilde{X}^1A'\ .$$

The molecules are bent in both states and in each case there is an increase in the bond angle in going from the ground to the excited state, in agreement with the predictions implicit in Fig. 7.

4. Thirteen Electrons: HO_2, HS_2, HNF. The first transition for this type of molecule should be of the type

$$\cdots\cdots (a's_A)(\bar{a}'')^2,\ \tilde{A}^2A' - \cdots\cdots (a's_A)^2(\bar{a}''),\ \tilde{X}^2A''$$

and should be at relatively long wavelengths. Indeed an absorption system of HNF has been found by Woodman (1970) with its 0-0 band near 5000 Å. The corresponding transitions for HO_2 and HS_2 have not yet been identified although recently there has been a suggestion that HO_2 has an absorption system in the infrared region of the spectrum. An absorption system of HS_2 is known (Porter 1950) but consists of diffuse bands and probably corresponds to a $\bar{\pi}-\pi$ type transition.

Summarizing the experimental results for HAB molecules, we note that electron configurations with no electrons in the $a's_A-\bar{\pi}$ orbital give electronic states in which the molecules are linear. With one or two electrons in the $a's_A-\bar{\pi}$ orbital the following angles have been found for HAB molecules in which A and B are both first-row atoms:

TABLE V
MOLECULAR ORBITALS AND GEOMETRIES FOR HAB MOLECULES[a]

		σ	σ	σ	π		$\bar{\pi}$				
		a'	a'	a'	a"	a'	$a's_A$	$\bar{a}$"	r_o(AH)	r_o(AB)	∠HAB
HCN	$\tilde{X}\ {}^1\Sigma^+$	2	2	2	2	2			1.064	1.156	180
(10)	$\tilde{A}\ {}^1A''$	2	2	2	1	2	1		1.140	1.297	125.0
HCP	$\tilde{X}\ {}^1\Sigma^+$	2	2	2	2	2			1.067	1.542	180
(10)	$\tilde{A}\ {}^1A''$	2	2	2	1	2	1		(1.14)	1.69	128
HCO	$\tilde{X}\ {}^2A'$	2	2	2	2	2	1		1.12_5	1.175	124.9_5[b]
(11)	$\tilde{A}\ {}^2A''(\Pi)$	2	2	2	2	2		1	1.064	1.186	180[b]
HNO	$\tilde{X}\ {}^1A'$	2	2	2	2	2	2		1.06_3	1.212	108.6
(12)	$\tilde{A}\ {}^1A''$	2	2	2	2	2	1	1	1.03_6	1.241	116.3
HPO	$\tilde{X}\ {}^1A'$	2	2	2	2	2	2		(1.433)	1.512	104.7
(12)											
HCF	$\tilde{X}\ {}^1A'$	2	2	2	2	2	2		(1.121)	1.314	101.8
(12)	$\tilde{A}\ {}^1A''$	2	2	2	2	2	1	1	(1.121)	1.297	127.2
HCCl	$\tilde{X}\ {}^1A'$	2	2	2	2	2	2		1.12	1.68_9	103.4
(12)	$\tilde{A}\ {}^1A''$	2	2	2	2	2	1	1			134
HSiCl	$\tilde{X}\ {}^1A'$	2	2	2	2	2	2		1.561	2.064	102.8
(12)	$\tilde{A}\ {}^1A''$	2	2	2	2	2	1	1	1.49_9	2.04_7	116.1
HNF	$\tilde{X}\ {}^2A''$	2	2	2	2	2	2	1	(1.06)	1.37	105[c]
(13)	$\tilde{A}\ {}^2A'$	2	2	2	2	2	1	2	(1.03)	1.34	125[c]

a. From Herzberg (1966) except where otherwise stated; b. Brown and Ramsay (1975); c. Woodman (1970). Numbers in parentheses denoted assumed values.

1 electron			2 electrons		
HCN	$\tilde{A}^1A''$	125.0°	HNO	$\tilde{X}^1A'$	108.6°
HCO	$\tilde{X}^2A'$	124.95°	HCF	$\tilde{X}^1A'$	101.8°
HNO	$\tilde{A}^1A''$	116.3°	HNF	$\tilde{X}^2A''$	105°
HCF	$\tilde{A}^1A''$	127.2°			
HNF	$\tilde{A}^2A'$	125°			

The values lie within the ranges 116-128° and 101-109° respectively. When A or B are second-row atoms the values are somewhat different but the same general trends apply.

III.C. AB_2 and BAC Molecules

Twelve molecular orbitals may be constructed for a BAC molecule from the s- and p-atomic orbitals in the valence shells of the A, B, and C atoms. We shall consider first, however, the symmetrical AB_2 molecule in both linear and bent (90°) configurations. The coordinate axes are the same as those shown in Fig. 4 for AH_2 molecules. It is convenient to consider the orbitals of the B atoms as a group, the possible group orbitals being formed by the in-phase and out-of-phase addition of the atomic orbitals.

For a linear AB_2 molecule the molecular orbitals may be approximately described as follows:

(1),(2) Lone pair s-atomic orbitals on each of the B atoms. Strictly speaking we should consider in-phase and out-of-phase combinations of these orbitals giving σ_g- and σ_u-molecular orbitals. The energies of these orbitals would not be expected to change appreciably with B—A—B angle.

(3) A σ_g-orbital bonding the A and B atoms formed by overlapping the s-orbital of A with the $(p_y + p_y)$ group orbital of B_2.

(4) A σ_u-orbital bonding the A and B atoms, formed from the p_y-orbital of A and the $(p_y - p_y)$ group orbital of B_2.

(5),(6) A π_u-orbital formed by the in-phase overlap of the $p\pi$-orbitals on the A and B atoms. If B is more electronegative than A this orbital will be more localized on the B than on the A atoms.

(7),(8) A π_g-orbital formed by the out-of-phase overlap of the $p\pi$-atomic orbitals on the B atoms. It has zero amplitude on the A atom and consequently is nonbonding between A and B.

(9),(10) A $\bar{\pi}_u$-orbital which is antibonding between A and B and is formed by the out-of-phase overlap of the $p\pi$-orbitals of A with the in-phase combinations of the corresponding orbitals of the B atoms. If B is more electronegative than A, this orbital will be more localized on the A than on the B atoms.

(11),(12) A $\bar{\sigma}_g$- and a $\bar{\sigma}_u$-orbital which are both antibonding between A and B and which are formed by the out-of-phase overlap of s_A and $(p_y + p_y)_{B_2}$ and of $(p_y)_A$ and $(p_y - p_y)_{B_2}$, respectively.

The above order will, in general, correspond to the order of decreasing binding energy; in some molecules, however, the $\bar{\sigma}_g$-orbital may lie below the $\bar{\pi}_u$-orbital.

For a nonlinear AB_2 molecule with an angle of 90° we shall assume as before that only the p-orbitals of the A atom are involved in bonding and that there is no mixing with the s-orbitals of A. The p-orbitals on the A and B atoms will initially be considered to lie either along or perpendicular to the AB bond directions. The p-orbitals on the B atoms lying along the bond directions can be added both in- and out-of-phase to give group orbitals which will be denoted by $(p + p)_1$ and $(p - p)_1$, and which have symmetry species a_1 and b_2, respectively. Similarly the p-orbitals of the A atom lying in the BAB plane may be replaced by equivalent orbitals p_z and p_y having the symmetry species a_1 and b_2 respectively. The molecular orbitals for the bent (90°) molecule can now be described briefly as follows and will be correlated with the corresponding orbitals of the linear molecule:

(1),(2) The lone-pair s-orbitals on the B atoms as discussed above. These give rise to an a_1- and a b_2-molecular orbital.

(3),(4) Two orbitals, a_1 and b_2, bonding the A and B atoms and formed by the overlapping of the p-orbitals of A and B lying along the bond directions. These orbitals may be written $(p_z)_A + (p + p)_1$ and $(p_y)_A + (p - p)_1$.

(5) A b_1-orbital formed by the in-phase overlap of the p_x-orbitals on the A and B atoms.

(6), (8) Two orbitals a_1 and b_2 formed by the in-phase and out-of-phase overlap of the p-orbitals on the B atoms, lying in the molecular plane and perpendicular to the bond directions. These orbitals will be written $(p + p)_2$ and $(p - p)_2$ and correlate with the π_u- and π_g-orbitals, respectively, in the linear molecule.

(7) An a_2-orbital formed by out-of-phase overlap of the p_x-orbitals on the B atoms. It has zero amplitude on the central A atom.

(9) A $\bar{b}_1$-orbital formed by overlapping the p_x-orbitals of the B atoms in-phase with the p_x-orbital of A out-of-phase.

(10) The lone-pair s-orbital on the A atom. This a_1-orbital is localized on the A atom and correlates with the $\bar{\pi}_u$-orbital of the linear molecule.

(11),(12) Two orbitals $\bar{a}_1$ and $\bar{b}_2$, which are antibonding between A and B and which are formed from the out-of-phase combinations $(p_z)_A - (p + p)_1$ and $(p_y)_A - (p - p)_1$ respectively [cf. (3) and (4)].

The orbitals for the linear and bent (90°) molecules are summarized in Table VI together with the correlations between the orbitals. The changes in the binding energies of the orbitals

TABLE VI

MOLECULAR ORBITALS FOR LINEAR AND BENT (90°) AB_2 MOLECULES[a]

Linear			Bent		Variation of binding energy with decrease of angle	
A	B_2		A	B_2	1st Effect	2nd Effect
1. 0	+ (s + s),	σ_g →	0	+ (s + s), a_1	No effect	No effect
2. 0	+ (s - s),	σ_u →	0	+ (s - s), b_2	No effect	No effect
3. s	+ (p_y+p_y),	σ_g →	p_z	+ $(p + p)_1$, a_1	Decrease	Increase
4. p_y	+ (p_y-p_y),	σ_u →	p_y	+ $(p - p)_1$, b_2	No effect	Decrease
5. p_x	+ (p_x+p_x) }	π_u →	p_x	+ (p_x+p_x), b_1	No effect	Increase
6. p_z	+ (p_z+p_z) },	π_u →	0	+ $(p + p)_2$, a_1	Decrease	Increase
7. 0	+ (p_x-p_x) }	π_g →	0	+ (p_x-p_x), a_2	No effect	Decrease
8. 0	+ (p_z-p_z) },	π_g →	0	+ $(p - p)_2$, b_2	No effect	Decrease
9. p_x	- (p_x+p_x) }	$\bar{\pi}_u$ →	p_x	- (p_x+p_x), $\bar{b}_1$	No effect	Increase
10. p_z	- (p_z+p_z) },	$\bar{\pi}_u$ →	s	+ 0, a_1	Increase	Increase
11. s	- (p_y+p_y),	$\bar{\sigma}_g$ →	p_z	- (p + p), $\bar{a}_1$	Decrease	Increase
12. p_y	- (p_y-p_y),	$\bar{\sigma}_u$ →	p_y	- $(p - p)_1$, $\bar{b}_2$	No effect	Decrease

[a]The symbol + denotes in-phase overlap and - denotes out-of-phase overlap.

with change of B—A—B angle may be estimated using the two principles given by Walsh and are also given in Table VI. An orbital diagram summarizing these results is given in Fig. 8. For BAC molecules the distinction between a_1 and b_2, and between a_2 and b_1 disappears, and we are left with two symmetry classes only, viz., a' and a", respectively. For convenience the primes and double primes have been added to the species symbols given in Fig. 8 following the procedure used by Walsh.

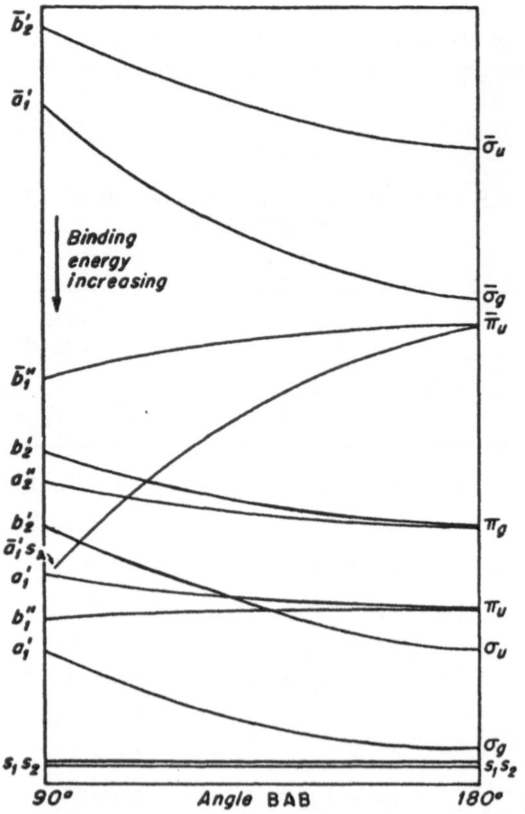

Fig. 8. Orbital diagram for AB_2 molecules (from A.D. Walsh 1953).

The orbital with the strongest tendency to produce a bent molecule is the $a_1's_A-\bar{\pi}_u$ orbital. Other orbitals with weaker tendencies to favor bent configurations are the $b_1''-\pi_u$ and $\bar{b}_1''-\bar{\pi}_u$ orbitals. Experimentally it is found that molecules always have linear structures until electrons occupy the $a_1's_A$ orbital. Then the value of the BAB (or BAC) angle is a sensitive function of the number of electrons in this orbital. The molecular geometries for AB_2 molecules are summarized in Table VII; no data are given for BAC molecules since to determine the individual bond lengths it is necessary to carry out an isotopic substitution and this has usually not been done.

In the following, g and u labels are included on the assumption that AB_2 molecules are being discussed; for BAC molecules these labels should be deleted. Also it should be noted that the order of the bonding σ_u and π_u orbitals sometimes changes from molecule to molecule.

1. Twelve Electrons: C_3 . The ground state of the C_3 molecule is expected to be a $^1\Sigma_g^+$-state with the configuration $(s_1)^2(s_2)^2(\sigma_g)^2(\sigma_u)^2(\pi_u)^4$. The lowest excited states should have the following electron configurations and term types:

$$\cdot\cdot\ (\sigma_g)^2(\sigma_u)^2(\pi_u)^3(\pi_g);\ {}^{1,3}\Delta_u,\ {}^{1,3}\Sigma_u^-,\ {}^{1,3}\Sigma_u^+$$

$$\cdot\cdot\ (\sigma_g)^2(\sigma_u)(\pi_u)^4(\pi_g);\ {}^1\Pi_u,\ {}^3\Pi_u.$$

The 4050 Å band system of C_3 may very reasonably be assigned to the transition $^1\Pi_u-{}^1\Sigma_g^+$. This transition involves the promotion of an electron from a weakly bonding σ_u orbital to a nonbonding π_g orbital and should be accompanied by a small increase in bond length, in agreement with observation [r'(CC) = 1.305 Å., r"(CC) = 1.277 Å.].

2. Thirteen Electrons: CNC and CCN . These two species have been observed in flash photolysis studies. The observed transitions arise from a $\pi_g \leftarrow \sigma_u$ excitation which gives the following states:

$$\cdot\cdot\ (\sigma_g)^2(\pi_u)^4(\sigma_u)(\pi_g)^2,\ {}^2\Sigma_u^+,\ {}^2\Sigma_u^-,\ {}^2\Delta_u\ \text{—}\ \cdot\cdot\ (\sigma_g)^2(\pi_u)^4(\sigma_u)^2(\pi_g),\ \tilde{X}^2\Pi_g.$$

All three excited states have been observed for CCN but so far only the $^2\Sigma_u^-$ and $^2\Delta_u$ excited states for CNC have been found. There is a small increase in the bond length in going from the ground to the excited states.

3. Fourteen Electrons: NCN and CCO . Both molecules have a $^3\Pi-{}^3\Sigma^-$ band system which arises from the transition

$$\cdot\cdot(\sigma_g)^2(\pi_u)^4(\sigma_u)(\pi_g)^3,\ \tilde{A}^3\Pi_u\ \text{—}\ \cdot\cdot(\sigma_g)^2(\pi_u)^4(\sigma_u)^2(\pi_g)^2,\ \tilde{X}^3\Sigma_g^-\ .$$

In addition a $^1\Pi_u-{}^1\Delta_g$ band system for NCN has been observed and can be assigned to the same excitation $(\pi_g \leftarrow \sigma_u)$.

4. Fifteen Electrons: CO_2^+, CS_2^+, BO_2, N_3, NCO, NCS, N_2O^+, COS^+. All molecules in this class have a $^2\Pi_i$ ground state arising from the configuration

$$\cdot\cdot\ (\sigma_g)^2(\pi_u)^4(\sigma_u)^2(\pi_g)^3,\ \tilde{X}^2\Pi_g(\text{inv})$$

TABLE VII

MOLECULAR ORBITALS AND GEOMETRIES FOR AB_2 MOLECULES[a]

		σ_g	π_u	σ_u	π_g	r_o(AB)	∠BAB
C_3 (12)	$\tilde{X}\ ^1\Sigma_g^+$	2	4	2		1.277	180°
	$\tilde{A}\ ^1\Pi_u$	2	4	1	1	1.305	180°
CNC (13)	$\tilde{X}\ ^2\Pi_g$	2	4	2	1	1.245	180°
	$\tilde{A}\ ^2\Delta_u$	2	4	1	2	1.249	180°
	$\tilde{B}\ ^2\Sigma_u^-$	2	4	1	2	1.259	180°
NCN (14)	$\tilde{X}\ ^3\Sigma_g^-$	2	4	2	2	1.232	180°
	$\tilde{A}\ ^3\Pi_u$	2	4	1	3	1.233	180°
BO_2 (15)	$\tilde{X}\ ^2\Pi_g$	2	4	2	3	1.265	180°
	$\tilde{A}\ ^2\Pi_u$	2	3	2	4	1.302	180°
	$\tilde{B}\ ^2\Sigma_u^+$	2	4	1	4	1.273	180°
N_3 (15)	$\tilde{X}\ ^2\Pi_g$	2	4	2	3	1.181	180°
	$\tilde{B}\ ^2\Sigma_u^+$	2	4	1	4	1.180	180°
CO_2^+ (15)	$\tilde{X}\ ^2\Pi_g$	2	4	2	3	1.177	180°
	$\tilde{A}\ ^2\Pi_u$	2	3	2	4	1.228	180°
	$\tilde{B}\ ^2\Sigma_u^+$	2	4	1	4	1.180	180°
CS_2^+ (15)	$\tilde{X}\ ^2\Pi_g$	2	4	2	3	1.554	180°
	$\tilde{B}\ ^2\Sigma_u^+$	2	4	1	4	1.564	180°

TABLE VII (continued)

		σ_g	π_u	σ_u	π_g		$\bar{\pi}_u$			
		a_1	b_1 a_1	b_2	a_2	b_2	$a_1 s_A$	b_1	r_o(AB)	∠BAB
CO_2 (16)	$\tilde{X}$ $^1\Sigma_g^+$	2	4	2	4				1.162	180°
	$\tilde{A}$ $^1B_2$	2	4	2	2	1	1		1.246	122°
CS_2 (16)	$\tilde{X}$ $^1\Sigma_g^+$	2	4	2	4				1.554	180°
	$\tilde{a}$ $^3A_2$	2	4	2	1	2	1		1.64	135.8°
NO_2 (17)	$\tilde{X}$ $^2A_1$	2	4	2	2	2	1		1.193	134.1°
	$\tilde{A}$ $^2B_1$	2	4	2	2	2		1		180°
	$\tilde{B}$ $^2B_2$	2	4	2	2	1	2		1.314	121.0°
SO_2 (18)	$\tilde{X}$ $^1A_1$	2	4	2	2	2	2		1.432	119.5°
	$\tilde{a}$ $^3B_1$	2	4	2	2	2	1	1	1.494	126.1°
	$^1A_2$	2	4	2	2	1	2	1	1.53	99°[b]
	$^1B_2$	2	4	2	1	2	2	1	1.560	104.3°[c]
CF_2 (18)	$\tilde{X}$ $^1A_1$	2	4	2	2	2	2		1.300	104.9°[d]
	$\tilde{A}$ $^1B_1$	2	4	2	2	2	1	1	1.32	122.3°[d]
ClO_2 (19)	$\tilde{X}$ $^2B_1$	2	4	2	2	2	2	1	1.473	117.6°
	$^2A_2$	2	4	2	1	2	2	2	1.619	107.0°[e]

a. From Herzberg (1966) except where otherwise stated.

b. Hamada and Merer (1974).

c. Brand and Srikameswaren (1972).

d. Mathews (1967).

e. Brand, Redding and Richardson (1970).

and two excited states arising from the configurations

$$\cdot\cdot\ (\sigma_g)^2(\pi_u)^4(\sigma_u)(\pi_u)^4,\ {}^2\Sigma_u^+$$

$$\cdot\cdot\ (\sigma_g)^2(\pi_u)^3(\sigma_u)^2(\pi_g)^4,\ {}^2\Pi_u\ \text{(inv)}\,.$$

The transitions ${}^2\Sigma_u^+-\tilde{X}^2\Pi_g$ and ${}^2\Pi_u-\tilde{X}^2\Pi_g$ are both allowed, but the transition between the two excited states is forbidden ($u \nleftrightarrow u$) except for unsymmetrical BAC molecules. The energies of the excited states are usually in the region of 30,000 cm^{-1} (±10,000 cm^{-1}). For BO_2, CO_2^+ and NCS the ${}^2\Sigma^+$ state lies above the ${}^2\Pi$ while the converse is true for NCO. For N_3, CS_2^+ and N_2O^+ only the ${}^2\Sigma^+$ state has been found; for COS^+ only the ${}^2\Pi$ state is known.

5. <u>Sixteen Electrons CO_2, CS_2, COS, N_2O</u> . All the molecules are linear in their ${}^1\Sigma_g^+$ ground states which arise from the configuration

$$\cdot\cdot\ (\sigma_g)^2(\pi_u)^4(\sigma_u)^2(\pi_g)^4,\ \tilde{X}^1\Sigma_g^+\,.$$

For CO_2 the first singlet excited state is bent ($\angle$OCO = 122°) and for CS_2 the angle in the first excited triplet state is 135.8°. The electron configurations are given in Table VII. Both excited states have one electron in the $a_1 s_A-\bar{\pi}_u$ orbital, the orbital with the strongest tendency to produce a bent molecule.

6. <u>Seventeen Electrons: NO_2</u>. NO_2 is nonlinear in its ground ${}^2A_1$ state. The first excited state arises from the promotion of an electron from the $a_1 s_A-\bar{\pi}_u$ orbital to the $\bar{b}_1-\bar{\pi}_u$ orbital and has a linear equilibrium structure

$$\cdot\cdot\ (a_2)^2(b_2)^2(a_1 s_A)^0(\bar{b}_1),\ \tilde{A}^2B_1\ —\ \cdot\cdot\ (a_2)^2(b_2)^2(a_1 s_A),\ \tilde{X}^2A_1\,.$$

Other excited states arise from the configurations

$$\cdot\cdot\ (a_2)^2(b_2)(a_1 s_A)^2,\ \tilde{B}^2B_2$$

$$\cdot\cdot\ (a_2)(b_2)^2(a_1 s_A)^2,\ {}^2A_2\,.$$

The $\tilde{B}^2B_2$ excited state is responsible for absorption in the 2500 Å region and has a smaller angle (121°) than the ground state (134.1°) in agreement with expectations. The transition to the ${}^2A_2$ excited state from the ground state is forbidden.

7. Eighteen Electrons: O_3, SO_2, S_2O, CF_2, SiF_2, FNO, ClNO. The electron configurations for the ground state and for the first few excited states are

$$\cdots (a_2)^2(b_2)^2(a_1s_A)^2 \quad , \tilde{X}^1A_1$$
$$\cdots (a_2)^2(b_2)^2(a_1s_A)(\bar{b}_1) \quad , {}^1B_1,\ {}^3B_1$$
$$\cdots (a_2)^2(b_2)(a_1s_A)^2(\bar{b}_1) \quad , {}^1A_2,\ {}^3A_2$$
$$\cdots (a_2)(b_2)^2(a_1s_A)^2(\bar{b}_1) \quad , {}^1B_2,\ {}^3B_2 .$$

The angles for these molecules in their ground states are:

O_3	SO_2	S_2O	CF_2	SiF_2	FNO	ClNO
116.8°	119.5°	118.0°	104.9°	101.0°	110°	104.2°

and are all considerably smaller than the angle in the ground state of NO_2.

Molecular geometries are known for the $^3B_1$, $^1A_2$ and $^1B_2$ excited states of SO_2; data are also available for the $^1B_1$ excited state of CF_2 (Table VII).

8. Nineteen Electrons: ClO_2, NF_2 . The ground state and first three excited states have the following electron configurations and term types:

$$\cdots (a_2)^2(b_2)^2(a_1s_A)^2(\bar{b}_1) \quad , \tilde{X}^2B_1$$
$$\cdots (a_2)^2(b_2)^2(a_1s_A)(\bar{b}_1)^2 \quad , {}^2A_1$$
$$\cdots (a_2)^2(b_2)(a_1s_A)^2(\bar{b}_1)^2 \quad , {}^2B_2$$
$$\cdots (a_2)(b_2)^2(a_1s_A)^2(\bar{b}_1)^2 \quad , {}^2A_2 .$$

The visible bands of ClO_2 have been assigned to the transition $^2A_2$-$\tilde{X}^2B_1$ (Brand, Redding and Richardson 1970). The bond length is considerably longer in the excited state which is consistent with a $\bar{\pi}\leftarrow\pi$ excitation, and the angle is smaller (107.0° ← 117.6°).

It is interesting now to summarize the data for AB_2 molecules and to correlate them with the orbital diagram given in Fig. 8. First we will consider the electronic states of molecules with one and two electrons in the a_1s_A-$\bar{\pi}_u$ orbital:

1 electron				2 electrons			
CO_2	$\tilde{A}$	$^1B_2$	122°	NO_2	$\tilde{B}$	$^2B_2$	121.0°
NO_2	$\tilde{X}$	$^2A_1$	134.1°	CF_2	$\tilde{X}$	$^1A_1$	104.9°
CF_2	$\tilde{A}$	$^1B_1$	122.3°	SO_2	$\tilde{X}$	$^1A_1$	119.5°
CS_2	$\tilde{a}$	$^3A_2$	135.8°			$^1A_2$	99°
SO_2	$\tilde{a}$	$^3B_1$	126.1°			$^1B_2$	104.3°
				ClO_2	$\tilde{X}$	$^2B_1$	117.6°
						$^2A_2$	107.0°

The ranges are not as closely defined as for AH_2 and HAB molecules but, nevertheless, for one and two electrons in the a_1s_A-$\bar{\pi}_u$ orbital the angles lie in the ranges 122-136° and 99-121° respectively.

Next we will consider the changes of bond length which accompany various types of excitation:

	$\pi_g \leftarrow \sigma_u$	$\pi_g \leftarrow \pi_u$		$\bar{\pi}_u \leftarrow \pi_g$
C_3	+0.028 Å		CO_2	+0.084 Å
CNC	+0.004 Å		CS_2	+0.086 Å
	+0.014 Å		NO_2	+0.121 Å
NCN	+0.001 Å		SO_2	+0.098 Å
N_3	-0.001 Å			+0.128 Å
BO_2	+0.008 Å	+0.037 Å		
CO_2^+	+0.003 Å	+0.051 Å		
CS_2^+	+0.010 Å			

The change in bond length accompanying a $\pi_g \leftarrow \sigma_u$ excitation (non-bonding ← weakly bonding) is small (-0.001 to +0.028 Å). For a $\pi_g \leftarrow \pi_u$ excitation (non-bonding ← bonding) the change is larger (0.037 to 0.051 Å). For a $\bar{\pi}_u \leftarrow \pi_g$ excitation (antibonding ← non-bonding) the change is still larger (0.084 to 0.128 Å). All these results are consistent with the predictions of simple theory.

REFERENCES

Berthou, J.M., Pascat, B., Guenebaut, H. and Ramsay, D.A., 1972. Can. J. Phys. 50, 2265.

Brand, J.C.D., Redding, R.W. and Richardson, A.W., 1970. J. Mol. Spectry. 34, 399.

Brand, J.C.D. and Srikameswaren, K., 1972. Chem. Phys. Letters 15, 310.

Brown, J.M. and Ramsay, D.A., 1975. Can. J. Phys. In Press.

Condon, E.U., 1928. Phys. Rev. 32, 858.

Dieke, G.H. and Kistiakowsky, G.B., 1934. Phys. Rev. 45, 4.

Dressler, K. and Ramsay, D.A.,1959. Phil. Trans. Roy. Soc. A251, 553.

Dubois, I., 1968. Can. J. Phys. 46, 2485.

Duxbury, G., Horani, M. and Rostas, J., 1972. Proc. Roy. Soc. (London) A331, 109.

Franck, J., 1925. Trans. Faraday Soc. 21, 536.

Hamada, Y. and Merer, A.J., 1974. Can. J. Phys. 52, 1443.

Herzberg, G. and Teller, E., 1933. Z. Physik. Chem. (Leipzig) B21, 410.

Herzberg, G. and Innes, K.K., 1957. Can. J. Phys. 35, 842.

Herzberg, G., 1966. Electronic Spectra of Polyatomic Molecules, D. Van Nostrand Co. Inc., Princeton, N.J., U.S.A.

Herzberg, G. and Johns, J.W.C., 1971. J. Chem. Phys. 54, 2276.

Ingold, C.K. and King, G.W., 1953. J. Chem. Soc. pp. 2702-2755.

Innes, K.K., 1954. J. Chem. Phys. 22, 863.

King, G.W., Hainer, R.M. and Cross, P.C., 1943. J. Chem. Phys. 11, 27.

Lew, H. and Heiber, I., 1973. J. Chem. Phys. 58, 1246.

Mathews, C.W., 1967. Can. J. Phys. 45, 2355.

Porter, G., 1950. Disc. Faraday Soc. 9, 60.

Renner, R., 1934. Z. Physik. 92, 172.

Walsh, A.D., 1953. J. Chem. Soc. pp. 2260-2331.

Wasserman, E., Yager, W.A. and Kuck, V.J., 1970. Chem. Phys. Letters 7, 409.

Woodman, C.M., 1970. J. Mol. Spectry. 33, 311.

LASER EXCITATION OF OPTICAL SPECTRA

D.A. Ramsay

National Research Council of Canada

Ottawa, Ontario, Canada

ABSTRACT

A brief description is given of the characteristics and types of lasers available. The use of lasers for the excitation of molecular spectra, especially resonance fluorescence spectra, is illustrated with examples. The technique of saturation spectroscopy, or "inverse Lamb-dip" spectroscopy, as a means for achieving ultrahigh resolution is discussed. The application of high-powered lasers to the observation of two-photon spectra, both with and without Doppler broadening, is also discussed.

I. CHARACTERISTICS AND TYPES OF LASERS

The main characteristics of optical lasers are:

(i) high monochromaticity
(ii) high power
(iii) coherence.

Both CW and pulsed lasers are available for most of the optical region of the spectrum. We will consider a few specific examples.

I.A. CW Lasers

He-Ne, A^+ and Kr^+ lasers are readily available with powers ranging from milliwatts to a few tens of watts. An argon ion laser with an output of 1 watt in the 4800 Å line produces $\sim 2.5 \times 10^{18}$ photons/sec in a line with a width of ~ 0.01 cm^{-1}. Such lasers are essentially fixed frequency lasers although they can sometimes be tuned over very narrow frequency ranges.

Dye lasers have the advantage that they can be tuned over much larger wavelength ranges. The most commonly used CW dye laser is the Rhodamine-6G dye laser pumped by an argon ion laser. The useful wavelength range is from 5600 to 6300 Å and typical output powers are ∿20% of the pumping powers. The normal band width of the output radiation is ∿0.1 cm^{-1} but with an etalon inside the cavity the band width can be reduced to ∿0.001 cm^{-1} (30 MHz).

I.B. Pulsed Lasers

The first optical laser was a flash lamp-pumped ruby laser and such lasers are still in common use. The output is essentially fixed frequency (6943 Å) but a small degree of tuning (±5 Å) can be achieved by varying the temperature of the ruby. Many other solid state lasers are available and pulses up to 100 MW can be obtained for short durations, e.g. a few nanoseconds. The outputs can also be frequency-doubled by using suitable materials, e.g. KH_2PO_4.

Pulsed dye lasers are available throughout the visible region of the spectrum (4000-7000 Å). The pump source may either be a flash lamp with a low repetition rate (e.g. 1/sec-1/min) or a pulsed N_2 laser with repetition rates up to 10^2/sec. With a 300 kW N_2 laser operating at 20 pulses/sec, a dye laser would typically have output pulses of 5-20 kW lasting for a few nanoseconds and having a band width of ∿0.1 Å. By using an etalon in the cavity the band width can be reduced to ∿0.001 Å with some sacrifice in peak power.

II. EXCITATION OF MOLECULAR SPECTRA

Two examples will be considered, both taken from work in our own laboratory.

II.A. Band Spectra

The emission spectrum of glyoxal can be readily excited using the various lines of a 1W argon ion laser. The emission contains bands of the first singlet-singlet system (^1A_u-^1A_g) and of the corresponding triplet-singlet system (^3A_u-^1A_g) (Fig. 1). The relative intensities of the bands depend on the exciting line used. With the 4880 Å laser line some new bands were excited and could not be fitted into the above systems. These new bands were subsequently shown to be produced by <u>cis</u> glyoxal which is present in glyoxal at room temperature to the extent of a few percent (Holzer and Ramsay 1970, Currie and Ramsay 1971, Dong and Ramsay 1973).

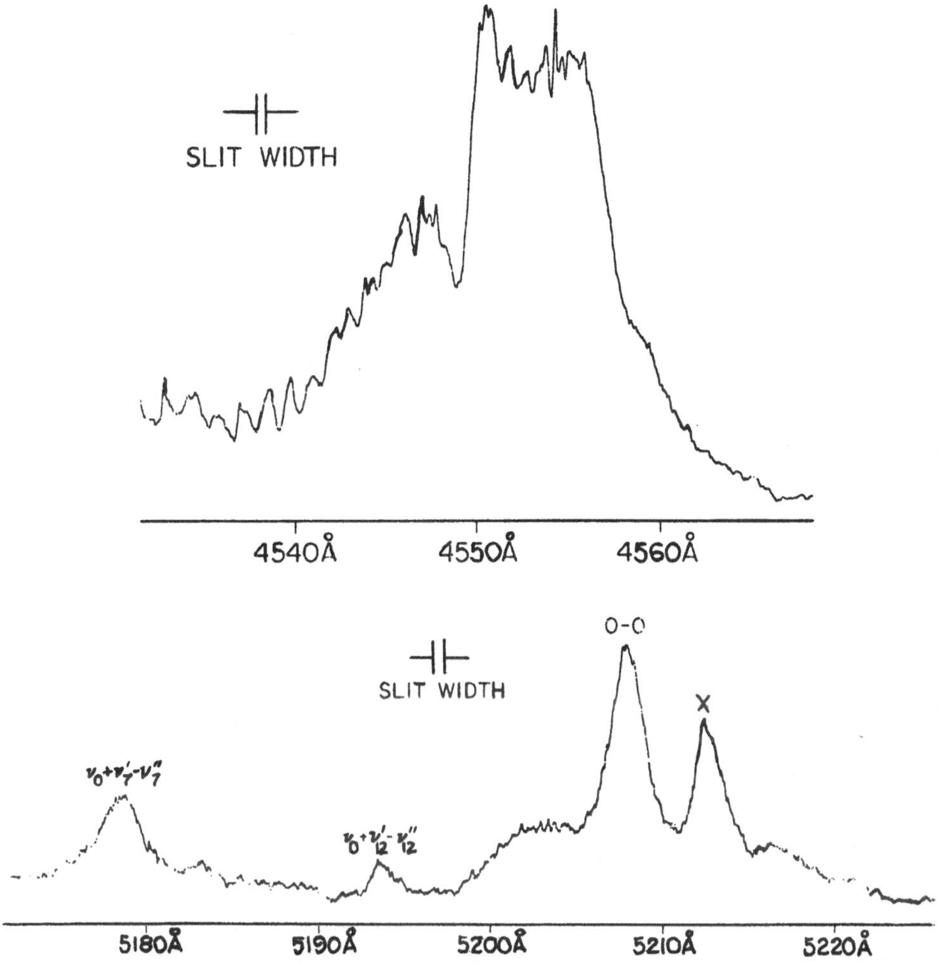

Fig. 1. The upper trace shows the 0-0 band of the ^1A_u-^1A_g system of glyoxal excited by the 4765 Å line of an argon ion laser.

The lower trace shows bands of the ^3A_u-^1A_g system of glyoxal excited by the 4880 Å line of an argon ion laser. The band marked with an X is a band of cis glyoxal.

The 4880 Å line overlaps the 0-0 band of the first singlet-singlet system ($^1B_1$-$^1A_1$) of cis glyoxal which is thereby selectively excited. In this way cis glyoxal was first identified experimentally.

The vibrational assignments for the bands of trans glyoxal were straightforward, since the absorption spectrum had been studied in considerable detail and furthermore all the ground state frequencies were well-known. However, new information was obtained for some of the vibrational frequencies of the cis isomer.

II.B. Resonance Fluorescence

The emission spectrum of the NH_2 free radical can be readily excited by means of a tunable dye laser. Kroll (1975) prepared NH_2 radicals in a flow system by reacting H atoms with N_2H_4 and observed the excitation spectrum by tuning a Rhodamine-6G dye laser through the rotational lines of the absorption spectrum (Fig. 2). When the laser was tuned to a single absorption line the emission spectrum was predominantly a resonance fluorescence spectrum, i.e. the lines observed had a common upper level connected to different rotational and vibrational levels in the lower state. In this way the 100, 110, 020 and 040 levels of the ground state were identified for the first time and preliminary rotational constants determined for these levels.

In addition to the resonance fluorescence spectrum, other lines were observed with increasing relative intensity as the pressure increased. Such lines originate in additional rotational levels of the excited state populated by collisional processes. In this way some information on rotational selection rules for collision processes was obtained.

III. SATURATION SPECTROSCOPY

The factor which limits the resolution which can be achieved in conventional high resolution spectroscopy of simple gases, is usually the Doppler width of an individual rotational line. This limit depends on the molecular weight and temperature of the gas, but Doppler widths for lines in the visible region of the spectrum are usually $\sim 0.05\ cm^{-1}$. Laser lines, however, can be considerably narrow, e.g. $\sim 0.001\ cm^{-1}$.

Let us consider the interaction of a laser beam with an individual line in the spectrum of a gas. If the frequency of the laser, ν_L, is equal to the center frequency of the absorption line, ν_o, then laser power can be absorbed only by those molecules which are moving transversely to the laser beam. If the laser frequency is slightly detuned from the center frequency, then laser power can be absorbed only by those molecules which have a velocity component v along the laser direction (and away from the laser), such that

$$\nu_L = \nu_o \left(1 - \frac{v}{c}\right). \tag{1}$$

If the laser beam is sufficiently intense, then the population of

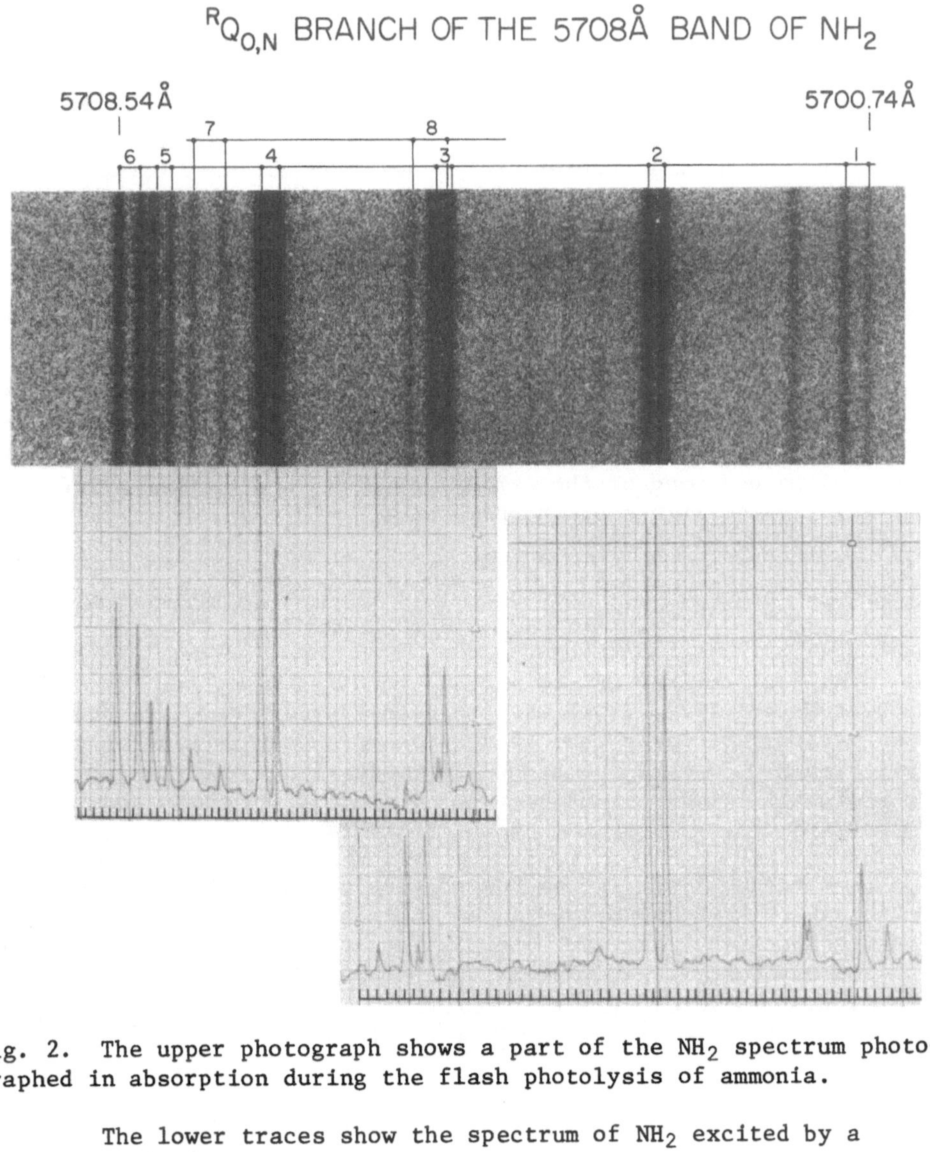

Fig. 2. The upper photograph shows a part of the NH_2 spectrum photographed in absorption during the flash photolysis of ammonia.

The lower traces show the spectrum of NH_2 excited by a tunable dye laser.

the group of molecules with velocity component v will be depleted in the lower state and increased in the upper state.

Consider now the experimental arrangement shown in Fig. 3. Suppose that the intense laser beam interacts with molecules which have a velocity component +v in the direction of the laser beam. Then the weaker reflected beam will interact with a different group of molecules having the same velocity component v but moving in the opposite direction. If the frequency of the laser is varied over the width of the absorption line, then the reflected beam can be used to monitor the profile of the absorption line. A normal Doppler profile will be obtained except at the center of the line where a "dip" in the absorption will be recorded, since the population of the group of molecules with $v = 0$ was depleted by the incident beam. Center frequencies of lines can therefore be determined with precisions depending on the widths of the laser lines.

As an example of the increased resolution which can be obtained using the technique of saturated absorption ('inverse Lamb-dip') spectroscopy, Hanes, Lapierre, Bunker and Shotton (1971) have resolved the 21 nuclear hyperfine components of the P(33) line in the 6-3 band of the B←X system of I_2 and have measured the relative positions of the components with a precision of 0.36 MHz (0.000012 cm^{-1}).

IV. TWO-PHOTON SPECTROSCOPY

The basic theory of two-photon processes was given by Göppert-Meyer (1929, 1931) more than forty years ago, yet only comparatively recently have these processes been studied in the optical region of the spectrum. The main reason was the lack of sufficiently intense light sources to observe the phenomena until the advent of the laser in the early 1960's.

The probability I_{ik} for two photons of the same frequency ω to cause a transition between an initial state i and a final state k is given by

$$I_{ik} \propto \left| \sum_j \frac{\langle i|M|j\rangle\langle j|M|k\rangle}{\omega-\omega_{ij}} \right|^2, \qquad (2)$$

where M is the dipole moment operator, ω_{ij} is the frequency of an "intermediate" transition, and the summation extends over all other states j. The probability is considerably enhanced if there are strongly allowed single-photon transitions between the intermediate state j and the initial and final states i and k, and also if $\omega \sim \omega_{ij}$. The selection rules are different from those for single-photon processes and the possibility exists for observing new excited states by two-photon absorption. For example, in a molecule with a center of symmetry the selection rules for two-photon transitions (g ↔ g, u ↔ u, g ↮ u) are the reverse of those for single-photon transitions.

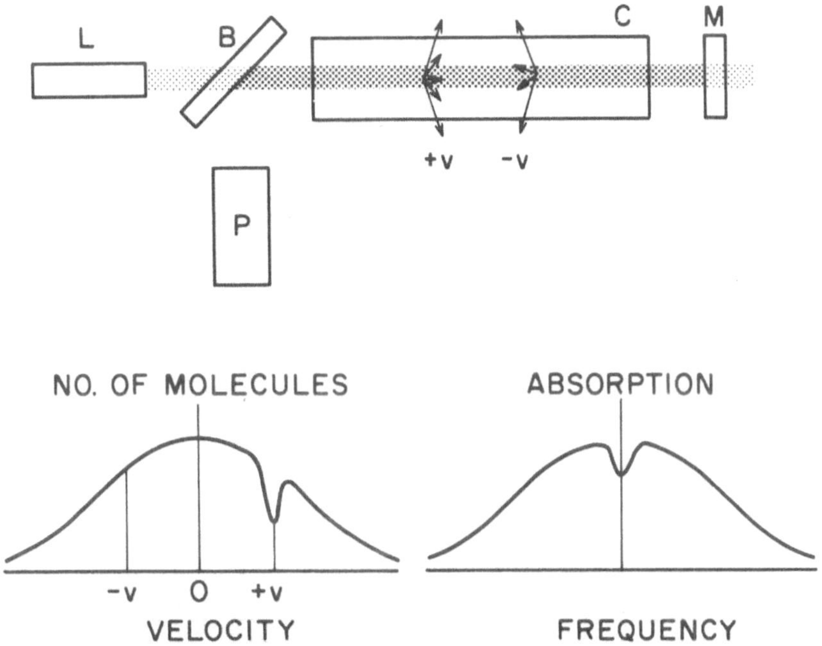

Fig. 3. Apparatus for observing "inverse-Lamb dip" effects: L-laser, B-beam splitting mirror, C-cell containing absorbing gas, M-partially reflecting mirror, P-photodetector. The velocity vectors of molecules interacting with the forward beam (+v) and with the weaker returning beam (-v) are indicated.

Bottom left - Velocity profile showing the molecules which interact with the intense incident beam (+v) and the weaker reflected beam (-v).

Bottom right- Absorption profile obtained with the reflected beam, showing the inverse Lamb-dip at the center frequency.

IV.A. Detection of Two-Photon Processes

One method for detecting two-photon absorption spectra involves monitoring the emission from the final state. In the apparatus shown in Fig. 4 light pulses from a dye laser pumped by a pulsed N_2 laser, are focussed into a cell containing the absorbing molecule. Emission in the ultraviolet is monitored using a photomultiplier with filters to eliminate scattered light from the dye laser and from the pulsed N_2 laser. The spectrum is scanned by rotating the grating in the laser cavity. Using this type of apparatus Hochstrasser, Sung and Wessel (1973) have observed two-photon spectra in benzene both in the crystal and vapor and have observed eight out of the ten ungerade fundamentals in the $^1B_{2u}$ excited state. Bray, Hochstrasser and Wessel (1974) have also observed rotational fine structure in the two-photon excitation spectrum of the $A^2\Sigma^+$-$X^2\Pi$ transition of NO and have observed two new branches, O ($\Delta J=-2$) and S($\Delta J=+2$), which are permitted by the two-photon

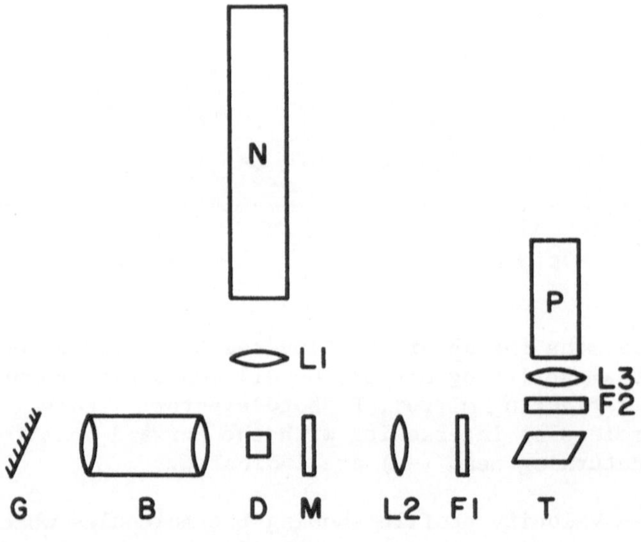

Fig. 4. Apparatus for observing two-photon excitation spectra: N-nitrogen laser, L1, L2 and L3-lenses, G-plane grating, B-beam expander, D-dye cell, M-output mirror, F1 and F2-filters, T-two photon cell, P-photomultiplier.

selection rules. We have observed a two-photon excitation spectrum in thiophosgene ($CSCl_2$) in the vapor (Hillier and Ramsay 1975). In all cases it has been verified that the intensity of the two-photon spectrum varies as the square of the intensity of the incident radiation.

Two photon absorption spectra have also been observed using a high-powered ruby laser (up to 10 MW) to provide one of the photons and a xenon arc lamp to provide the other (Eisenthal, Dowley and Peticolas 1968, Monson and McClain 1970, 1972). The spectra were plotted point by point and show only low resolution.

IV.B. High Resolution Spectra without Doppler Broadening

The 4d-3s transition in Na vapor has been studied by Hänsch, Harvey, Meisel and Schawlow (1974) using a single mode CW tunable dye laser with 30-50 mW power. The transition is forbidden for a single-photon electric dipole process but is allowed for double-photon absorption.

The apparatus is similar to the one in Fig. 4 except that the dye cell is pumped continuously by an argon ion laser. When the dye laser is tuned to half the transition frequency (λ=5787.32 Å), two-photon absorption occurs and is monitored by emission at 3302 Å corresponding to the second step in the 4d→4p→3s cascade process. The two-photon absorption is particularly strong in this system since a 3p intermediate state lies close in energy to the exciting radiation and the transitions 3p-3s and 4d-3p are both strongly allowed.

When a concave mirror is placed behind the absorption cell and is accurately aligned so that the return beam overlaps the incident beam in a narrow waist inside the absorption cell, there is a dramatic increase in the signal. If the frequency of the laser is scanned slowly, four sharp resonances are observed with widths ∿30 MHz. These resonances correspond to the hyperfine components of the 3s state; the hyperfine splittings in the 4d state are much smaller. Under these experimental conditions, the two photons are taken one from each beam and the effects of Doppler motion cancel. Furthermore all the molecules in the beam contribute towards the signal. If the two beams are misaligned then the two photons are taken from one of the beams only and a much weaker Doppler-broadened profile is observed.

It appears that this technique for overcoming the effects of Doppler broadening will have valuable application in high resolution molecular spectroscopy.

REFERENCES

Bray, R. G., Hochstrasser, R. M. and Wessel, J. E. 1974. Chem. Phys. Letters 27, 167.
Currie, G. N. and Ramsay, D. A. 1971. Can. J. Phys. 49, 317.
Dong, R. Y. and Ramsay, D. A. 1973. Can. J. Phys. 51, 1491.
Eisenthal, K. B., Dowley, M. W. and Peticolas, W. L. 1968. Phys. Rev. Letters 20, 93.
Göppert-Meyer, M. 1929. Naturwissensch 17, 929.
________ 1931. Ann. Physik 9, 273.
Hanes, G. R., Lapierre, J., Bunker, P. R. and Shotton, K. C. 1971. J. Mol. Spectry. 39, 506.
Hänsch, T. W., Harvey, K. C., Meisel, G. and Schawlow, A. L. 1974. Optics Comm. 11, 50.
Hillier, R. M. and Ramsay, D. A. 1975. Unpublished results.
Hochstrasser, R. M., Sung, H-N. and Wessel, J. E. 1973. J. Am. Chem. Soc. 95, 8179.
Holzer, W. and Ramsay, D. A. 1970. Can. J. Phys. 48, 1759.
Kroll, M. 1975. J. Chem. Phys. In press.
Monson, P. R. and McClain, W. M. 1970. J. Chem. Phys. 53, 29.
________ 1972. J. Chem. Phys. 56, 4817.

TECHNIQUES OF FLASH PHOTOLYSIS

S. Claesson

Institute of Physical Chemistry, Uppsala University

Uppsala, Sweden

ABSTRACT

In this article an elementary discussion is given of the design and properties of a flash-photolysis apparatus. Special emphasis is placed on simple and reliable performance. The design of discharge circuitry, flash lamps and other experimental aspects are referred to as a means to get good data from simple experiments. Properties of a flash photolysis apparatus are illustrated by seven units built in Uppsala ranging from 500J to 200,000J in energy and lamp voltages between 1kV and 100kV. Some simple statistical means for the interpretation of experimental data are also given.

I. INTRODUCTION

The technique of flash photolysis was introduced by Norrish and Porter (1) in Cambridge and it rapidly became an indispensable tool in chemistry and they were awarded the Nobel Prize in 1967 for this work. Independently but somewhat later the technique was also developed by Herzberg in Ottawa who received the Nobel Prize in 1971 for his work in molecular spectroscopy. Furthermore Professor Porter, together with M.A. West,has recently published a comprehensive article on flash photolysis (2),covering about 100 pages,which also gives extensive references to the scientific literature.

Therefore there is no need to present a longer paper here to describe the technique of flash photolysis in detail but rather to point out certain facts which seem to be of central interest and special importance to scientists who wish to apply experimentally the technique of flash photolysis.

Flash photolysis apparatuses are at present commercially available from several companies. However, different problems may require different experimental approaches, particularly regarding space for complicated sample cells for highly sensitive and reactive chemical systems (low or high temperature, high vacuum or high pressure, etc.).

In many cases it may therefore be more convenient to design and build a flash photolysis unit for the special research work planned. In fact it is not difficult to design such an apparatus for the problems at hand and the cost is small compared to the other pieces of equipment necessary, such as oscilloscopes, monochromators, fast photomultipliers with their electronic systems, and so on.

This discussion will be based on our work in Uppsala over the last 20 years where we have designed flash photolysis units in the energy range from a few joules to 400,000 joules. This work was made possible through generous gifts of capacitors from the ASEA company through Dr. H. Liander, former Vice President of that company. Most of our work has been done with equipment in the energy range 3000 to 30,000 joules where the duration of the flash (at half intensity) is of the order 5 μs and 15 μs, respectively. These specific properties will be discussed at the end of this lecture.

II. DISCHARGE CIRCUITS

The flash is formed by discharging a capacitor of proper size through a quartz tube filled with a suitable gas at an intermediate pressure (5-50 mmHg) and provided with electrodes at both ends.

The circuit can therefore be looked upon as a capacitor with capacitance C (microfarads) in series with a resistor R (ohms) formed by the resistance of the leads and the lamp during discharge, and also in series with a self-inductance L (microhenries) caused by the self-inductance in the lamp and the leads. For such a situation it is well known that the energy is

$$E = \frac{1}{2} CV^2 \quad , \tag{1}$$

where V is the voltage to which the capacitor is charged. If the resistance is small the circuit will oscillate during discharge with a ringing frequency such that the period is

$$T = 2\pi\sqrt{LC} \quad . \tag{2}$$

As the resistance of the circuit is increased the oscillations will be more and more damped and for a certain resistance

$$R_C = 2\sqrt{\frac{L}{C}} \quad , \qquad (3)$$

the damping will be critical which means that the voltage decays exponentially. (The lamp resistance varies during discharge but for the present discussion where we are mainly interested in orders of magnitude it can be regarded as constant and proportional to the length of the lamp and inversely proportional to its cross section. The final adjustment of a circuit will always be done experimentally.)

It is obviously necessary to have the lamp critically damped as that corresponds to the shortest decay of the flash. If underdamped it will oscillate and the flash will show a fluctuating tail; if overdamped it will decay slower than in the critically damped case.

In order to get a fast discharge the time T must also be made as small as possible. Therefore L·C should be minimized. However, in order to keep the energy sufficiently large it is then necessary to use a high voltage V in order to make C small. For practical reasons it is inconvenient to us a voltage much greater than 25kV and we regard 50kV as an absolute maximum. At higher voltages corona effects become troublesome. As a typical set of values occurring in practice we can take C = 6μF, V = 20,000 volts, $E=(3\times10^{-6})\times(4\times10^{8})$ = 1200J, $L\sim 1\mu H$ giving $R_C = 2\sqrt{1/6} = 0.8$ ohms. Typical values for flash lamp resistance are 0.1 - 0.5Ω and therefore a small resistance has to be added in series with the lamp to avoid oscillations which would create a long tail in the flash making it unsuitable for fast work.

However, if the resistance to be added is large in comparison with that of the lamp itself, it will dissipate most of the energy and the flash will be weak. It is therefore imperative to do the utmost to decrease the self-inductance as much as possible in order to minimize R_C. That is done by using a coaxial design throughout the circuit. Coaxial cables for high currents are easily made from ordinary plastic insulated copper wires covered with extra plastic tubing on the outside which in turn is placed inside an ordinary copper braid which is used as the outside (ground) connector. Furthermore to decrease the self-inductance, two identical lamps should be mounted parallel to each other (bifilar arrangement) and the electrodes at one end connected. Two capacitors (or groups of capacitors) should be connected to each other and grounded. Their other electrodes (high voltage) are connected through the coaxial cables, one to each lamp and the outer braids grounded at the capacitors' ground and connected to each other close to the lamps. The high voltage generated should charge the capacitors to +V and -V respectively. If each capacitor is C (microfarads), the combined capacity is $\frac{C}{2}$; with the voltage across 2V this gives

$$E = \frac{1}{2} \cdot \frac{C}{2} (2V)^2 = CV^2 .$$

We have gained three advantages:

1) the capacitance seen by the lamps is only $\frac{C}{2}$ at a voltage 2V,
2) voltage towards ground is only V instead of 2V,
3) the two lamps are electrically in series and must therefore fire simultaneously.

The common end of the lamp-pair is always kept at ground potential through a high ohmic resistor(50kΩ or higher)made from a large number of radio resistors connected in series and placed in a tube filled with paraffin oil (not solid paraffin which can crack,leading to undesired effects) . Similar high-ohmic resistors are used to monitor the voltage of the capacitors connected to microammeters at the grounded end.

This basic design has been used since we built our first flash photolysis apparatus in 1955 and further improvements in design have had only marginal effects, except for very fast flashes below 1 μs.

If groups of capacitors are used each should have its own coaxial cable which decreases the self-inductance as $1/L = 1/L_1 + 1/L_2 + \ldots$

II.A. Series Resistors for Flash Lamps

The rather small resistors, 0.1-0.2Ω, which occasionally have to be added in series with the flash lamps to achieve critical damping are sometimes difficult to design. In many cases a carbon rod or a metal wire with a high melting point are satisfactory. However, for very fast, small flash lamps (≤1μs) these resistors must fit into the coaxial design with little space for resistors. A short, thin metal wire will explode and a long, heavier wire becomes too bulky.

Most semiconductors, carbon films and metal films were both expensive and difficult to obtain. However, when designing the fast-flash Apparatus VII with a flash duration of 0.7 μs an almost ideal resistor was found. It consists of two metal electrodes dipping into a solution of NaCl and water. By varying electrode area, electrode separation and salt concentration a proper resistance was easily found by trial and error by looking at the flash profile on the oscilloscope when these parameters were varied. At these high current densities and voltages it was not practical to try to calculate the dimensions and concentrations from normal conductivity values in the literature.

II.B. Flash Lamp Design

Flash lamps of many different types have been described in the literature. The most convenient and simplest is a straight quartz tube with an electrode at each end. Earlier the electrodes were either cemented or fused into the quartz tube; today almost everybody uses O-ring seals. We use tungsten rods for electrodes, which reach about 5 cm into the lamps, thus avoiding direct illumination of the O-rings during the flash. During the flash the gas inside the lamp gets very hot and some quartz also evaporates. It is therefore convenient to have an expansion volume (glass-flask) outside the electrode holder, connected to the lamp by a narrow hole (2-3 mm diameter). In this volume the quartz vapor will condense and the lamps will remain clear. This expansion volume also helps to reduce the shock-wave in the lamp during discharge.

Different gases have been suggested for filling the lamps. We prefer to use oxygen (15mm Hg). This gas is produced from the quartz during the flash anyway and the triggering characteristics would change with other gas fillings. At higher energies the difference in light output is not too dependent on the gas; however, for low energies it might be advantageous to use a noble gas.

We trigger our lamps by discharging a small capacitor through a pulse transformer which is connected to the two electrodes kept at ground potential by a 50 kΩ resistor. To avoid discharge of the large capacitors through the pulse transformer once the triggering has occurred, a very small high voltage capacitor is placed between the pulse transformer and the lamp electrodes. It is made with small pieces of aluminum foil isolated by sheets polymer film and immersed in paraffin oil. The peak voltage of the small trigger pulse is normally about twice as high as that of the lamp driving voltage from the main capacitors. Photographs of a typical O-ring seal and expansion vessel for the flash lamp are shown in Figs. 1A and 1B, respectively.

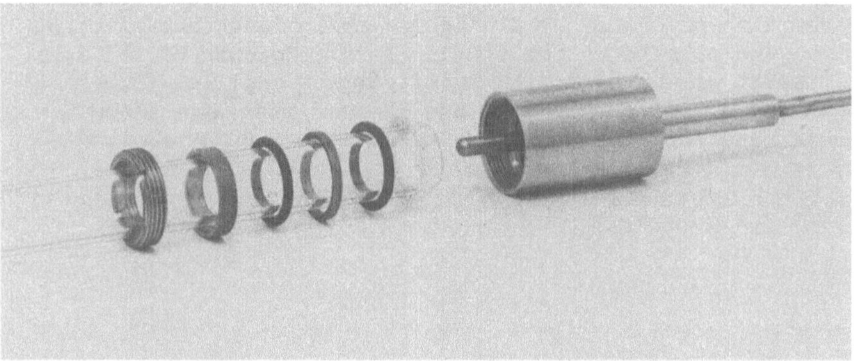

Fig. 1A. O-ring seal for flash lamp.

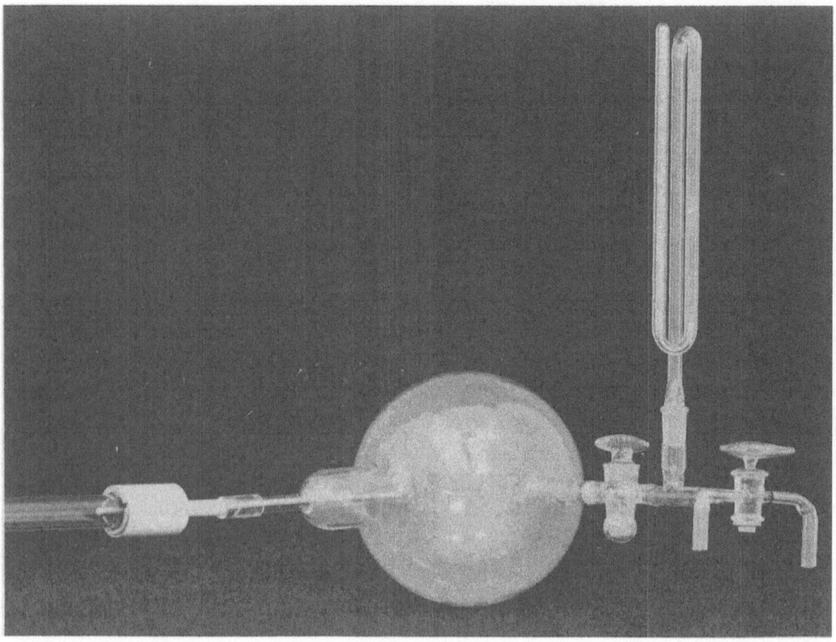

Fig. 1B. Expansion vessel for oxygen-filled flash lamp showing some deposit from quartz vapor.

III. EXAMPLES OF FLASH PHOTOLYSIS APPARATUSES

A series of typical apparatuses developed in Uppsala according to the above mentioned principles will be shown in the next figures and a few remarks made about them.

Apparatus I, Fig. 2A, was very slow because of too high a self-inductance. As shown in Fig.2B, it has now been rebuilt into a square-pulse generator by adding low-ohmic inductances to the circuit. The pulse duration time is 1 ms. Because of the long pulse time it can be used with small,compact capillary flash lamps up to 10,000 J. Figs. 3A and 3B show such apparatuses, which are used for flash-photolysis studies of human and animal skin.

Fig. 2A. Apparatus I, 1370μF,7kV, 34,000J, τ = 100μs (3).

Fig. 2B. Apparatus I rebuilt. τ = 1 ms.

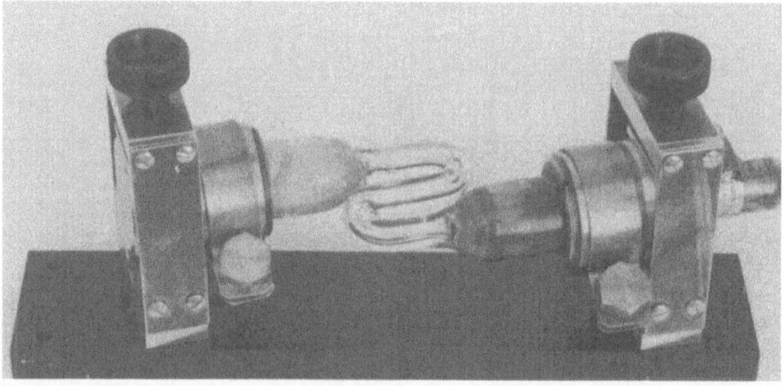

Fig. 3A. Apparatus of type used in flash-photolysis studies of human and animal skin.

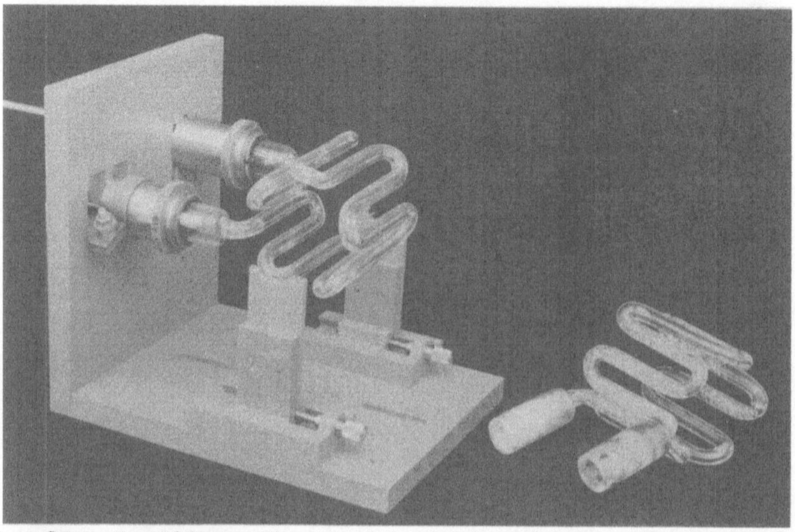

Fig. 3B. Capillary flash lamp showing the typical microscopic cracking. This develops some hours after a flash and disappears when flashed again.

Apparatus II (coaxial), with a capacitance of 150μF, voltage of 10 kV, and τ = 5-20μsec, is shown in Figs. 4A and 4B. For its size this apparatus was fast at that time, but it was difficult to achieve critical damping in such a large coaxial discharge. Nowadays smaller flash units of this type are very good for pumping dye-lasers.

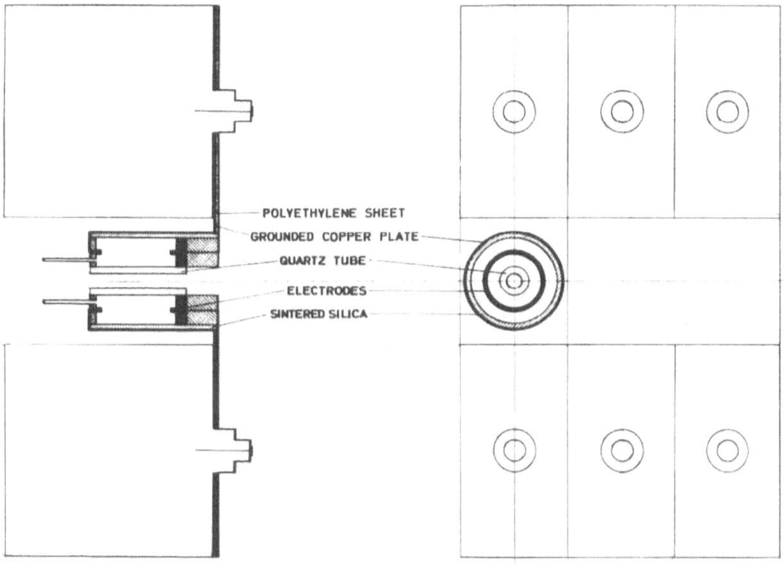

Fig. 4A. Cross section of flash lamp in Apparatus II.

Fig. 4B. Apparatus II, coaxial, 150µF, 10kV, 7500J, τ = 5-20µs (4).

Apparatus III, Figs. 5A, 5B, and 5C, was first used for spark illumination around capillary tubes (Fig. 5B). The circuit was then very underdamped. From oscillograms (Fig. 5C, 100µs per large division) it is seen that the circuit was underdamped and the shortest wavelengths were produced during the first oscillations when the voltage was high. This capacitor bank has also been used with very long lamps (600 cm) and as a pulse generator.

Fig. 5A. Apparatus III.

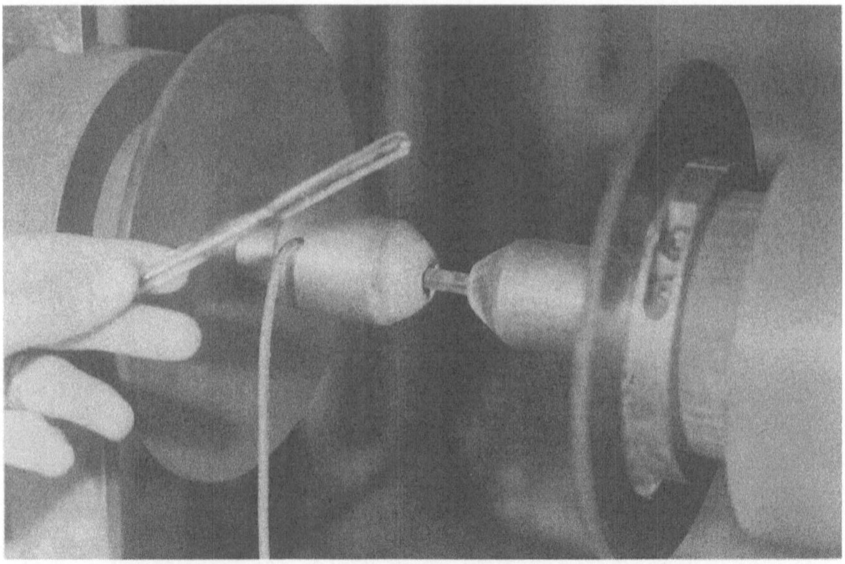

Fig. 5B. Apparatus III used for spark illumination around capillary tubes.

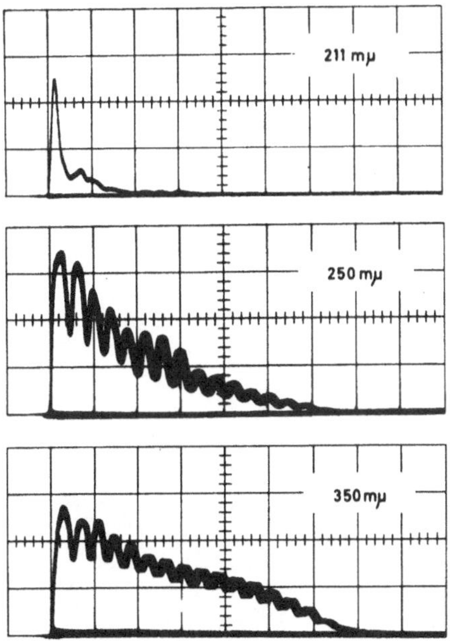

Fig. 5C. Oscillograms showing pulse duration of Apparatus III (100 μs per large division).

Apparatus V is shown in Figs. 6A and 6B. The coaxial cables are of equal length, one for each capacitor and connected in parallel. In Fig. 6B, the two lamps are again seen in parallel; the ballast vessel and the high-voltage resistors made from radio resistors are in series and placed in oil-filled tubes. Observe the coaxial caps over the porcelain insulators on top of the capacitors. Proper damping is of very great importance. Introducing a carbon resistance to accomplish proper damping improves the "1/e time" for a 3 kJ flash from 13 to 8μsec, without seriously affecting the peak intensity, and with an almost constant "1/2 time" of 6μsec. For the critically damped circuit, the flash has approximately an exponential decay (with a half-life of about 4.5μsec) down to about 1% of maximum light intensity. Increasing the discharged energy above a certain limit (about 2kJ for the underdamped and 3kJ for the critically damped circuit) leads to only a small increase in peak light intensity. For very fast reactions the optimum working condition is at the onset of this "saturation".

Fig. 6A. Apparatus V.

Fig. 6B. Apparatus V, showing two lamps in parallel, the ballast vessel and high-voltage resistors in series and placed in oil-filled tubes.

Fig. 7A. Apparatus VI, 2·24 μF, $\pm$ 50kV, 60,000J (1964, unpublished). τ = 13μs for 60 cm lamps and 24 μs for 200 cm lamps.

Apparatus VI is depicted in Figs. 7A and 7B. With this very fast apparatus only the long lamps will take all the energy without exploding. The following may be seen in Fig. 7B: the parallel arrangement of the equal-length coaxial cables; the conductors (insulated inside) around the procelain insulators of the individual 96 capacitor units (0.5 μF, 55 kV); and the 50 kΩ resistor and the small high voltage blocking capacitor in oil-filled tubes (on the top of the box to the right with the pulse transformer). This unit and Apparatus V have been used continuously

Fig. 7B. Apparatus VI, showing the parallel arrangement of coaxial cables, and the conductors around the porcelain insulators of the individual capacitor units (0.5μF, 55kV).

for more than 10 years, and the only service necessary has been to change the quartz tubes of the lamps a few times a year when they become too sputtered with tungsten from the electrodes.

In Apparatus VII (Figs. 8A and 8B), coplanar insulated metal sheets have been used instead of cables to reduce inductance to a minimum. This apparatus is critically damped by means of an ammonium chloride solution. The length of the lamps can easily be changed from 5 cm to 30 cm. Only the longer lamps take the full energy. To trigger, the pressure in the lamp is reduced by opening a valve to an evacuated flask. The resistor with salt solution is clearly seen in Fig. 8B where the tube used to reduce the pressure has been removed for clarity.

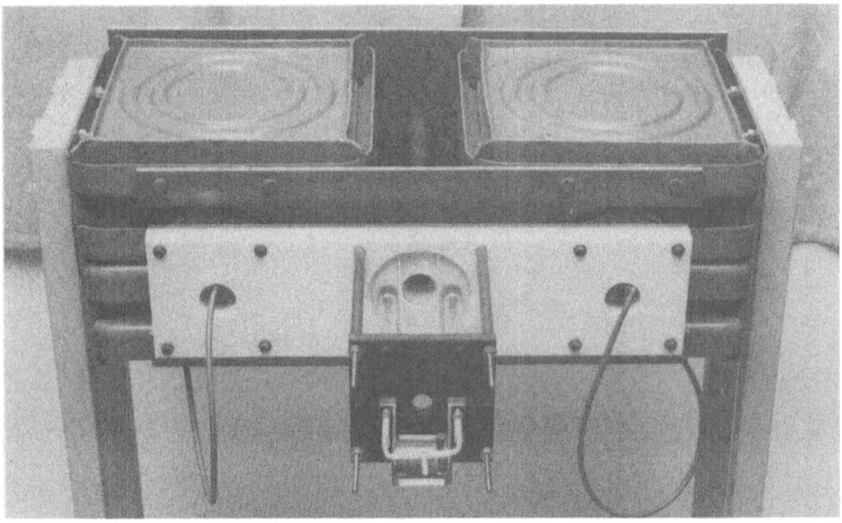

Fig. 8A. Apparatus VII, 2·0.4 μF, ± 35kV, 500J, τ=0.7 μs (1971, unpublished).

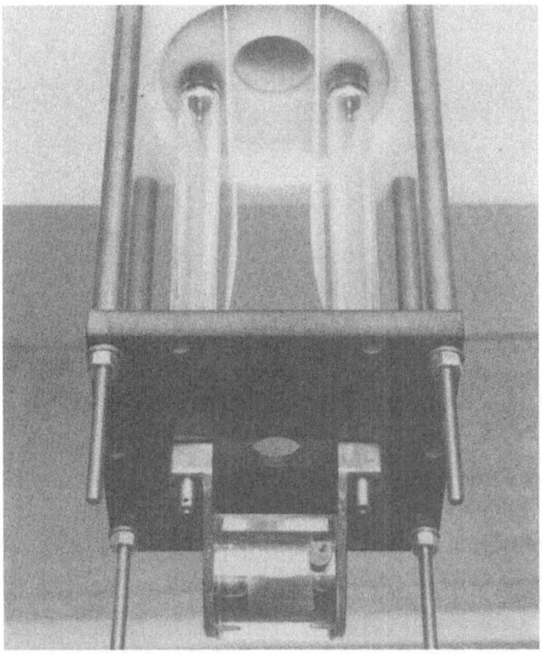

Fig. 8B. Apparatus VII showing resistor with salt solution. The tube used to reduce the pressure has been removed for clarity.

IV. KINETIC STUDIES BY MEANS OF FLASH PHOTOLYSIS

High-resolution kinetic studies of molecules in the gas phase are primarily done by photographic techniques using a very short pulse from a small flash lamp or a dye-laser. However in this section the discussion will be limited to kinetic studies of solutions. In this case the principal technique is almost always the same.

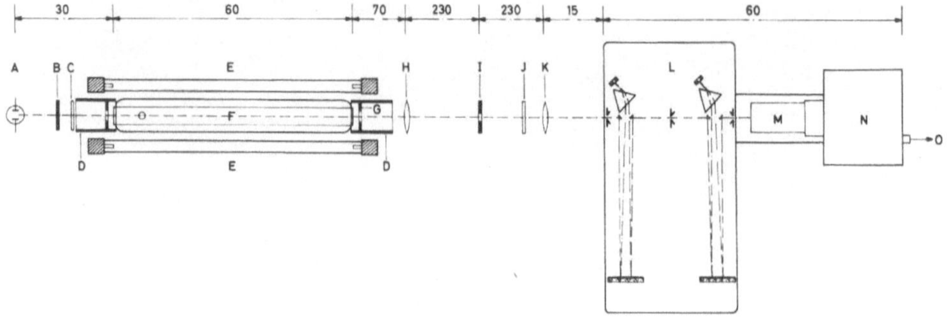

Fig. 9. Schematic of apparatus for kinetic studies of solutions.

As shown in Fig. 9, the solution to be studied is placed between the flash lamps in the cell, F. This cell is surrounded by a mantle with a filter solution to select only a certain wavelength region of the light emitted from the flash lamps. The monitoring light source, A, of constant intensity is focused on the slit of the monochromator, L, through a series of lenses and optical stops in such a way that stray light is minimized. The monitoring wavelength is selected on the monochromator (or polychromator) and the transmitted light intensity is obtained on an oscilloscope screen via photomultiplier(s) and cathode follower(s). From these readings the changes in optical density in the solution following flash can be calculated. What are normally recorded, however, are the changes in transmission vs. time. Some remarks will be made about experimental details in this regard.

The monitoring light source must be intense and as constant as possible. If that is not the case the signal/noise ratio will be low, requiring a high time constant of the photomultiplier system in order to get an accurate reading on the oscilloscope, making fast measurements impossible. Most workers therefore use high intensity xenon arcs but some simple alternative light sources exist which quite often are more convenient and give more light. One is a pulsed tungsten lamp. The intensity as a function of wavelength for a typical lamp (10V, 12 amp) is shown in Fig. 10 curve (D). If now this lamp is pulsed by increasing the voltage to 15 volts for a few milliseconds, which is normally all that is required for a measurement, the intensity increases about fivefold to curve (C) (note logarithmic scale). Such a sharp high current pulse can easily be obtained from an ordinary car battery by means of a mercury switch of the type used in ordinary water thermostats, and a simple pulse generator to trigger the switch. Another possibility is to use an ordinary capillary mercury lamp run on the 50 c/s line voltage. If the triggering of the flash experiment is synchronized with the line voltage so that the mercury lamp is at peak voltage during the flash experiment, the light output is at maximum and very stable for a long time (remember that $\cos x = 1 - x^2/2$ for small x). The curve (A) is recorded for a mercury lamp used in such a way and curve (B) for a normal xenon arc. It is seen that

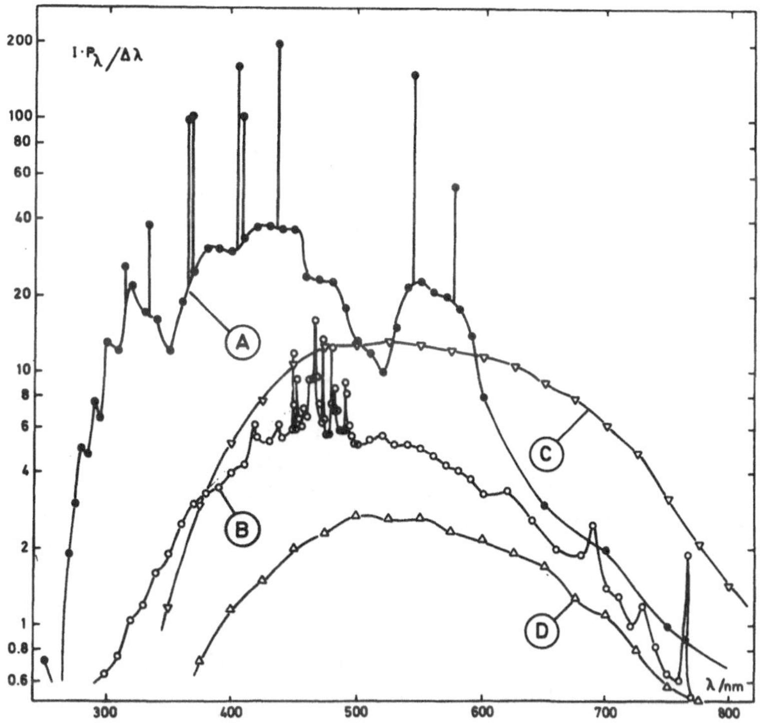

Fig. 10. Plot of Intensity vs. Wavelength for:
(A) mercury lamp with triggering synchronized with line voltage;
(B) normal xenon arc; (C) pulsed tungsten lamp;
(D) tungsten lamp (10V, 12 amp).

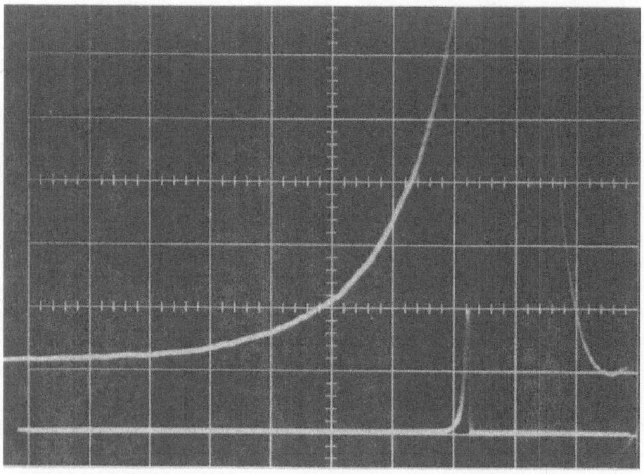

Fig. 11. Dual beam oscilloscope trace with delayed output pulse. The position of the base line is shown by the slow sweep and the change in optical transmission by the fast sweep.

already the intensity of the background of the mercury lamp compares well with the intensities of the xenon arc and the lines are about 20 times brighter (Finnström and Hunt).

Triggering and location of the base line often create difficulties. Oscilloscopes which can give delayed output pulses are indispensable in this kind of work and dual beam scopes are extremely helpful to locate base lines. Fig. 11 illustrates this. Before flash the two sweeps, one slow with low sensitivity, were placed exactly 10 mm above each other on the screen. Then the slow sweep was triggered by the trigger pulse to the flash lamp (delay 3 divisions, time increases from right to left). The fast sweep is triggered by the flash from the lamp (pickup by a few turns of copper wire at the end of a coaxial cable placed close to the lamps is better than photoelectric pickup). Now one can follow the rapid change in optical transmission at high magnification and at the same time know the position of the base line from the slow sweep.

V. CALCULATION OF RATE PROCESSES WHICH ARE AS FAST AS THE FLASH DECAY

This will be described with the help of a system studied recently by Dr. Finnström in our laboratory as a model where the nature of the components is immaterial to the discussion. The following mechanism was suggested and proved:

$$TH^+ + h\nu \rightarrow X \quad I_a\phi \qquad (4a)$$

$$X \rightarrow B \quad k_1 \qquad (4b)$$

$$B + B \rightarrow \text{products} \; 2k_2 \, . \qquad (4c)$$

This mechanism gives the following two rate expressions,

$$\frac{dX}{dt} = \phi I_a - k_1 X \qquad (5a)$$

$$\frac{dB}{dt} = k_1 X - 2k_2 B^2 \quad . \qquad (5b)$$

If the excitation is homogeneous throughout the cell, I_a can be expressed as

$$I_a = \text{constant} \times I_o \quad , \qquad (6)$$

according to Lambert-Beer's law, where I_o is the incident light intensity. It has also been assumed that the degree of excitation is low, say less than 5% in order that the concentration of the ground state be not much affected.

Equation (5a) is a linear first-order differential equation with the solution

$$X = c' e^{-k_1 t} \int_o^t I_o(\tau) e^{k_1 \tau} d\tau, \qquad (7)$$

where c' is a constant, which has been introduced to allow $I_o(t)$ to be measured in relative units. If the expression for X is substituted into (5b) the rate of change of B can be written as

$$\frac{dB}{dt} = -2k_2 B^2 + c' \, k_1 e^{-k_1 t} \int_o^t I_o(\tau) e^{k_1 \tau} d\tau \quad . \qquad (8)$$

This differential equation is of the Riccati type and can not in its general form be solved by quadratures for B, but can naturally be solved numerically, if the time dependence of the flash profile is known.

For our flash-photolysis unit we found the following time dependence of I_o:

$$I_o(t) = \begin{cases} \text{point determined}, & t \leq a \\ I_o(a) e^{-k_o (t-a)}, & t \geq a, \end{cases} \qquad (9)$$

with $a = 8 \mu s$ and $k_o = 1.28 \times 10^5 \; s^{-1}$. It is thus obvious that the integral in (8) can be evaluated for $t > a$. For $t < a$, (8) is left unchanged, but for $t \geq a$ we have

$$\frac{dB}{dt} = -2k_2B^2 + c'k_1e^{-k_1t} \int_0^a I_o(t)e^{k_1t}\,dt + $$
$$+ c'\frac{k_1I_o(a)}{k_o-k_1}\left[e^{-k_1(t-a)} - e^{-k_o(t-a)}\right];\ \begin{matrix} t\geq a \\ k_1\neq k_o \end{matrix} \quad . \tag{10}$$

Although the value of the rate constant of the reaction $X \rightarrow B$ does not make it necessary to introduce the complete rate expression, the examination of the two methods brings up the question of how well an approximation of the time-dependence of the rise of the flash profile could give a satisfactory result.

Two cases were investigated aside from the exact method outlined above. They are both more convenient to apply because they avoid the point determination of the rise. In the first case (I) the rise of the flash intensity was taken to be infinitely steep and the decay was exponential with the condition that the total light output should be unchanged. In the second case (II) a linear rise to maximum intensity was adopted, followed by an exponential decay.

VI. NUMERICAL ANALYSIS OF EXPERIMENTAL DATA

Such treatments are always very difficult if the rate constants do not differ sufficiently. A simple statistical approach to this problem,which was worked out by Dr. Johnsen in our laboratory, will be illustrated for the case of a combined first- and second-order reaction.

The rate of disappearance of the (light) absorbing species will be proportional to its concentration, C, and the square of its concentration. Thus, $-dC/dt = k_1C + k_2C^2$, which leads to $d(C^{-1})/dt = k_1(C^{-1} + k_2/k_1)$. Upon integration we obtain $C^{-1} = k_2/k_1 + (C_o^{-1} + k_2/k_1)\exp(k_1t)$, where C_o is the concentration at T = 0.

In terms of the absorbance, A, we obtain $A^{-1} = -k_2'/k_1 + (A_o^{-1} + k_2'/k_1)\exp(k_1t)$. In this case, k_2' is equal to k_2 divided by the molar extinction coefficient and the path length. The equation has the general form $Y = a + b\exp(c\cdot x)$ and our objective is to determine the best estimates of a, b and c from our experimentally obtained x, y data. In the treatment of experimental data we most often assume a normal (random) distribution of errors and therefore use a least-squares procedure to obtain the best estimates of the experimental parameters.

By this we mean that we choose values of a, b and c so that the sum of the square of the differences between y_i and their estimated values will be a minimum:

$$SS = \sum_i [y_i - a - b \exp(c \cdot x_i)]^2 \quad . \qquad (11)$$

The solution of this problem would be exceedingly time consuming without the use of a computer. Sophisticated numerical methods and programming techniques, however, have a tendency to separate the experimentalist from his data. The technique described here has been designed for our table-top calculator with less than 300-word memory capacity. The program consists, essentially, of a weighted, linear, least-squares routine and a searching routine. Both of these routines are noticeably lacking in mystery and easily programmed by a programming novice.

Our approach to the solution of this problem has been as follows. First we make an educated guess for the value of c. Then we use this value of c to transform all of the x's to z's, where $z = \exp(c \cdot x)$. We can now determine the best values of a and b corresponding to this chosen value of c and the x,y data. Finally we calculate SS corresponding to the guessed value of c. By repeating this procedure with different guesses of c we will find that SS will vary with c in a parabolic fashion. At the minimum of this parabola we find the best estimate of c and the corresponding best estimates of a and b.

Our search for the minimum of SS as a function of c is greatly aided by a calculation of the derivative of SS with respect to c. From equation (11) we see that the partial derivative of SS with respect to c will be $-2 \cdot b \sum_i x_i \exp(c \cdot x_i)[y_i - a - b \cdot \exp(c \cdot x_i)]$. If our guess for c is too small this derivative will be negative, and if our guess is too large, then the derivative will be positive. Our strategy is to guess a value of c, determine the sign of the derivative of SS (with respect to c) and then increase or decrease our guess in c by 10%. If the sign of the derivative of SS changes for the next guess, then we know that these two estimates of c bracket the best estimate. Our next guess in c will be midway between these two c values. The new interval containing the best estimate of c will be bounded by this latest c estimate and that one of the two previous c estimates which has an SS derivative with its sign opposite to that of the midpoint c estimate. That is to say: the "best c interval" is always bounded by values of c whose SS derivatives have opposite signs. We continue to halve this interval until it becomes smaller than a predetermined size.

Our problem would now be solved if we could be satisfied with our initial assumption of random errors in y. In our current problem y is A^{-1}. In our spectroscopic measurements we assume random errors in transmittance, T. T is related to A by the equation

$$A = -\log_{10}T. \quad (12)$$

A small error in T is related to a small error in A by $dA = -(\log_{10}e)T^{-1}dT$, and a small error in A^{-1} will be $d(A^{-1}) = -A^{-2}\,dA = (\log_{10}e)(A^2T)^{-1}dT$. A slight rearrangement of this equation yields $dT = (\ln 10)(A^2T)d(A^{-1})$. The best fit to our experimental data therefore will require that we minimize the sum of squares of the deviations of the T's from their estimated values, rather than the sum of squares of the deviations of the A^{-1}'s from their estimated values. If we square the last equation, $(dT)^2 = (\ln^2 10)A^4T^2(dA^{-1})^2$, we find that we can achieve our goal by statistically weighting our A^{-1} values with $(\ln^2 10)A^4T^2$. Thus for the present analysis SS will take the form

$$SS = \sum_i (\ln^2 10)A_i^4 T_i^2\,[A_i^{-1} - a - b\cdot\exp(c\cdot t_i)]^2.$$

REFERENCES

1. R.G.W. Norrish and G. Porter, Nature 164, 658 (1949).

2. G. Porter and M.A. West, in Techniques of Chemistry VI, 367 (1974).

3. S. Claesson and L. Lindqvist, Arkiv Kemi 11, 535 (1957).

4. S. Claesson and L. Lindqvist, Arkiv Kemi 12, 1 (1957).

SOME FAST REACTIONS IN GASES STUDIED BY FLASH PHOTOLYSIS AND KINETIC SPECTROSCOPY

R. G. W. Norrish

Cambridge University

Cambridge, England

This Nobel lecture was originally delivered December 11, 1967. The author acknowledges the Nobel Foundation for permission to reprint it.

I. INTRODUCTION

Realisation that free radicals and atoms take part in chemical reactions has focussed attention on the process of photochemistry which is not only paramount in the geochemistry of the upper atmosphere but is also basic to many reactions of organic chemistry involving free radicals and the triplet state; this realisation also has led to the development of gas lasers, and to the exploration in detail of the intimate anatomy of reactions of pyrolysis, combustion and explosion.

Classical photochemistry emerged in 1908 with the understanding by Stark of the distinction between the primary and secondary photochemical processes, of which the former is the immediate result of the absorption of a light quantum by a molecule or atom and the latter the subsequent "dark" reactions initiated by the products of the former (1). Into this simple pattern it has been possible to fit the whole gamut of photochemical phenomena - fluorescence, phosphorescence, photolytic and photosynthetic processes, photocatalytic and photosensitised reactions. Determination of quantum yields led to the distinction of endoactinic and exoactinic reactions: the former, being endothermic in character, draw their energy requirement from the absorbed quantum and rarely exceed an overall quantum yield of 2; the latter, being exothermic, release their "pent up" energy by photochemical initiation and are usually of the nature of chain reactions, with high quantum yields, and sometimes explosive characteristics. For example, the dissociation of hydrogen iodide into its elements is 2200 cal endothermic and its quantum yield is limited to two (2) while the synthesis of hydrogen chloride from its elements is exothermic to the extent of 22,000 cal and may have a quantum yield as high as 10^6 (3).

It was indeed the study of these two reactions that first led to the conclusion that the primary reaction may involve photolysis of the reactant into atoms (and later free radicals). In the former case we have:

$$HI + h\nu = H + I$$

in the latter:

$$Cl_2 + h\nu = Cl + Cl$$

followed by the well known H_2 - Cl_2 chain reaction. We owe much to Bodenstein, Warburg and Nernst (4) by whose early work the reality of the participation of atoms in chemical reactions was made apparent and the concept of the chain reaction established. Following this, the reactions of H atoms generated by an electric discharge through hydrogen gas were established by R. W. Wood (5) and by Bonhoeffer (6), and the production of free alkyl radicals by the pyrolysis of metal alkyls proved unequivocally by Paneth (7).

Simultaneously the growth of the study of the band spectra of gaseous molecular species in particular by Frank (8) and V. Henri (9) clarified the quantum mechanisms of the processes of thermal dissociation, photodissociation and predissociation, indicating the production of free radicals and atoms in both ground and electronically excited states. It may justly be claimed that from the marriage of photokinetics with spectroscopy there resulted a new insight into the mechanism of chemical reactions; the part played by atoms, free radicals and excited species as transient intermediates became abundantly apparent. The reactions of these transients however, which together make up the overall process of conversion of reactants to final products are so fast that they can neither be observed nor isolated by classical means, and their nature and participation could until recently only be deduced from the circumstantial evidence of reaction kinetics, quantum yields, and the spectroscopic characteristics of the reactants.

It therefore became of importance, if further progress was to be made, to endeavour to obtain objective evidence of the presence of short-lived transients both in thermal and photochemical reactions. Using continuous sources of the highest attainable intensity (e.g. a 10 kW high pressure mercury arc) the author and his collaborators in 1946 attempted to obtain evidence by spectroscopic means of a stationary concentration of intermediates in such reactions as the photolysis and photon oxidation of ketene without success. In no case could any absorption spectrum which could be attributed to reacting transients be observed in the reacting system and it became apparent that their reactivity was so great that no sufficient stationary concentration for detection by the means then available could be achieved.

II. FLASH PHOTOLYSIS AND KINETIC SPECTROSCOPY

It was the realisation that enormously greater "instantaneous" light intensities could be obtained from a powerful light flash than from a conventional light source, and that such a flash need not be of greater duration than the half life of the elusive transients that led Porter and me to study the results of applying such flashes to suitable responsive photochemical systems (10). Using an electric discharge from a condenser bank through inert gas, dissipating about 10,000 joules, it was immediately found that the resulting light flashes of about 2-3 milliseconds duration were able to create large measures of photodecomposition in reactants such as nitrogen peroxide, chlorine, ketene, acetone and diacetyl, amounting to 100% in some cases. It was obvious that momentarily there must be very high concentrations of free radicals or atoms in such reacting systems which by suitable means should be detectable by absorption spectroscopy. This was first achieved by Porter (11) who using a second less powerful flash triggered mechanically by the method of Oldenberg (12) at specific short intervals after the first was able to observe the complete dissociation of chlorine by the disappearance of the Cl_2 absorption spectrum and its return over a period of milliseconds as the atoms recombined.

The modern method of flash photolysis developed from this uses an electronic technique by which the first flash (photoflash) is caused photo-electrically to trigger the second flash (specflash) at specific short intervals measured in microseconds and milliseconds (13). The photoflash is generated by discharging a capacity of the order of 40 μF at 10 kV through an inert gas such as krypton or xenon contained in a quartz tube generally 50 cm in length and 1 cm in diameter. The reaction vessel is a quartz tube of similar dimension with plane quartz end plates lying close to and parallel to the photoflash tube. The specflash lamp consisting of a quartz capillary tube about 10 cm in length is placed "end on" to the reaction vessel, and has a plane quartz end plate so that by means of a lens and limiting

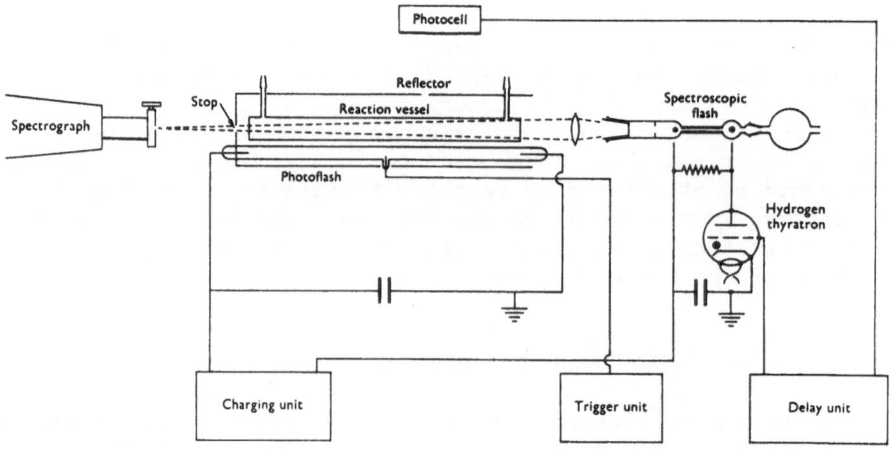

Fig. 1. Diagram of flash photolysis apparatus.

stop, a beam of light can pass longitudinally through the reaction vessel to a suitable spectrometer to register the absorption spectrum of the reacting system at any specific interval after the photoflash (Fig. 1). The discharge is made as before through inert gas.

The energy dissipated by the discharge of a condenser is given in joules by the relationship, $E = (\frac{1}{2}) CV^2$, where the capacity is measured in microfarads, and the potential difference in kilovolts. For a given energy, the duration of the flash is shorter the smaller C and the greater V; the self-inductance of the circuit must be kept as low as possible. For the photoflash, a convenient energy dissipation is 2000 J derived from the discharge of 40 μF at 10 kV. The half life of the light flash is about 10 μsec. For the specflash a discharge of 100 J is generally used, obtained by discharging 2 μF at 10 kV; its half life is of the order 2 μsec. The pressure of the gas in both lamps is of the order 5-10 cm Hg. The reaction vessel may be double-walled for the introduction of gaseous or liquid colour filters in the annular space. Both it and the photolysis lamp are surrounded by a tubular reflector coated on the inside with magnesium oxide, and when necessary the whole can be mounted in a tubular electric furnace. A general description of the apparatus which throughout our work has had several minor modifications is given in detail by Norrish, Porter and Thrush (13); the technique in use at present, represents a compromise among all the factors affecting its operation. Improvements have been effected by using highly transparent "spectrosil" quartz which transmits down to 1600 A, end plates of lithium fluoride, and vacuum spectrographs for the detection of transients whose absorption spectra lie in the far ultraviolet. Of great importance for the future is the reduction in the periods of the photoflash and specflash to achieve greater time resolution, and the development of highly transparent materials for construction of apparatus suitable for shorter wave photolysis than is at present available.

It may readily be calculated that the "instantaneous" dissipation of only 1 joule of energy (i.e. about 0.05% of the total output of the photoflash) by 150 ml of gas at 1.0 mm pressure will raise the temperature of the reactant by about 5000°C for there is not time for cooling during the short period of the flash. Thus the early results of flash decomposition are more properly regarded as flash pyrolysis than flash photolysis, and unless steps are taken to neutralise this rise in temperature by diluting the system with a large excess of inert gas, we cannot expect to study the photochemical effects divorced from thermal complications. This however is readily done: by the introduction of inert gas at pressures of 100 to 500 times that of the reactant, the temperature rise can be kept. below 10°C, which for practical purposes may be regarded as iosthermal, while for reactions in solution of course, there is no problem. On the other hand we may take advantage of flash heating in undiluted systems to administer an adiabatic shock which for many purposes is superior and certainly simpler than the technique of shock wave kinetics. This arises from the fact that, by flash-heating, the whole system is instantaneously and nearly homogeneously heated to high temperatures, making possible the detection of the transient products of pyrolysis and growth and decay of intermediates in the chain reactions leading to explosion in suitable systems. Indeed, it is the homogeneity of the explosive processes which makes it possible to observe in absorption the unexcited radicals taking part; we have in fact in a reaction vessel 1/2 metre in length a "flame front" virtually 1/2 metre thick which is very different from the thin element propagating an explosive wave. This is important because it makes possible for the first time the observation of the reactions of unexcited species leading to and taking part in explosion as well as the electronically excited species to which we were limited in the past.

Thus there are two ways in which we can employ the techniques of kinetic spectroscopy and flash photolysis: the ISOTHERMAL method and the ADIABATIC method. The field of their application is almost unlimited; I must content myself with general remarks, and three specific examples.

The first objective of flash photolysis, namely to observe the growth decay of radical species by kinetic spectroscopy has been achieved; following the first demonstration of the dissociation of chlorine, the spectrum of the ClO radical was first seen in absorption on flashing a mixture of chlorine and oxygen. Its origin was ascribed to the almost complete dissociation of chlorine, and to the reaction to be expected from the chlorine atom in an atmosphere of oxygen. It was possible to show that the sequence of reactions:

$$Cl_2 + h\nu = Cl + Cl$$

$$Cl + O_2 = ClOO$$

$$ClOO + Cl = ClO + ClO$$

$$ClO + ClO = Cl_2 + O_2,$$

in which the final state of the system is the same as the first, provides a complete basis for explaining the reaction. The study of this reaction (14) constituted an early success in the application of flash photolysis to chemical kinetics.

Dr. Hussain has collected references to some sixty simple free radicals and atoms which have been discerned and characterised in absorption, either by isothermal flash photolysis or by adiabatic flash pyrolysis and explosion. Prominent among them are CH, CH_2, CH_3, NH, NH_2, OH, HCO, HNO, CN, CS, ClO, BrO, IO, NCl, NCl_2, PH, PH_2, PO, PN, SH, SO, SiO, TeO, TeH, W, Te, Sn, Hg, Fe, Mn and also highly vibrating states of several molecular species, such as O_2. The collection of this information is the first step towards identifying the nature of radical reactions observed by kinetic spectroscopy. To illustrate this we shall now consider two examples of the application of the isothermal technique, the first involving the primary photolysis of nitrosyl halides, the second the secondary reactions associated with the photolysis of nitrogen dioxide, chlorine dioxide and ozone.

III. VIBRATIONAL EXCITATION BY PRIMARY REACTION. THE FLASH PHOTOLYSIS OF NITROSYL HALIDES--VIBRATIONAL RELAXATION

The sequence of spectra shown in Fig. 2 show the course of the photolytic dissociation of nitrosyl chloride, typical also of nitrosyl bromide, which takes place in the region of 2600 Å (15).

A study of a large number of plates showed that the primary product, NO, is highly vibrationally excited in the ground state comprising all levels from $v'' = 11$ to $v'' = 0$. All these were observed in absorption in the β, γ, δ, and ε spectra of NO; the rotational temperature of the molecule was, however, unaffected. By using NO as a light filter surrounding the reaction vessel it was proved that these excited species do not have their origin in the secondary excitation of NO molecules and after consideration of all possibilities it was concluded that they are in fact the product of the primary photolysis of the halide, NOX:

$$NOX + h\nu = NO^* + X.$$

It was found that the relative "instantaneous" population of the higher levels of nitric oxide increased as the halide pressure decreased and that at first the level $v'' = 1$ was barely detectable. The decay of the higher excited levels was however extremely rapid and increased with the pressure of the halide yielding ultimately the level $v'' = 1$ which accumulated and was virtually the only excited level detectable after the photoflash. It was in fact established that the rate of decay is determined by the pressure of the unchanged nitrosyl halide, and that on the other hand, the effect of inert gases was not detectable.

The rapidity of decay of NO* and the specific effect of the parent NOX suggest that near-resonance transfer processes are operating the deactivation, as indeed is confirmed by the fact that the vibrational frequencies of NO in the range of levels $v = 11$ to $v = 1$ lie between 1900 and 1600 cm^{-1} while Burns & Bernstein (16) found 1800 cm^{-1}.

At this point, however, there arises an apparent anomaly. The observation of Pearse & Gaydon (17) showed that the first five levels in the ground state of NO can be populated by fluorescence as shown diagrammatically in Fig. 3. This fluorescence which consists of the banded $v = 0$ progression $A^2\Sigma^+ \leftarrow X^2\Pi$ of NO was also seen by Basco, Callear & Norrish (18) using the flash technique; yet by the same means they were unable to observe levels higher than $v = 1$ in absorption, with the exception of $v = 2$ <u>very</u> faintly, Fig. 4. It might be postulated that the higher levels are populated very weakly relatively to the first, but this is not so; Pearse & Gaydon from a measurement of the intensities of the fluorescent bands found the first five levels to be populated almost equally. Herein lies the problem: why is only the level $v = 1$ seen by kinetic spectroscopy and why do the higher levels $v = 2$, 3, 4 and 5 decay too rapidly to be observed in absorption when the same and higher

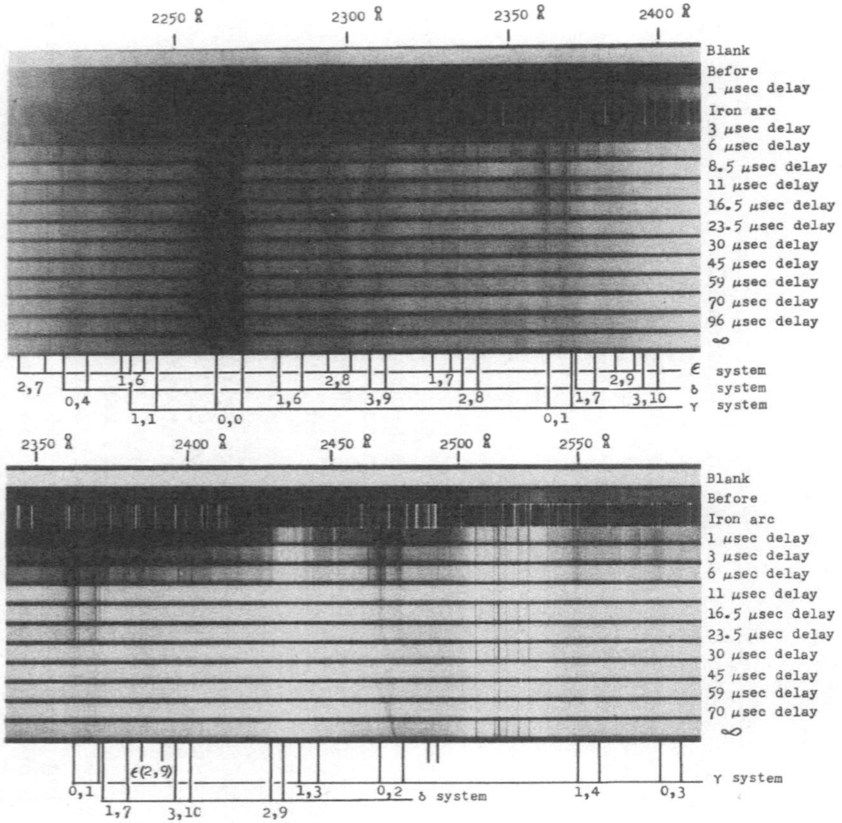

Fig. 2. Vibrationally excited NO from NOCl. Upper picture: pressure of NOCl = 1.0 mm Hg; pressure of N_2 = 375 mm Hg. Lower picture: pressure of NOCl = 2.0 mm Hg; pressure of N_4 = 420 mm Hg; flash energy 1600 J (15).

levels derived from the photolysis of the nitrosyl halides are readily detected and their decay, albeit rapid, easily followed in times measured in microseconds?

The solution to this apparent anomaly may be achieved by means of the two following hypotheses (19):

(1) The most favourable resonant collisions are between closely associated levels of the vibrating species, e.g.

$$NO_{\nu=n} + NO_{\nu=(n-2)} \rightarrow 2NO_{\nu=(n-1)}$$

and owing to change in frequency of levels due to anharmonicity, the most favourable of all will be obtained when the frequency levels differ by 2 as above.

(2) At the instant of production from the nitrosyl halide the HO* is formed in very high vibrating states: say, ν = 12, 11, or 10.

The vibrational energy of ν = 11 is 55 kcal, and since the bond strength of NO-CL is 38 kcal there is plenty of energy available from the light quantum (say, 98 kcal for 2800 Å) for this to occur. The same applies for NOBr. In consequence there is a gap between ν = 10 (say) and ν = 0 and in the absence of other

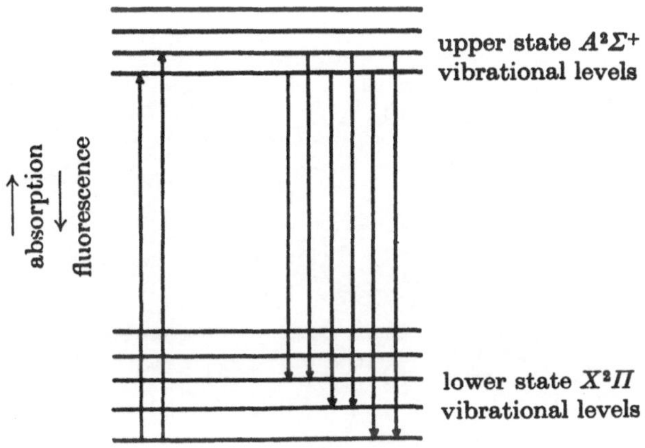

Fig. 3. Diagrammatic representation of population of vibrational levels of NO in the ground state by fluorescence.

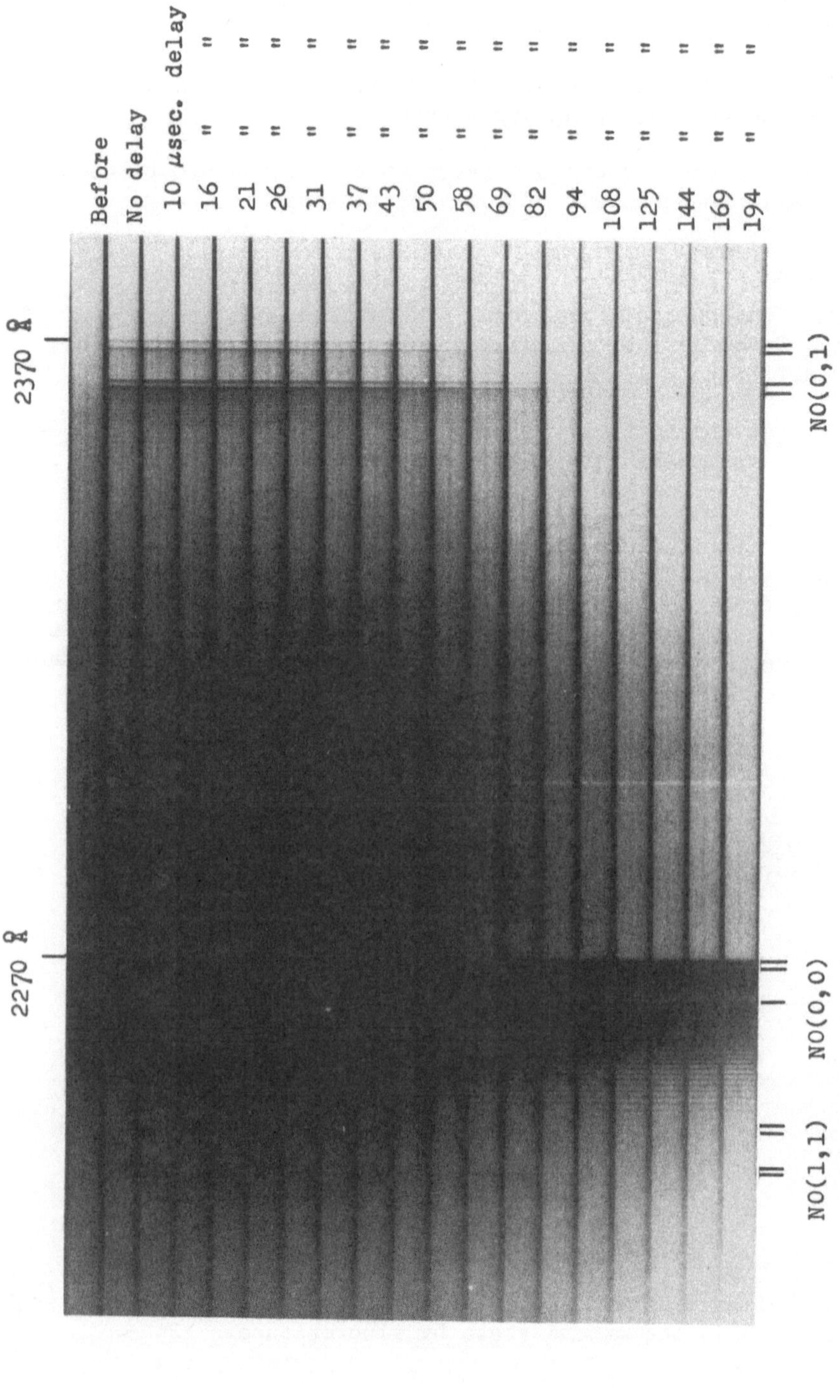

Fig. 4. NO $^2\Pi(v=1)$ produced by flash fluorescence of NO, showing decay. Pressure of NO = 5 mm Hg; pressure of N_2 = 600 mm Hg; flash energy 1600 J.

deactivating species (inert gases ineffective) the high vibrational levels cannot be relaxed. This of course is an ideal conception; lower levels will be built up by collisional deactivation by species such as the nitrosyl halides as we have seen, but it will be a relatively slow process compared with self-deactivation. As the lower levels are populated so will resonant self-quenching increase, but there will always be an irregular distribution which will cause a retardation, and further, since high levels are continually fed in by the flash, the irregular distribution will be preserved and all levels will be observed during its operation.

In contrast, when the first five vibrational levels of the ground state of NO are populated by fluorescence they are populated as we have seen above, nearly equally; thus the highly efficient process of self-quenching described in hypothesis (1) can take place as shown in Fig. 5, and all levels are deactivated to $\nu = 1$ when the resonant process must of necessity stop. The collapse of the pattern is so rapid that only the first level is seen to be overpopulated, and this can only be deactivated slowly by the inefficient process of collisional conversion to translational energy. If, however, we have a gap in the vibra-

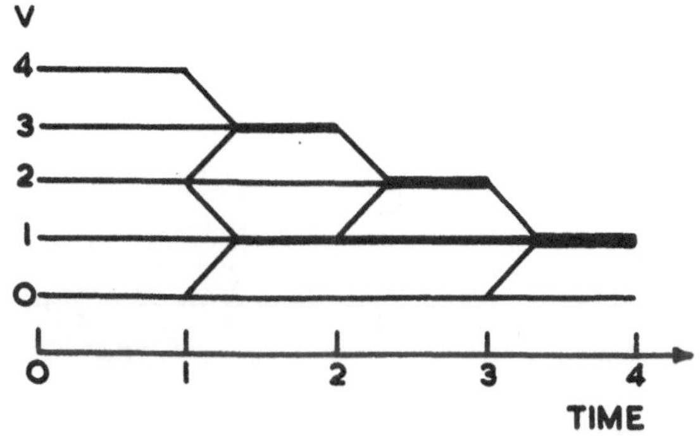

Fig. 5. Diagrammatic representation of relaxation of vibrational energy of NO by self-quenching.

tional distribution or a series of irregularities in the seauence of population of the pattern of vibrational levels, as with NO* derived from nitrosyl halides, the resonant deactivitation must be brought to a halt, or slowed down, shown for an ideal case in Fig. 6. In this case overpopulation of all higher levels is observed.

The overpopulation of the $NO_{\nu=1}$ level in the ground state by fluorescence (Fig. 4) makes possible the quantitative study of the relaxation reaction (18)

$$NO_{\nu=1} + M = NO_{\nu=0} + M^1.$$

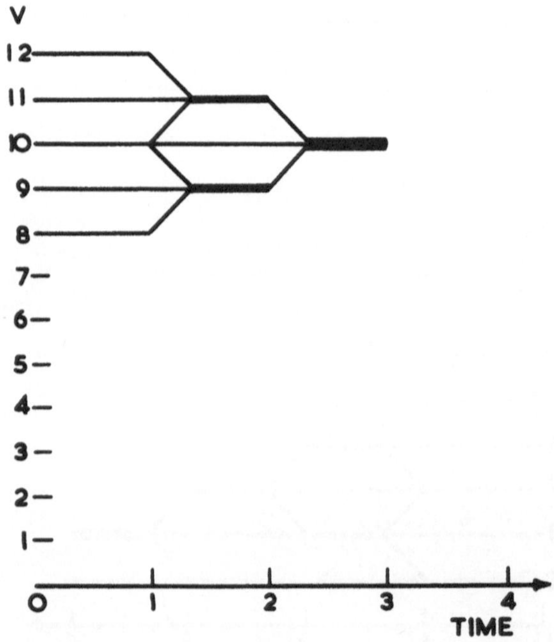

Fig. 6. Diagrammatic representation of relaxation of high levels of vibrational energy of NO, restricted by isolation (Ideal).

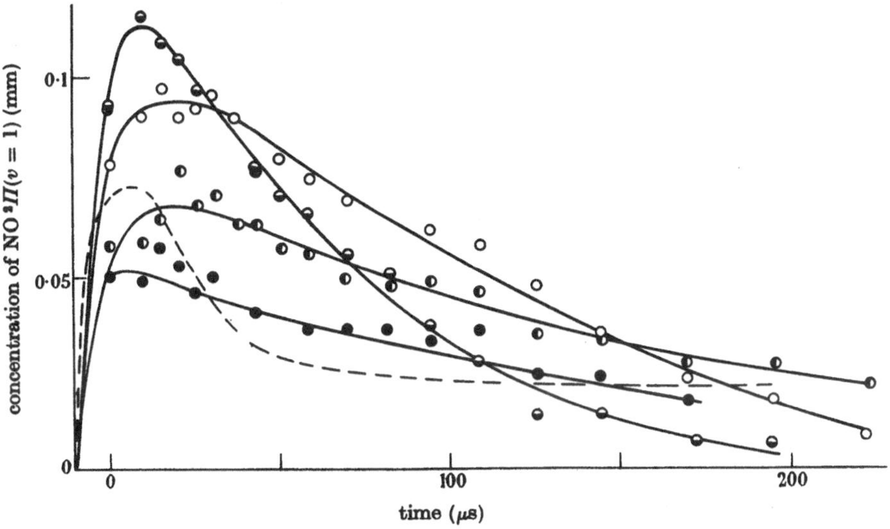

Fig. 7. Rise and Decay of NO $^2\Pi$ (ν=1) measured by plate photometry:
O 2 mm NO + 600 mm N_2·(1600 J)
● 2 mm NO + 220 mm N_2·(1600 J)
◐ 1 mm NO + 600 mm N_2·(1600 J)
◓ 5 mm NO + 600 mm N_2·(1600 J)
-- 50 mm NO + 467 mm N_2·(900 J)

This arises from the fact that the absolute concentration of $NO_{\nu=1}$ can be measured by plate photometry because the (0,1) band is visible spectroscopically in absorption with nitric oxide at atmospheric pressure, and since its concentration at equilibrium is given by

$$[NO^*] = [NO]e^{-\frac{h\nu}{kT}},$$

the photometric curves can be calibrated to give absolute concentrations by choosing one particular line in the band for measurement. In this way the curves shown in Fig. 7 were obtained. When plotted logarithmically they give good straight lines indicating first order decay from which the unimolecular constant k_3 can be obtained. $1/k_3$ is the mean lifetime τ of the excited species, and if this suffers Z collisions per second then P_1-0, the probability of energy transfer at one collision, is given by

$$P_{1-0} = \frac{1}{\tau Z} = \frac{k_3}{Z}.$$

k_3 can be split into two terms depending on relaxation by NO, and by any added gas M. Thus

$$k_3 = k_4(NO) = k_5(M) ,$$

and k_4 and k_5 may be calculated from the various values of k_3 derived from the curves of the type shown in Fig. 7 for nitrogen. The data shown in Table 1 show preliminary figures for the quenching probabilities of various added gases, the high value for water being probably due to chemical reaction.

TABLE 1

Molecule	NO	CO	H_2O	CO_2	N_2	Kr
P_{1-0}	3.55×10^{-4}	0.25×10^{-4}	7×10^{-3}	1.7×10^{-4}	4×10^{-7}	zero

Further studies (18) of relaxation by CO indicated unmistakably that the process occurs by resonant transfer of vibration:

$$NO_{\nu=1} + CO_{\nu=0} = NO_{\nu=0} + CO_{\nu=1}.$$

The concentration of CO* was measured by photometering the unresolved band of the fourth positive $A^1\Pi \rightarrow X^1\Sigma^+$ system which is visible in the spectrum of CO at atmospheric pressure and so can be used to measure in absolute terms the vibrational exchange between NO and CO shown in Fig. 8.

Studies of the photolysis of $(CN)_2$, CNBr and CNI generically represented as CNR by kinetic spectroscopy yield results similar to those described for NOBr and NOCl. These substances absorb at the short end of the quartz ultraviolet below 2300 Å and on flashing in the presence of inert gas yield vibrationally excited CN radicals up to $\nu = 6$ which are observed spectroscopically in absorption in the $\Delta\nu = 0 \pm 1$ and - 2 sequences of the violet ($B^2\Sigma - X^2\Sigma$) system at 3590, 3883, 4216 and 4660 Å. Decay sequences with time of CN* indicated the preferential production of CN* in the higher excited vibrational states and their decay by collision with CNR as with the analogous nitrosyl halide reactions, but owing to the very high extinction coefficient of the CN radical itself there was also detected, using colour filters, a high secondary population of CN^*_0 resulting from absorption by CN of light in the region 3500-4500 Å, far

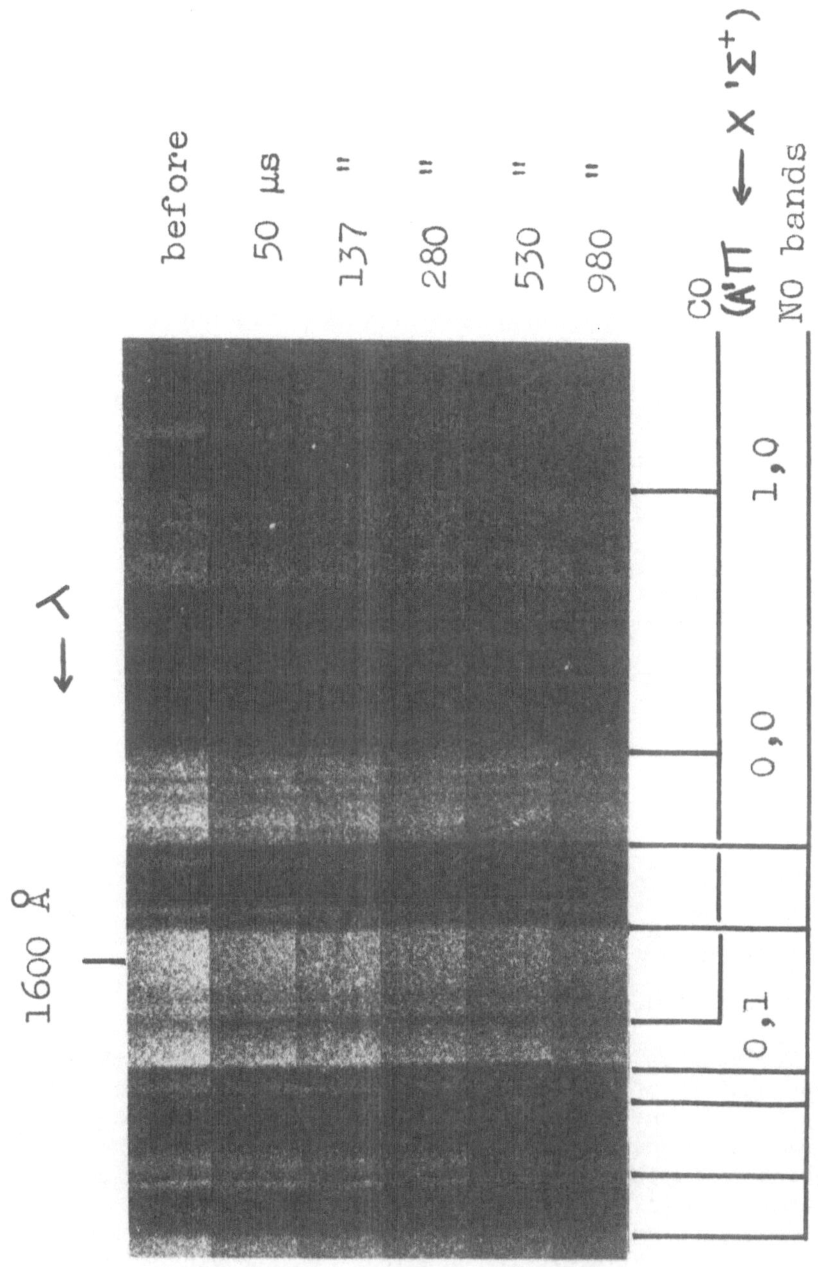

Fig. 8. Production of CO(v=1), ground state, by resonance with NO(v=1), ground state, (18).

outside the photolytic wavelengths of CNR. The process

$$CN \cdot X^2\Sigma(\nu = 0, 1, 2. . .) \rightleftarrows CN \cdot B^2\Sigma(\nu = 0, 1, 2. . .)$$

as indicated involves many reversible excitations during the flash, the reverse reaction taking place either by fluorescence or collision---but in the end only $\nu = 1$ persists as before.

IV. VIBRATIONAL EXCITATION BY SECONDARY REACTIONS

IV.A. The Reactions of Oxygen Atoms

The photolysis of nitrogen dioxide, chlorine dioxide and ozone studied by the techniques of classical photochemistry were all concluded to proceed by similar mechanisms, involving the primary generation of oxygen atoms, as follows (21):

Nitrogen dioxide:

$$NO_2 + h\nu = NO + O$$
$$O + NO_2 = NO + O_2;$$

Chlorine dioxide:

$$ClO_2 + h\nu = ClO + O$$
$$O + ClO_2 = ClO + O_2$$
$$ClO + ClO = Cl_2 + O_2;$$

Ozone:

$$O_3 + h\nu = O_2 + O$$
$$O + O_3 = O_2 + O_2^*$$
$$O_2^* + O_3 = O_2 + O_2 + O$$
$$O_2^* + M = O_2 + M^1 .$$

The quantum yield of the first two reactions in the near ultraviolet is of the order 2. In the case of O_3 it was measured to be up to 8 in the region of 2000-2500 A, but limited to 2 when photolysis occurs at the red end of the spectrum. Thus a chain reaction is indicated in the former case, which owing to the inherent simplicity of the system, must be propagated as shown by excited oxygen molecules, considered by the earlier workers to be an electronically excited species. On studying these reactions by isothermal kinetic means, the scheme of reactions shown above was confirmed, but in addition highly vibrating oxygen molecules in the ground state, cold rotationally and translationally, were produced

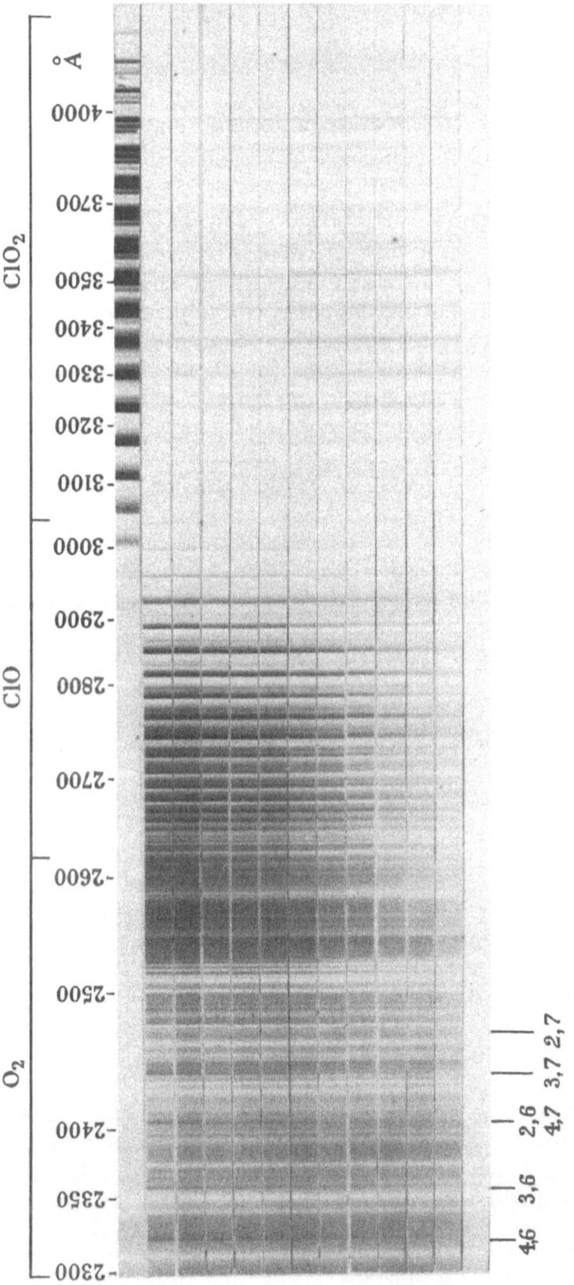

Fig. 9. Flash photolysis of ClO_2 (ClO_2 pressure, 0.5 mm Hg; N_2 pressure, 580 mm hg; flash energy 320 J) showing ClO and vibrationally excited O_2 (latter seen with difficulty owing to low dispersion) (22).

in each case. Thus the reactions

$$NO_2 + O = NO + O_2^*$$

$$ClO_2 + O = ClO + O_2^*$$

$$O_3 + O = O_2 + O_2^* ,$$

were indicated (22), (23). With NO_2, vibrational levels up to $\nu = 11$ were observed; with ClO_2, levels up to $\nu = 8$, and with O_3, levels up to $\nu = 17$-20. In each case more than half the exothermic energy of reaction appeared unequilibrated as vibrational energy of the oxygen molecule observed in absorption in the Schumann-Runge spectrum. Fig. 9 shows the flash photolysis of ClO_2 in which after flashing, the transient spectrum of the ClO radical is seen together with the absorption by highly vibrating oxygen molecules. The production of excited O_2^* is seen more clearly in Figs. 10 and 11 resulting from the photolysis of NO_2 and O_3, respectively.

These results led McGrath and Norrish (24) to the tentative generalisation that when an atom reacts with a polyatomic molecule, a large proportion of the exothermic energy of reaction is preferentially located initially as vibration in the newly formed bond, i.e.

$$A + BCD = AB^* + CD.$$

Qualitatively this seems reasonable since the main interaction must be visualised as between A and B, while the elimination of CD could well occur without much appreciable change in the interatomic distance between the parts C and D. The generalisation has now been widely confirmed. McGrath and Norrish (25) have shown by flash photolysis that the reactions

$$Cl + O_3 = ClO^* + O_2$$

$$Br + O_3 = BrO^* = O_2 ,$$

yield highly vibrating ClO and BrO with up to six quanta of vibration, while likewise, the reactions of ^1D oxygen atoms derived from ozone on reacting with a wide range of hydrides yield vibrationally excited OH (25), e.g.:

$$O + H_2O = OH^* + OH$$

$$O + NH_3 = OH^* + NH_2$$

$$O + H_2 = OH^* + H .$$

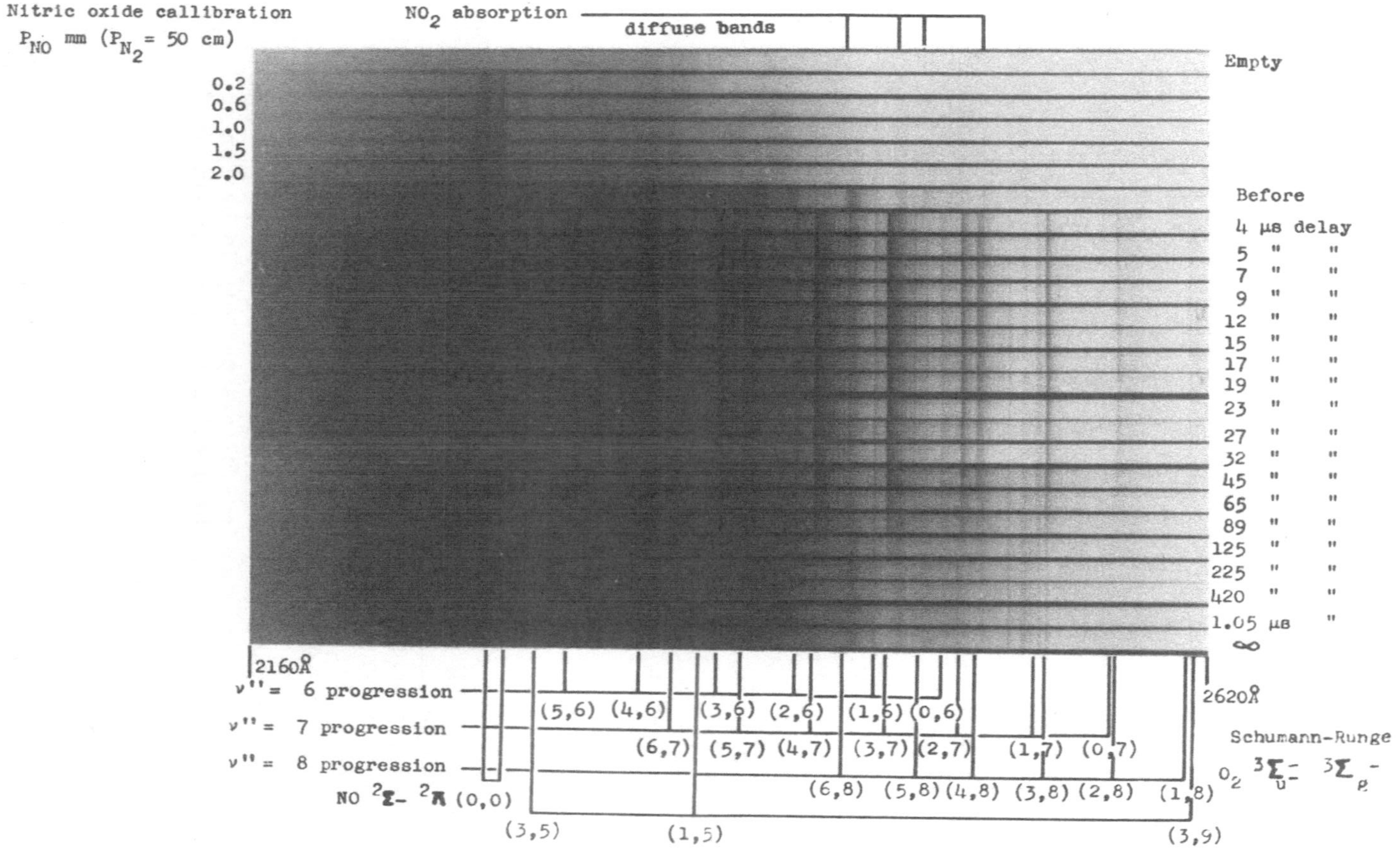

Fig. 10. Decay of vibrationally excited O_2 resulting from the flash photolysis of NO_2. NO_2 pressure, 2 mm Hg; N_2 pressure, 500 mm Hg; flash energy 2025 J (34).

Other examples, such as the reactions of hydrogen atoms observed by McKinley, Garvin and Boudart (27) and Cashion and Polyani (28),

$$H + O_3 = OH^* + O$$

$$H + Cl_2 = HCl^* + Cl$$

$$H + Br_2 = HBr^* + Br,$$

further confirm the correctness of our generalisation which invites detailed quantitative study, and must probably await greater time resolution in our technique before it can be achieved. For example, we cannot yet be sure whether the vibrationally excited products are produced ab initio in their highest vibrating state and relax subsequently, or whether a complete spectrum of vibrating states results directly as part of the reaction mechanism.

IV.B. The Photolysis of Ozone

The photolysis of ozone was first discerned as a chain reaction by Heidt and Forbes (21) and confirmed for pure ozone by Norrish and Wayne (29) who observed quantum yields up to 16 in the ultraviolet. The nature of the excited oxygen functioning as chain carrier would now appear to be identified as the vibrating molecule with more than 17 quanta of vibration. For the propagation of the chain the endothermic reaction

$$O_3 + O_2^* = O_2 + O_2 + O$$

requires 69 kcal, and this is supplied precisely by a molecule vibrating with more than 17 quanta. All those vibrating with less are visible by flash photolysis and decay by normal relaxation processes. Those with more react so rapidly with ozone molecules that they are not seen, except that they may be faintly discerned up to $\nu = 20$ as a consequence of competition between reaction and collisional deactivation. This conclusion is based upon the deduction that the oxygen atom is generated in the first electronically excited state, ^1D, lying 45 kcal above the ground state and that the chain reaction is propagated uniquely by ^1D oxygen atoms, because no chain reaction follows photolysis by "orange" light where the magnitude of the quantum is only sufficient for the generation of ^3P oxygen atoms.

The chemical proof that the oxygen atoms generated by the photolysis of ozone in ultraviolet light are in the ^1D state lies in the fact that when small quantities of water vapour are added to the system, the spectrum of vibrationally excited O_2 molecules is

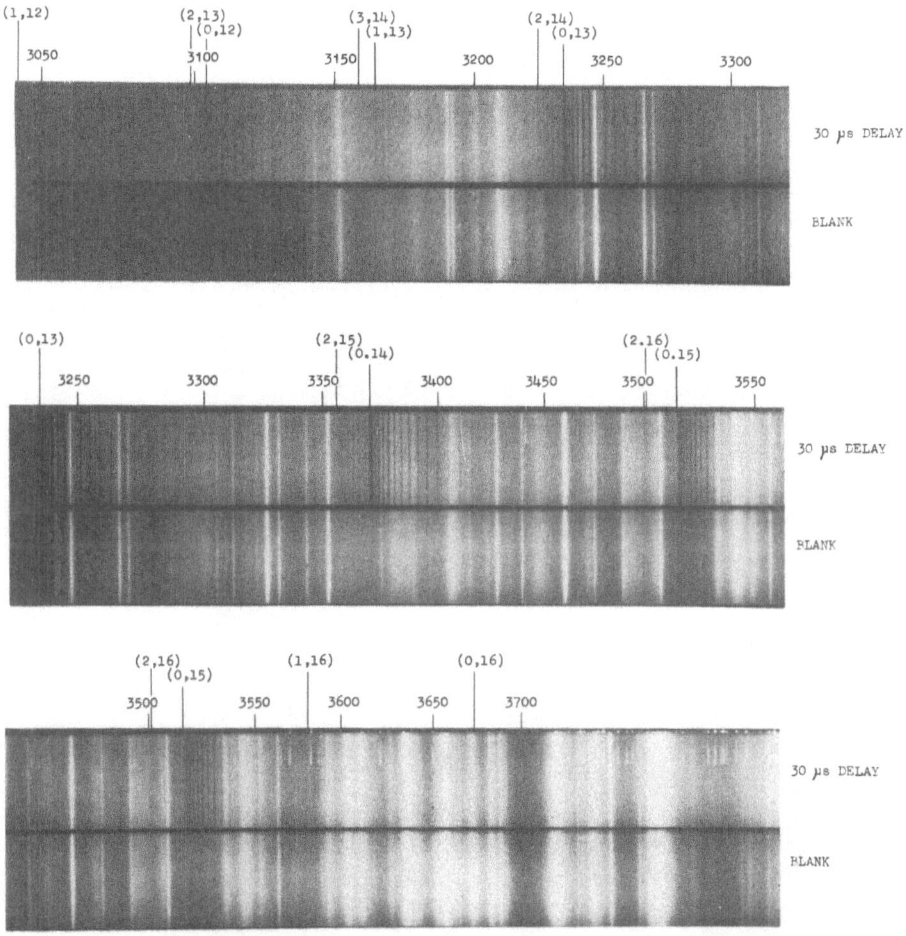

Fig. 11. Vibrationally excited O_2 produced by flash photolysis of ozone. O_3 pressure, 20 mm Hg; N_2 pressure, 800 mm Hg; flash energy 200 J (32).

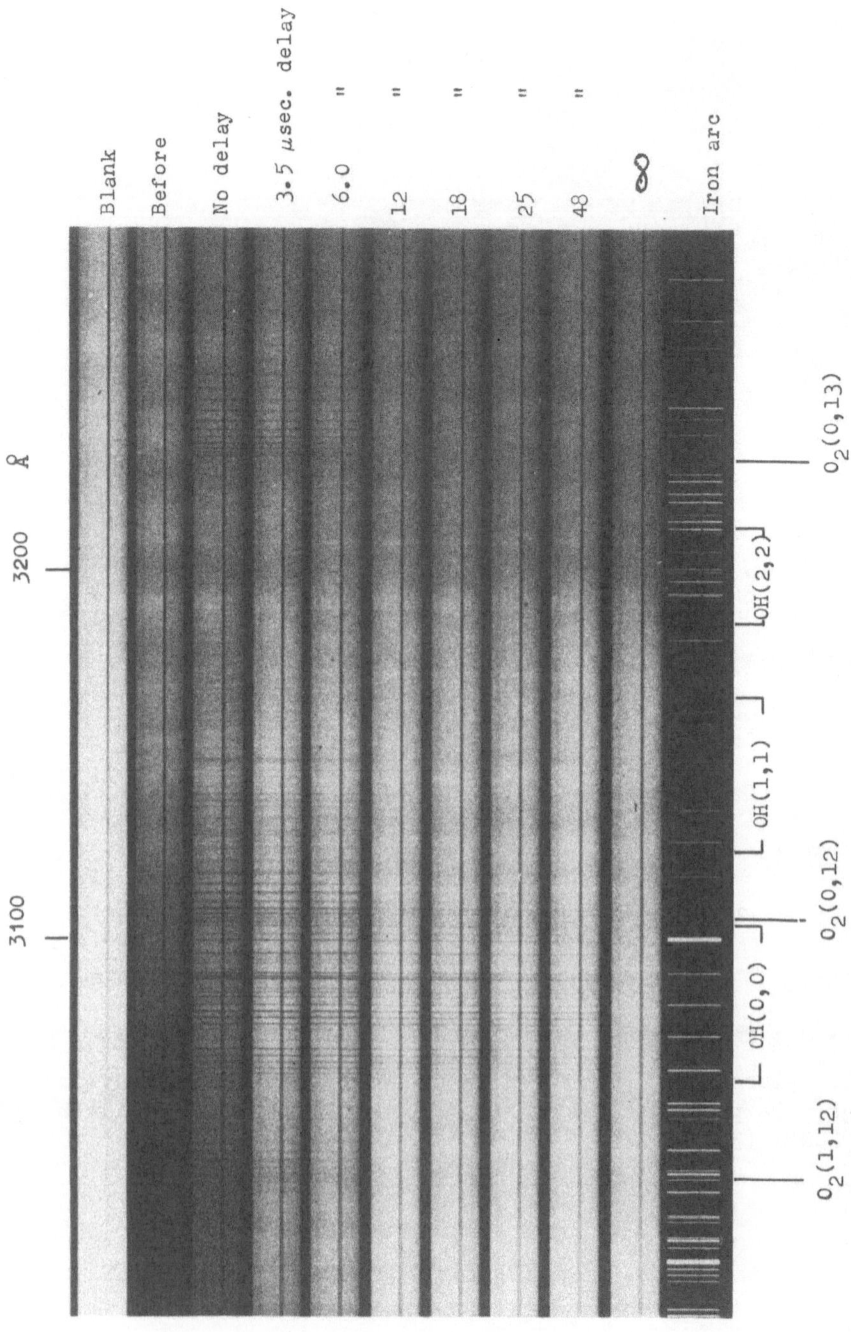

Fig. 12. Production of excited hydroxyl by reaction of $O(^1D)$ with water vapour. Pressure of ozone, 6mm Hg; pressure of water vapour, 4 mm Hg; pressure of nitrogen, 200 mm Hg; flash energy 1600 J (25).

progressively suppressed and replaced by the absorption spectrum of OH as seen in Fig. 12. This is to be correlated with the observation of Forbes and Heidt (39) that in "damp" ozone the quantum yield is increased to values as high as 130, as compared with their maximum of 8 for dry ozone, and in the light of our observation it may be concluded that an entirely new mechanism of chain propagation is substituted as a consequence of the successful competition of water with ozone for the oxygen atom, i.e.:

$$O \xrightarrow{+O_3} O_2 + O_2^*$$
$$O \xrightarrow{+H_2O} OH + OH^* .$$

This however can only take place if the O atom is excited to the ^1D state for the reaction of $O(^3P)$ with water is endothermics. We have

$$O(^3P) + H_2O = 2\ OH - 11\ kcal$$

$$O(^1D) + H_2O = OH + 34\ kcal .$$

In the presence of water, the chain reaction may be written

$$O_3 + hv = O_2 + O(^1D)$$

$$O(^1D) + H_2O = OH + OH$$

$$\left.\begin{aligned} OH + O_3 &= HO_2 + O_2 \\ HO_2 + O_3 &= OH + 2O_2 \end{aligned}\right\} \text{propagation ,}$$

followed by the chain ending by intercombination of radicals. This scheme satisfies the kinetic findings by Forbes & Heidt; it explains the appearance of the OH radical and demands the formation of the excited O atom. The reaction of O atoms with other hydrides referred to above is also equally dependent on the photolytic generation of $O(^1D)$ in the ultraviolet. It is significant that water has no effect on ozone photolysis in "orange" light where only ^3P oxygen atoms can be generated. The quantum yield remains unchanged at 2 in accord with the simple scheme (31):

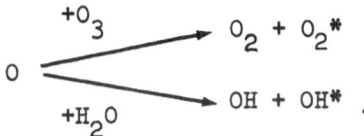

$$O_3 + h\nu = O_2 + O(^3P)$$

$$O(^3P) + O_3 = O_2 + O_2 ,$$

analogous to the photolysis of NO_2 and ClO_2.

It was shown by McGrath & Norrish (32) that the rate of decomposition of ozone by the secondary reactions subsequent to the flash is strongly affected by the addition of inert gases. Starting with 2.94 mm of O_3 and diluting with added gas to give a mixture of ratio of O_3/M = 1:163, the rate of disappearance of O_3 was determined by photometering the O_3 absorption in a series of spectra such as those shown in Fig. 13. In Fig. 14 are seen three typical curves showing ozone decay. From these curves could be measured the efficiencies of third bodies M in the back reaction

$$O(^1D) = O_2 + M = O_3 + M'.$$

When M is O_2 the ratio of O_2 to O_3 is 163:1 so it is hardly surprising that the above reaction predominates over the reaction

$$O(^1D) + O_3 = O_2 + O_2^* ,$$

to such an extent as to reverse all O_3 decomposition. With other added gases the relative efficiences of the molecules M for the three body recombination were determined as He = 1, A = 1, SF_6 = 1.5, CO_2 = 14, N_2 = 16, N_2O = 17.

The gases divide into two groups: (1) the inert gases and SF_6 and (2) N_2, CO_2 and N_2O. Group (1) gases exhibiting low efficiency are spherically symmetrical and chemically inert. Group (2) gases are much more efficient. It is possible with Group (2) that some form of chemical affinity is operative in forming intermediate transition species, and that a more facile energy transfer is possible due to readily stimulated vibrational modes. Further work along these lines may well prove rewarding.

V. APPLICATION OF THE ADIABATIC METHOD. THE STUDY OF EXPLOSIVE PROCESSES EXEMPLIFIED BY THE OXIDATION OF HYDRIDES

The gaseous oxidation of hydrides, including hydrocarbons occurs by exothermic processes which have the characteristics of chain reactions; that is to say they proceed by initiation, propagation, multiplication and extinction of reacting centres. The reactions are said to be autocatalytic and if the conditions are such that multiplication of propagating centres exceeds extinction, the process may develop to explosion. These conditions

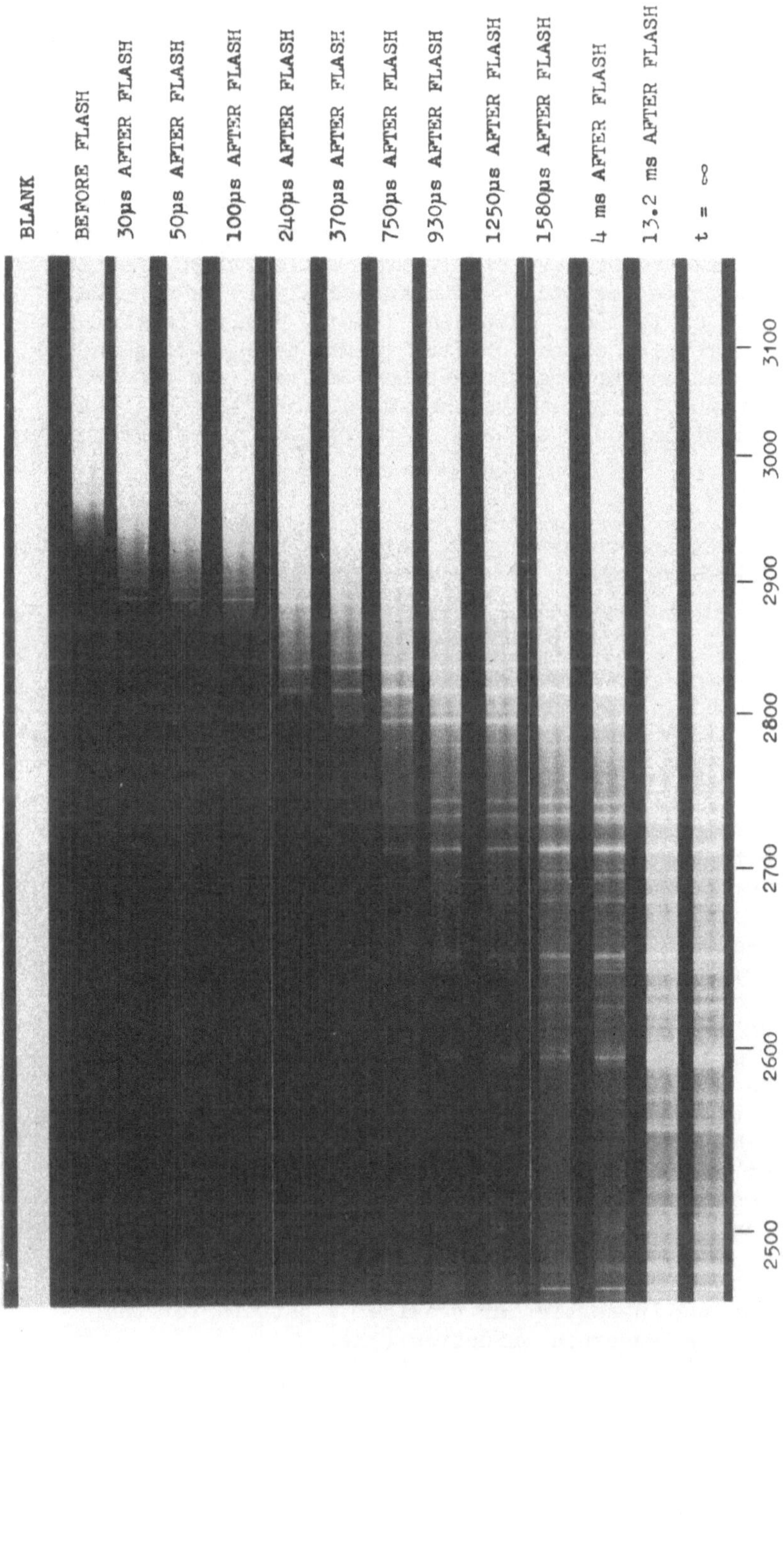

Fig. 13. Disappearance of ozone spectrum in time after flash. Flash energy 1280 J; O_2/N_2 mixture ratio 1:163: ozone pressure 2.93 mm Hg.

depend on the parameters of temperature, total pressure, relative concentrations of reactants, catalytic activity of the surface in initiating or terminating reaction chains, the geometry of the reaction vessel, and the activity of added catalysts and inhibitors. The variation of these parameters gives rise to sharp limits of explosion, and by judicious kinetic experiments the separate effect of each can be isolated and defined by keeping all but the one under examination constant.

The development of the slow reaction and the incidence of ignition are subject to an induction or incubation period during which autocatalysis occurs (initially exponentially) to a steady state or to explosive reaction. This autocatalysis is dependent on the magnitude of the "net branching factor," which is the result of the interplay of the physical parameters leading to multiplication and extinction of reaction centres. In the notation of Semenov, f represents the sum of the reactions leading to multiplication and g the sum of those leading to extinction:

$$(f - g) = \Phi,$$

where Φ is the net branching factor, which may obviously be positive or negative. The development of the reaction velocity (V) in time (t) is given by

$$V = Ae^{\Phi t},$$

where the pre-exponential term A varies only slowly and in a much less dramatic way than Φ, with changing kinetic conditions. When Φ is negative from the beginning a finite and small stationary reaction velocity is imposed. When Φ is positive, rapid exponential development of velocity to explosion may occur. This is the case with the reaction of hydrogen with oxygen which shows sharp explosive limits dependent on the parameters listed above. There exist cases, however, where Φ starting positive, may give rise to exponential development of the reaction in a big way, but owing to consumption of reactants or varying catalytic factors may become negative during the course of reaction which, as it were, starting hopefully towards explosive build-up is finally quenched to a stationary state and subsequent decline, by the failure of the net branching factor to remain positive. Such reactions are termed degenerate explosions by Semenov. They are distinguished by having a small but positive initial value of Φ and depend for branching on the reaction of a "precariously stable" intermediate which builds up as the reaction proceeds and which can be detected by kinetic and analytical observation. The recognition of degenerately branched chain reactions represents the culminating triumph in Semenov's interpretation of branching chain reactions and in particular provides a pattern for the understanding of hydrocarbon oxidation (33).

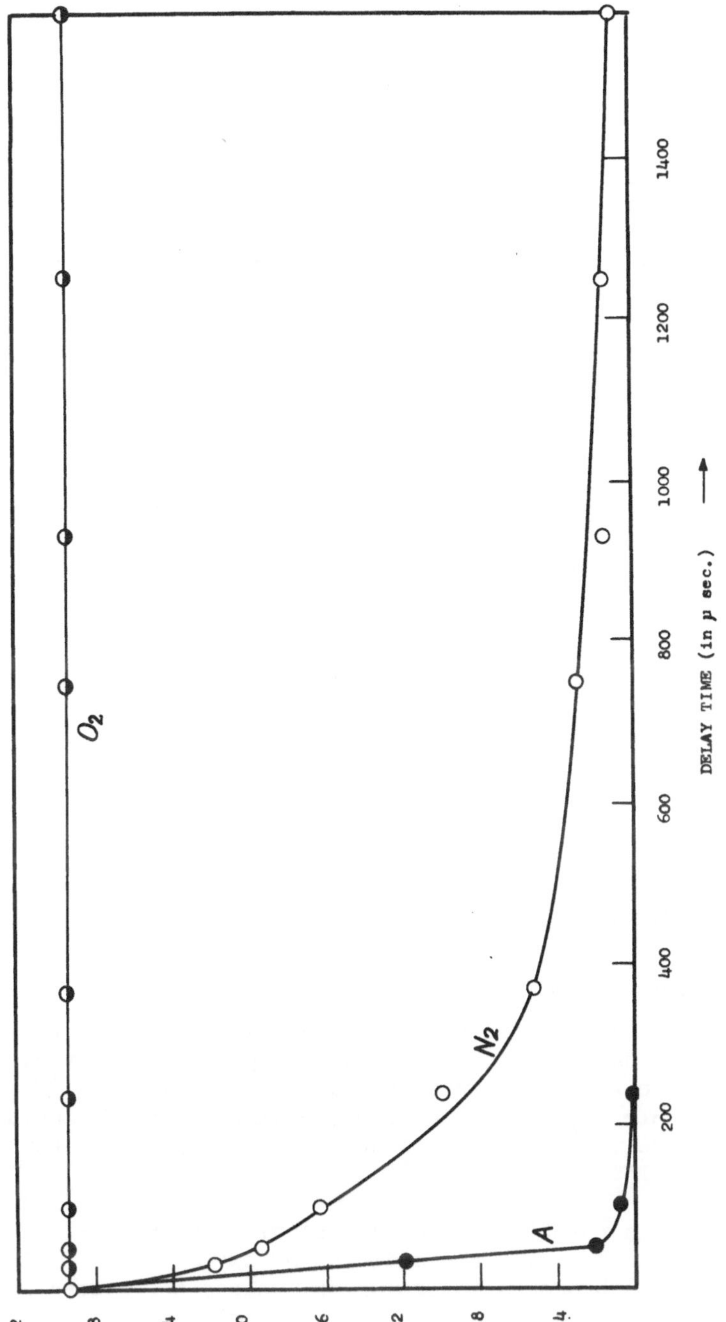

Fig. 14. Typical ozone decay curves for O_3/N_2, and O_3/O_2 mixtures. The mixture ratio in all cases is 1:163.

But while giving us the overall pattern of reaction, neither the experimental methods nor the mathematical conceptions were capable of exposing the intimate nature of the precise reactions involved. These were deduced in some instances from circumstantial evidence, with the gradual realisation that atoms and free radicals are more often than not concerned in the chain processes.

It has remained for flash photolysis and kinetic spectroscopy not only to confirm and amplify the general conclusions of the classical studies of chain reactions, but also to provide objective proof of the nature and reactions of the transient participating species. For this purpose we use the adiabatic method taking advantage of the free radicals produced by pyrolysis and photolysis for initiation, and flash heating to generate temperatures suitable to sustain the propagation and branching reactions upon which the autocatalytic chain reaction depends. We take for example the case of hydrides.

Since oxygen does not absorb energy from the photolytic flash under the condition imposed by the limitations of the transparency of quartz, it is fortunate therefore that many hydrides absorb sufficiently to provide the necessary pyrolysis for initiation. This is true for hydrogen sulphide, hydrogen telluride, ammonia, hydrazine, and phosphine, all of which photolyse isothermally and pyrolyse adiabatically by eliminating a hydrogen atom:

$$XH_n + hv = XH_{n=1} + H \ .$$

The growth and decay of the free radical $XH_{n=1}$ so generated can be followed by kinetic spectroscopy. Under pyrolytic conditions however when the concentration of free radicals generated may be high, the above reactions may be followed by further elimination of hydrogen from the free radical, e.g.

$$XH_{n=1} + XH_{n-1} = XH_n + XH_{n-2}.$$

This is true for example for ammonia (34) which under isothermal conditions gives only NH_2 but under adiabatic conditions yields NH radicals as well. We observe the same result with PH_3 (35), H_2S (36), and H_2Te (37), the last two yielding HS and S, and HTe and Te, respectively, even under isothermal conditions as shown for example in Fig. 15.

The pyrolytic reactions under our conditions are generally limited in extent, but with the addition of oxygen in sufficient quantity oxidation proceeds to explosion, unless the system is partially cooled by the addition of an inert diluent. Sufficient

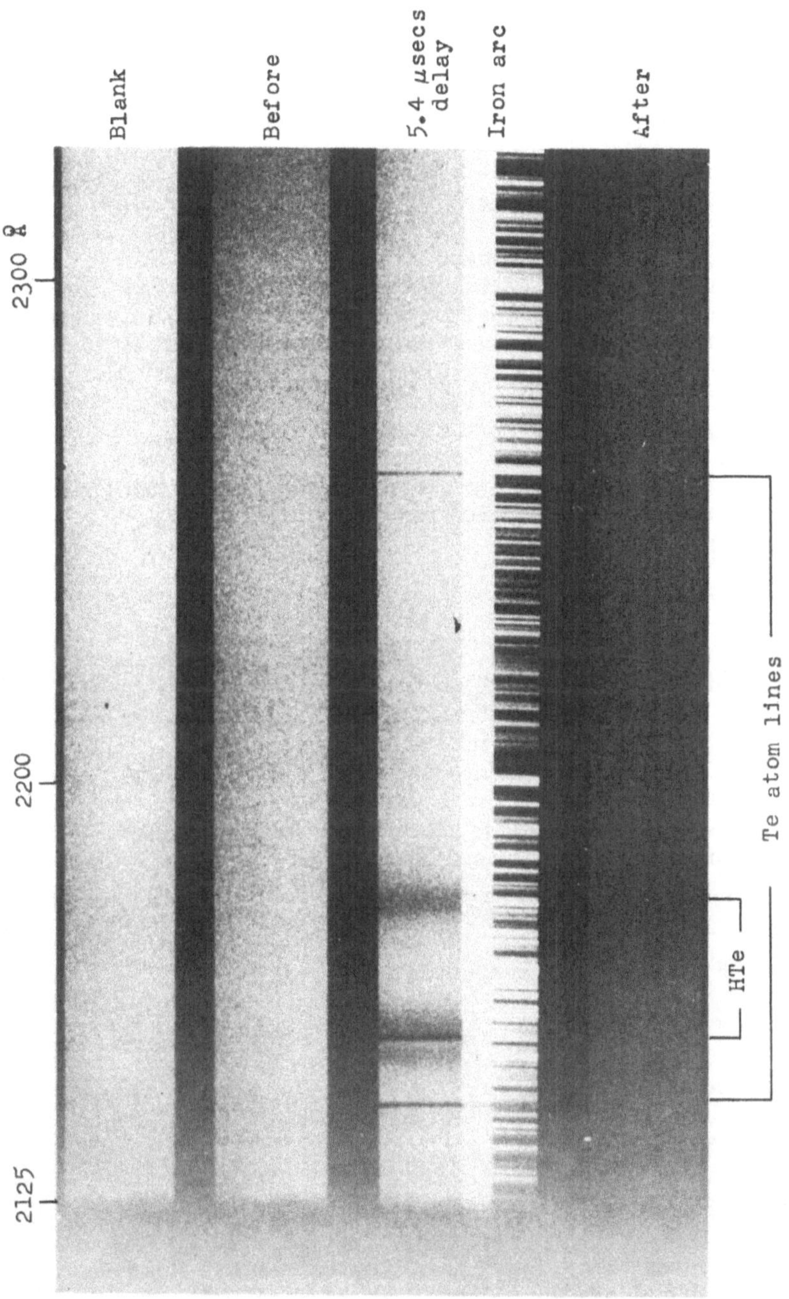

Fig. 15. Flash photolysis of tellurium hydride. Pressure of TeH_2, 0.25 mm Hg; pressure of N_2, TeH_2, 250 mm Hg; flash energy, 2500 J (37).

excess fuel or oxygen has the same effect. Under such conditions the oxidation proceeds by a quenched chain reaction, and is much more limited in extent.

The development of reaction from initiation to explosion involves an incubation period of less than a millisecond, and in oxygen-rich mixtures the onset of ignition is marked by a copious burst of hydroxyl radicals. It is clear that in all cases studied the hydroxyl radical acts as a chain carrier.

Hydrocarbons on the other hand do not in general absorb light transmitted by quartz (with the exception of highly unsaturated compounds) and the fuel oxygen mixture therefore does not respond to the flash. To initiate explosive reaction it is necessary to add a small quantity of sensitizer such as chlorine, nitrogen peroxide or alkyl nitrite. These, by absorbing strongly, raise the temperature of the system, and simultaneously photolyse and pyrolyse to give free atoms or radicals which act as initiators. Nitrogen peroxide for example absorbs strongly throughout the spectrum and yields oxygen atoms which give ready initiation (38):

$$NO_2 \rightarrow NO + O.$$

In Fig. 16 is shown a sequence of absorption spectra illustrating the explosion of a mixture of $2H_2 + O_2$ sensitized by nitrogen peroxide (39). The growth and decay of the OH radical is seen in the (0, 0) and (0, 1) bands of the transition $^2\Sigma^+ - {}^2\Pi$. This and the earlier study of the reaction of oxygen atoms with hydrogen by Norrish and Porter (38) go far towards confirming the scheme of oxidation of hydrogen proposed by Lewis and Von Elbe (40) of which the following are some constituent reactions:

$$\left.\begin{array}{l} OH + H_2 = H_2O + H \\ H + O_2 = OH + O \\ O + H_2 = OH + H \end{array}\right\} \text{propagation and branching}$$

$$\left.\begin{array}{l} OH + \text{surface} = \text{products} \\ H + O_2 + M = HO_2 + M' \end{array}\right\} \text{termination .}$$

The explosion of hydrocarbons sensitized by amyl nitrate was studied by Erhard & Norrish (41) and is illustrated in Figs. 17 and 18, which show the effect of adding tetraethyl lead to a mixture of hexane and oxygen. The first shows the ignition in the absence of the addendum with the rapid disappearance of the spectrum of the sensitizer on flashing, followed by an incubation

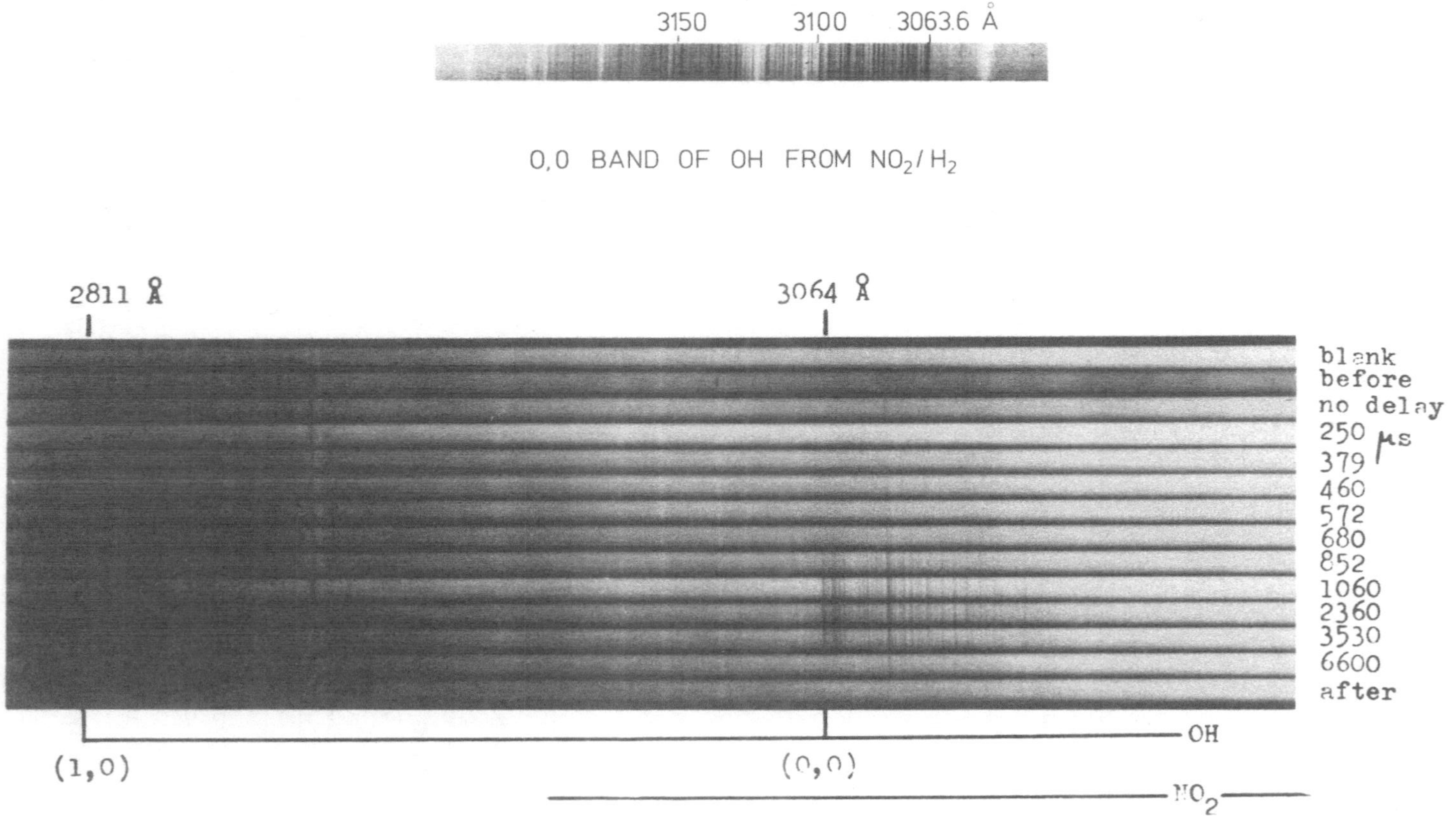

Fig. 16. (a) Photon-reaction of $NO_2 + H_2$ giving OH radical. Pressure of H_2, 2 mm Hg; pressure of NO_2, 2 mm Hg; no delay (38). (b) Explosion of $2H_2 + O_2$ sensitized by NO_2 showing the growth and decay of the OH radical. Pressure of NO_2, 2 mm Hg; pressure of $(2H_2+O_2)$ 15 mm Hg; pressure of N_2 30 mm Hg; flash energy 3300 J (39).

period of 875 μsec to the onset of explosion as marked by the sudden growth of the OH radical. The second shows the ignition under identical conditions in the presence of the addendum. It is seen that the incubation period is increased some three-fold to ∿ 2600 μsec, while during the growth to explosion the spectrum of gaseous lead oxide is strongly developed. At the point of ignition the PbO spectrum disappears completely and is replaced by the resonance spectrum of lead. Both the OH and the Pb spectra are very faintly visible before ignition. These and other experiments in which ignition was observed photoelectrically by the sudden growth of OH emission, throw light upon the mechanism of antiknock in the internal combustion engine which we conclude to be dependent on the moderating effect of Pb and PbO on the development of the autocatalytic growth to explosion.

Knock has been proved by Miller & Male (42, 43) to be due to the homogeneous detonation of the residual charge in the cylinder at the end of the ignition stroke. It is believed to be due to the generation of centers of autoignition (peroxides, aldehydes, etc.) due to adiabatic rise of temperature, which replaces the smooth explosion wave generated by the spark ignition. We have concluded that the tetraethyl lead clearly must operate in the gas phase and suggest that (1) it may remove the centers of autoignition by reduction - e.g. peroxides may be removed by

$$R.OOH + Pb\ (C_2H_5)_4 \rightarrow R + OH + PbO + \text{products}\ ,$$

and (2) moderate the liberation of energy as the reaction develops to explosion by the following reactions

$$OH + Pb \rightarrow PbOH \xrightarrow{+OH} Pb\ (OH_2) \xrightarrow{-H_2O} PbO$$

$$PbO + R \rightarrow RO + Pb.$$

Thus by alternate oxidation and reduction from lead to lead oxide and back again the atomic lead and lead oxide can intervene in chain propagation by the removal of OH, and so by shortening chains retard their development. With the onset of explosion, the PbO is instantaneously decomposed to atomic lead, which as the system cools is finally deposited on the surface of the reaction vessel.

The question as to whether moderation of the explosive process occurs in the gas phase, or by chain ending on heterogeneous particles of lead or lead oxide ("smoke") would appear to be answered by these results, since no "smoke" is observed during the course of the reaction, which is seen to be completely homogeneous. In contrast the addition of tetraethyl tin which has

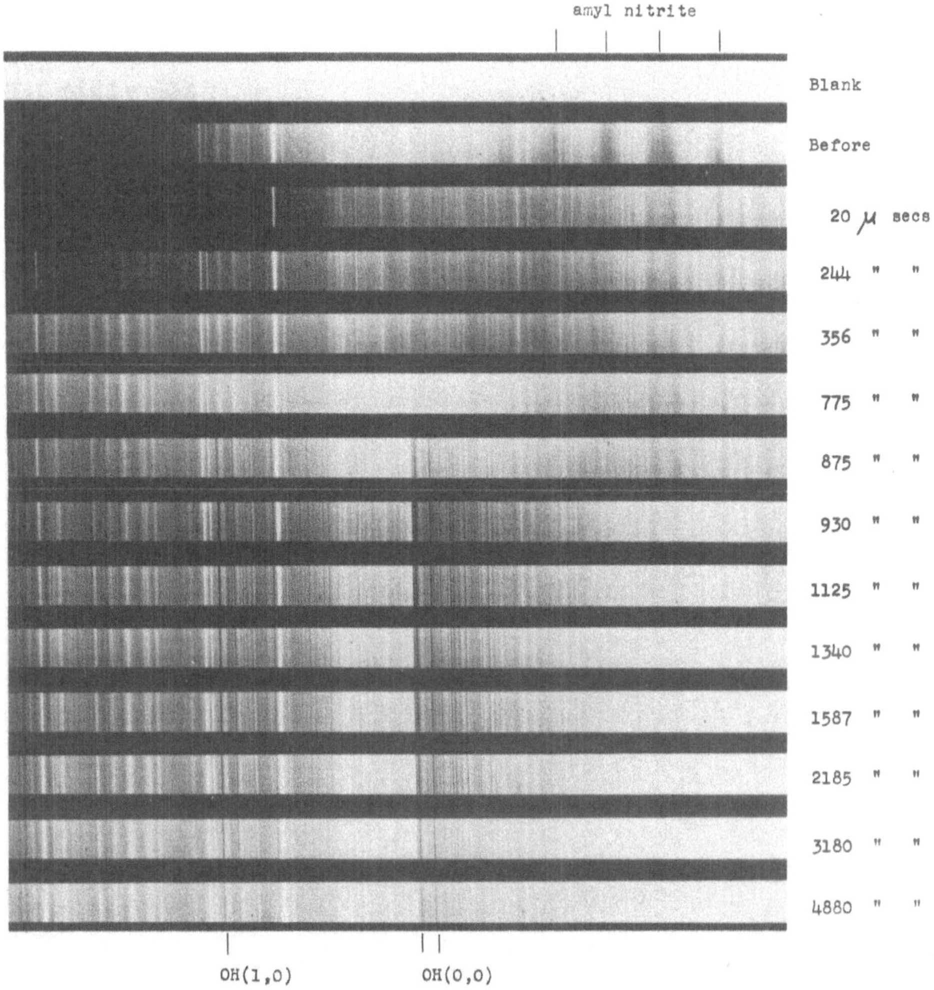

Fig. 17. Explosion of hexane and oxygen sensitized by amyl nitrite. Pressure of C_6H_{14}, 2 mm Hg; pressure of $C_5H_{11}ONO$, 2 mm Hg; pressure of O_2, 32.5 mm Hg; flash energy 2000 J (41).

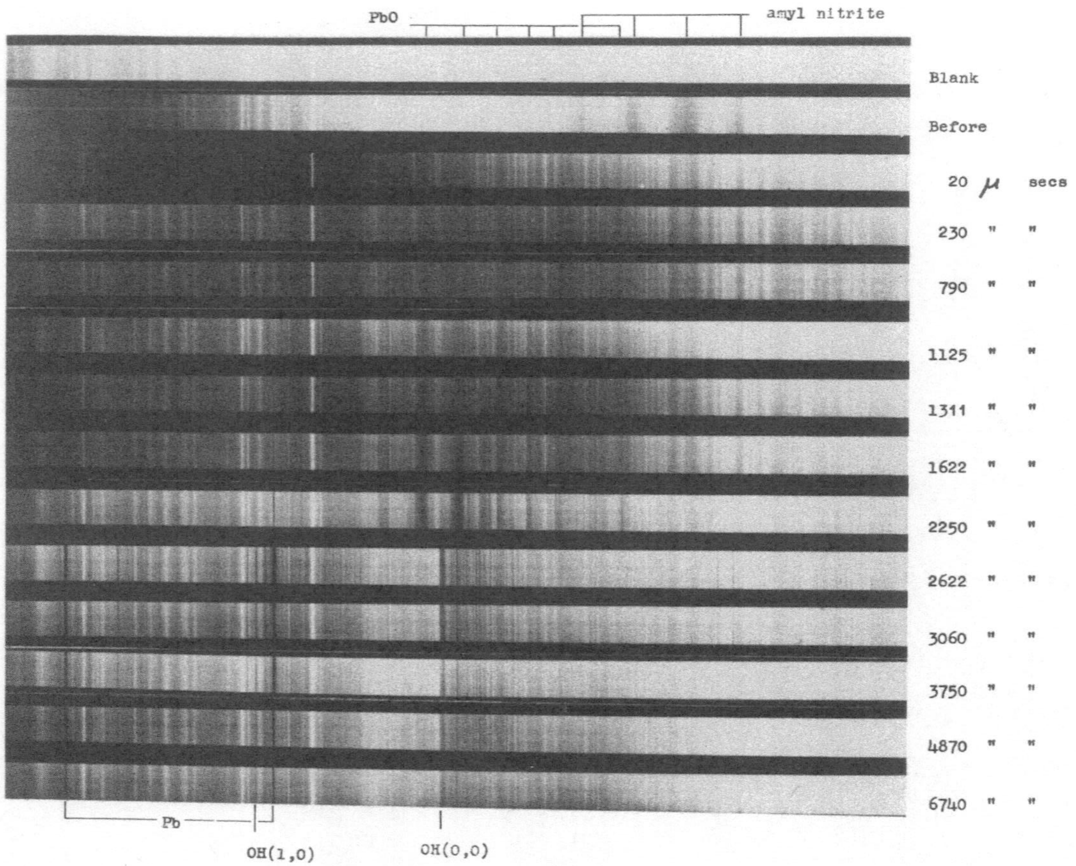

Fig. 18. Effect of tetraethyl lead on the hexane explosion. Pressure of C_6H_{14}, 2 mm Hg; pressure of $C_5H_{11}ONO$, 2 mm Hg; pressure of O_2, 32.5 mm Hg; pressure of tetraethyl lead, 0.2 mm Hg; flash energy 2000 J (41).

no antiknock action is accompanied by the copious formation of smoke. There is no sign of the production of gaseous SnO during the incubation period and the reactions leading to ignition (44). This is due to the lower volatility of SnO.

Many other studies of the affects of addenda on explosive reactions of hydrocarbons have been made by Callear & Norrish (44) with interesting results which cannot be discussed here. Reactions of this kind provide a plentiful source of free radicals and atoms derived from the addenda in high temperature reactions.

The growth and decay of free radicals as we pass through ignition is shown for the combustion of acetylene sensitized by NO_2 in Figs. 19 and 20. The curves were obtained by plate photometry of the various radical spectra seen in absorption at increasing intervals after initiation. They are also typical of curves obtained for ethylene and methane (46) and indicate the growth and decay of the observed radicals with time, though they cannot be compared in terms of absolute concentration since the extinction coefficients of the radicals are at present unknown. The combustion of acetylene and ethylene were shown by Bone (47) to depend on the stoichiometric equations

$$C_2H_2 + O_2 = 2CO = H_2$$

$$C_2H_4 + O_2 = 2CO + 2H_2 ,$$

according to which there is an apparent preferential burning of carbon. With oxygen in excess, water is formed, while in fuel-rich mixtures free carbon in the form of smoke is produced. These two conditions are very sharply distinguished on either side of the fuel-oxygen ratio of 1:1. This classical result is very clearly confirmed by the curves shown in Figs. 19 and 20. In the former we have an oxygen-rich system and the formation of water is indicated by the copious display of OH; in the latter (the fuel rich system) the OH is barely in evidence and the precursors of free carbon are observed in the CH, C_2, and C_3 radicals. The change from one type of display to the other takes place extremely sharply at the fuel-oxygen ratio of 1:1, nitrogen peroxide for this purpose being counted as oxygen.

The CN radical which is strongly in evidence in fuel-rich systems is derived from the sensitizer. It has been shown (48) that during the induction period of about 1/2 msec the temperature rises exponentially -- slowly at first and very sharply at the end. With the sudden appearance of the free radicals the explosive reaction is complete: we are witnessing in fact the after-burning of hydrogen in oxygen-rich mixtures, and the after-cracking of the fuel in fuel-rich mixtures. The only radical which can be seen during the induction period before ignition is the OH radical which

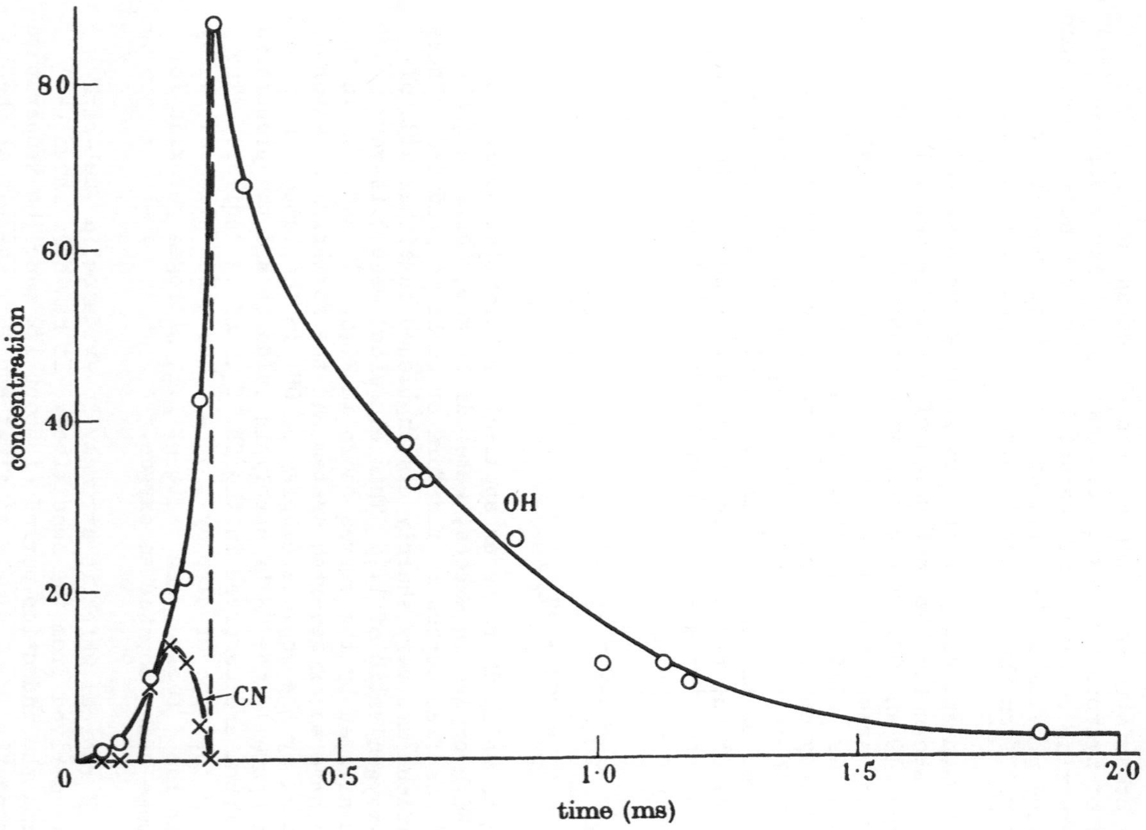

Fig. 19. Growth and decay of OH and CN radicals in oxygen-rich mixture of acetylene and oxygen (NO_2 counting as O_2). Pressure of C_2H_2, 10 mm Hg; pressure of O_2, 10 mm Hg; pressure of NO_2, 1.5 mm Hg (45).

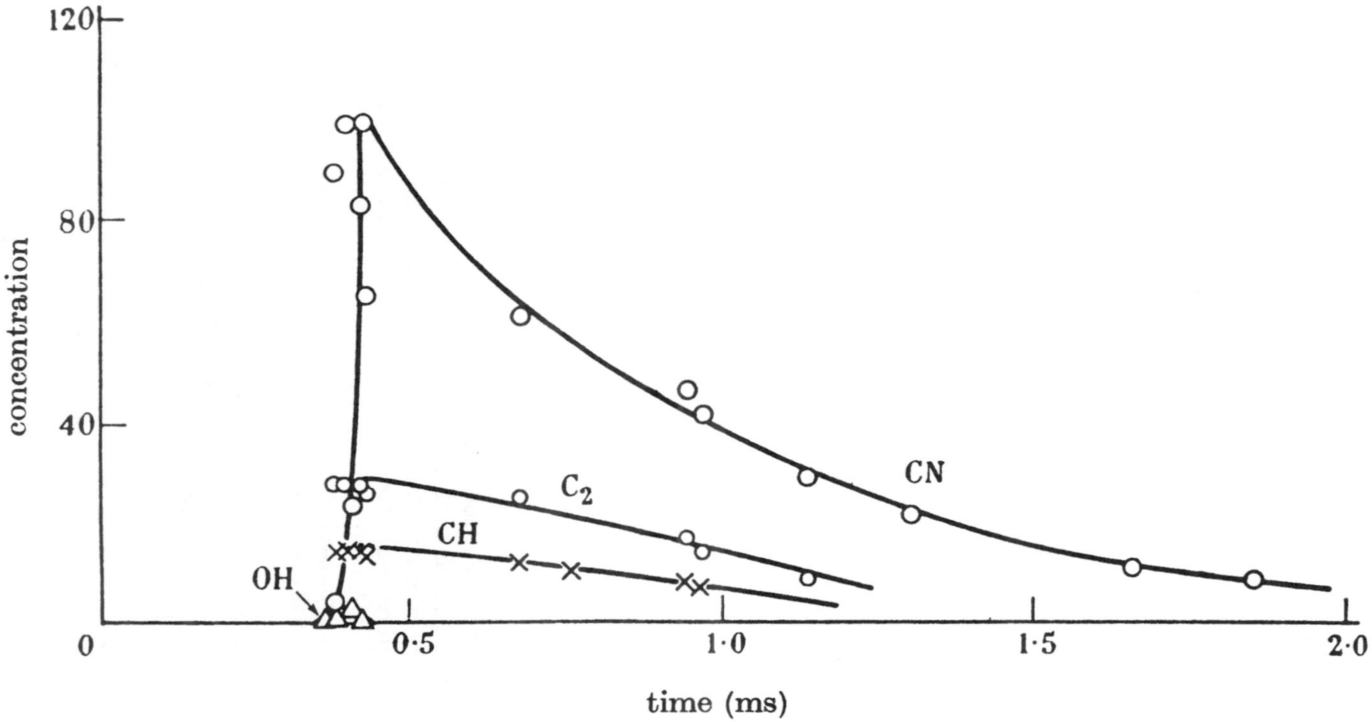

Fig. 20. Growth and decay of radicals in fuel-rich mixture of acetylene and oxygen. Ordinates of curves not comparable since extinction coefficients of radicals unknown. Pressure of C_2H_2, 13 mm Hg; pressure of O_2, 10 mm Hg; pressure of NO_2, 1.5 mm Hg (45).

grows in concentration as the reaction develops.

Further experiments with fuel-rich mixtures (46) indicated the growth and decay of a precursor of free carbon which followed closely the growth and decay of the carbon radicals C_2, C_3 and CH. It was possible to deduce the extinction coefficient of this carbon precursor at 3700 Å and to show that its high value is characteristic of aromatic polynuclear hydrocarbons. It may be suggested (49) that in the high temperature of the flame (>3000°C) cracking of some of the excess fuel occurs to yield free carbon atoms, which progressively "crystallise" through C_2 and C_3 to the "aromatic" structure of graphite. The confirmation or otherwise of this view must await further studies of the products of explosion by means of the vacuum spectrograph when we may hope to see the resonance line of carbon in absorption.

VI. A GENERAL MECHANISM FOR THE COMBUSTION OF HYDRIDES

As I have mentioned above we have noted that the hydroxyl radical is common to the ignition processes of all the hydrides so far examined. Where initiation occurs by the direct photolysis of the hydrides yielding an H atom, it may be generated by the reaction

$$H + O_2 = OH + O\ .$$

When initiation is by the photolysis of a sensitizer yielding oxygen atoms as with NO_2, OH may be derived from the reaction

$$O + XH_n = OH + XH_{n-1}.$$

During the incubation period the OH is observed gradually to increase, and the instant of ignition is marked in oxygen-rich mixtures by a very sudden and enormous increase in its concentration. By detailed comparison of the oxidation reactions of H_2S, NH_3, PH_3 and hydrocarbons it may be concluded that the pattern of chain propagation is the same in all cases, and represented by the general scheme

$$\left.\begin{array}{l} OH + XH_n = H_2O + XH_{n-1} \\ XH_{n-1}\,O_2 = XH_{n-2}\,O + OH \end{array}\right\}\ \text{propagation.}$$

Branching is dependent on the intermediate and may take place variously by any one of the following reactions

$$XH_{n-2}\,O + O_2 = XH_{n-2}\,O_2 + O$$

$$XH_{n-2}O + O_2 = XH_{n-3}O + HO_2$$

$$XH_{n-2}O = XH_{n-3}O + H .$$

The first reaction takes place in the autocatalysis of H_2S oxidation, in which SO (seen by kinetic spectroscopy) is the intermediate. The second occurs in the oxidation of methane which yields formaldehyde (readily detectable during the reaction of conventional methods of analysis). The third is exemplified by the oxidation of ammonia in which HNO is concluded to be the origin of chain branching.

TABLE 2

Hydride	Uniradical	Associated Intermediate
SH_2	SH	SO(36)
TeH_2	TeH	TeO(37)
NH_3	NH_2	HNO(34)
PH_3	PH_2	HPO(35)
N_2H_4	N_2H_3	NH_2NO(34)
CH_4	CH_3	H_2CO(50)
C_2H_4	CH_3	H_2CO(51)
B_2H_6	BH_2	HBO(52)

Table 2 shows the uniradical and the associated intermediate derived from a series of hydrides which take part in chain propagation and branching, in accordance with our conclusions based on comparative study both by classical kinetic methods and flash photolysis. In cases where the intermediate is moderately stable, as with SO from H_2S and H_2CO from CH_4, the overall oxidation exhibits the slow autocatalysis associated with degenerate branching. In other cases, with extremely unstable intermediates the branching factor may be high, and is reflected in kinetics which show very short incubation periods and sharp transition from very slow reaction to explosion.

All the uniradicals and associated intermediates are seen to be isoelectronic or electronically structurally similar. This and the uniform participation of the OH radical in all the chain propagation reactions would seem to provide a generalizing hypothesis of value and one which invites further experimental examination.

In connection with the continued study of the reactions of the OH radical, Horne & Norrish (53) have recently been able to measure quantitatively the kinetics of the reactions

$$OH + C_2H_6 \xrightarrow{k_4} C_2H_5 + H_2O$$

$(\log_{10}k_4 = (11.1 \pm 0.7) - (3600 \pm 600)/2.303RT \text{ l mol}^{-1}\text{sec}^{-1})$

and

$$OH + CH_4 \xrightarrow{k_5} CH_3 + H_2O$$

$(\log_{10}k_5 = 10.7 - 5000/2.303RT.\text{ l mol}^{-1}\text{sec}^{-1}$, approximately)

by kinetic spectroscopy.

The OH radicals were generated by flashing water vapour in highly transparent quartz and comparing their rates of decay in the presence of inert gases and hydrocarbons. Further measurements of this kind with other hydrides will be of value to the continued study of the combustion of hydrocarbons along the lines indicated above. They are also of course of importance in consideration of the reactions involved in the evolution of planetary atmospheres, as are many other reactions studied by kinetic spectroscopy such as the photochemistry of NO, and of ozone and the reactions of the oxygen atom described above.

The examples which I have cited give, I hope, some indication of the breadth of application of methods based on flash photolysis in the study of gas reactions. Other results of importance involve the discovery of new absorption spectra of chlorine and bromine by Briggs and Norrish (54), and the detection of population inversion such as is observed in the study by Donovan & Husain (55) of spin orbit relaxation of the metastable iodine atom I $(5^2P_{1/2})$ produced in the photolysis of CF_3I:

$$I(5^2P_{1/2}) \rightleftarrows I(5^2P_{3/2}).$$

Population inversion in favour of high vibrational levels is as we have seen also observed in NO and CN, and has also been studied effectively by Polanyi and his co-workers for atomic reactions such as

$$H + Cl_2 = HCl^* + Cl$$

$$Cl + Hl = HCl^* + I.$$

All these reactions form the basis of potential gas laser action and are being effectively studied in this connection.

The opportunity for the application of the methods of flash photolysis to chemical kinetics, not only in the gas phase, but also to the study of photochemical reactions in solution is very great, and is increasing steadily as improvement in technique gives greater time resolution, and ever increasing accessibility to reactions of the "vacuum ultraviolet."

In conclusion I give thanks to those who have collaborated and contributed to the work described in this lecture, many of whom are continuing to direct and develop it with distinction.

REFERENCES

1. J. Stark, Physik Z. 9, 889, 894 (1908).

2. E. Warburg, Sber. press Akad. Wiss. Phys. Math. K. 1., p. 314 (1918); ibid., p. 300 (1916).

3. M. Bodenstein and W. Dux, A. Phys. Chem. 85, 297 (1913).

4. See R. G. W. Norrish, Bakerian Lectures, Proc. Roy. Soc. A301, 1 (1967).

5. R. W. Wood, Phil. Mag. VI, 44, 538 (1922).

6. K. F. Bonhoeffer, Z. Phys. Chem. 113, 199 (1924).

7. F. Paneth and W. Hofeditz, Ber. deut. Chem. Ges. 62, 1335 (1929); also F. Paneth and W. Lautsch, ibid. 64, 2702.

8. J. Franck, Trans. Faraday Soc. 21, 536 (1926).

9. V. Henri, Comp. Rend. Paris 177, 1037 (1923).

10. R. G. W. Norrish and G. Porter, Nature 164, 658 (1949).

11. G. Porter, Proc. Roy. Soc. A. 200, 284 (1950).

12. O. Oldenberg, J. Chem. Phys. 3, 266 (1935); also ibid. 2, 713 (1934).

13. R. G. W. Norrish, G. Porter and B. A. Thrush, Proc. Roy. Soc. A. 216, 165 (1955).

14. G. Porter and F. J. Wright, Discuss. Faraday Soc. 14, 23 (1953)

15. N. Basco and R. G. W. Norrish, Proc. Roy. Soc. A. 268, 291 (1962).

16. W. G. Burns and H. J. Bernstein, J. Chem. Phys. 18, 1669 (1950).

17. R. W. B. Pearse and A. G. Gaydon, Identification of Molecular Spectra , Chapman and Hall, London (1950).

18. N. Basco, A. B. Callear and R. G. W. Norrish, Proc. Roy. Soc. A. 260, 293 (1961); ibid. A. 269, 180 (1962).

19. R. G. W. Norrish, The study of energy transfer in atoms and molecules by photochemical methods, 12th Solvay Conference, Brussels: The Transference of Energy in Gases, Interscience Publishers, New York, p. 99 (1964).

20. N. Basco and R. G. W. Norrish, Proc. Roy. Soc. A. 268, 291 (1962).

21. R. G. W. Norrish, J. Chem. Soc., 1158 (1929); J. W. T. Spinks and J. M. Porter, J. Am. Chem. Soc. 56, 264 (1934); G. Kistiakowsky. Z. Physik Chem. 117, 337 (1925); J. Heidt and G. S. Forbes, J. Am. Chem. Soc. 56, 264 (1934).

22. F. J. Lipscomb, R. G. W. Norrish and B. A. Thrush, Proc. Roy. Soc. A. 233, 455 (1956).

23. McGrath, W. D. and R. G. W. Norrish, Proc. Roy. Soc. A. 142, 265 (1957); N. Basco and R. G. W. Norrish, Proc. Roy. Soc. A. 260, 293 (1960); also Discuss. Faraday Soc. 33, 99 (1960).

24. W. D. McGrath and R. G. W. Norrish, Z. Phys. Chem. 15, 245 (1958); W. D. McGrath and R. G. W. Norrish, Proc. Roy. Soc. A. 254, 317 (1960).

25. N. Basco and R. G. W. Norrish, Proc. Roy. Soc. A. 260, 293 (1961); W. D. McGrath and R. G. W. Norrish, see reference (24).

26. N. Basco and R. G. W. Norrish, see reference (24).

27. J. D. McKinley, D. Garvin and M. J. Boudart, J. Chem. Phys. 23, 784 (1955).

28. J. K. Cashion and J. C. Polanyi, J. Chem. Phys. 29, 455 (1958); ibid. 30, 1097 (1959); ibid. 30 316 (1959); J. C. Polanyi, ibid. 25, 784 (1955), also Chemistry in Britain 2, 151 (1966).

29. R. G. W. Norrish and R. P. Wayne, Proc. Roy. Soc. A. 288, 200, 361 (1965).

30. G. S. Forbes and L. J. Heidt, J. Am. Chem. Soc. 56, 1671 (1934).

31. G. Kistiakowski, Z. Physik. Chem. 117, 337 (1925).

32. W. D. McGrath and R. G. W. Norrish, Proc. Roy. Soc. A. 242, 265 (1957).

33. N. N. Semenov, Chemical Kinetic and Chain Reactions, Oxford University Press (1935).

34. D. Hussin and R. G. W. Norrish, Proc. Roy. Soc. A. 273, 145 (1963).

35. R. G. W. Norrish and G. A. Oldershaw, Proc. Roy. Soc. A. 262, 1 (1961).

36. R. G. W. Norrish and G. A. Oldershaw, Proc. Roy. Soc. A. 240, 293 (1957).

37. R. G. W. Norrish and M. Osborne, to be published.

38. R. G. W. Norrish and G. Porter, Proc. Roy. Soc. A. 210, 439 (1952).

39. J. F. Nicholas and R. G. W. Norrish, to be published.

40. B. Lewis and G. Von Elbe, Combustion, Flames and Explosions of Gases, Academic Press, New York (1951).

41. K. Erhard and R. G. W. Norrish, Proc. Roy. Soc. A. 234, 178 (1956).

42. S. A. E. Miller, Quart. Trans. 1, 98 (1947).

43. T. Male, "Third Symposium on Flame and Combustion Phenomena," Williams and Wilkins, p. 271 (1949).

44. A. P. Callear and R. G. W. Norrish, Proc. Roy. Soc. A. 259, 304 (1960).

45. R. G. W. Norrish, G. Porter and B. A. Thrush, Proc. Roy. Soc. A. 216, 165 (1953).

46. R. G. W. Norrish, G. Porter and B. A. Thrush, Proc. Roy. Soc. A. 227, 423 (1955).

47. W. A. Bone, Proc. Roy. Soc. A. 137, 243 (1932).

48. R. G. W. Norrish, Conférence Plénière dans la Section des Mélanges Gazeux au XVIe Congrès de Chimie Pure et Appliquée,

49. R. G. W. Norrish, see reference 48.

50. R. G. W. Norrish, Revue de l'Institut Francais du Petrole 7, 288 (1949).

51. A. Harding and R. G. W. Norrish, Proc. Roy. Soc. A. 212, 291 (1952).

52. M. D. Carabine and R. G. W. Norrish, Proc. Roy. Soc. A. 296, 1 (1967).

53. D. Horne and R. G. W. Norrish, Nature 215, 1373 (1967).

54. A. G. Briggs and R. G. W. Norrish, Proc. Roy. Soc. A. 276, 57 (1963).

55. R. J. Donovan and D. Husain, Trans. Faraday Soc. 62, 11 and 1050 (1962).

ELECTRONIC AND VIBRATIONAL ENERGY TRANSFER

B. A. Thrush

Department of Physical Chemistry
University of Cambridge
Cambridge, England

ABSTRACT

Most of the information currently available as to the properties and mechanism of energy transfer processes involving small molecules has come from the study of electronic and vibrational relaxation processes. The general properties and kinetics of these processes of energy redistribution are discussed in terms of the mechanisms of energy transfer.

INTRODUCTION

The mechanisms by which molecules exchange energy with each other between electronic, vibrational, rotational and translational degrees of freedom are best studied spectroscopically. Much investigation remains to be done in this area and very little is yet known about many topics ; notably the role of rotation in energy transfer processes, is far from fully understood.

This review attempts to summarise the general conclusions which have been reached in studies of the relaxation of electronic and vibrational energy in small molecules.

II. VIBRATION - TRANSLATION TRANSFER

Measurements of the interconversion of vibrational and translational energy were originally made by ultrasonic dispersion; that is the increase in the velocity of sound with increasing frequency due to the rise in $\gamma = C_p/C_v$, as the vibrational energy can no longer follow the rapid oscillations in the translational (and rotational) energy. For this reason, vibrational energy transfer is normally discussed in terms of a relaxation time, τ, which is the time taken for the non-equilibrium element in an energy distribution to decay to 1/e of its initial value or in terms of the number of molecular collisions needed to achieve this. For the loss of the single quantum of vibrational energy in a molecule (which is the normal situation in measurements of sound dispersion), the number of collisions is given by

$$Z_{1,0} = 1/P_{1,0} , \tag{1}$$

where $P_{1,0}$ is the corresponding collisional probability. If Z is the total number of collisions undergone by a molecule per unit time then

$$\tau = 1/Z\, P_{1,0} . \tag{2}$$

The conversion of vibrational energy to translational energy depends on the perturbation of the molecular vibration during a collision. Landau and Teller (1) who considered the problem classically showed that the probability is highest for a steep interaction potential and a high relative velocity of the colliding particles where the interaction has a short duration. They established that the probability of vibrational relaxation decreased exponentially with the ratio of the collision duration to the vibrational period. The quantum mechanical formulation was first given by Zener (2) using a distorted wave treatment. This was developed by Jackson and Mott (3) who applied the distorted wave treatment to the analogous problem of particles colliding with an elastic surface.

Schwartz, Slawsky and Herzfeld (4) brought the theory into a form where it could be compared with experimental measurements on gases. For the relaxation of a single quantum($\Delta E = h\nu$) in a diatomic molecule with atomic masses m_b and m_c in collision with a rigid molecules of mass m_a where the repulsive potential of the form exp(-x/l), the probability of relaxation is found to be (4)

$$P_{1,0} = \frac{4\pi^2\mu^2 l^2 [\Delta E]}{\hbar^2} \cdot \frac{(m_B^2 + m_C^2)}{m_B m_C (m_B + m_C)} \exp\left(\frac{-4\pi l\,|\Delta E|}{\hbar(V_0 + V_1)}\right) , \tag{3}$$

where μ is the reduced mass of the colliding species A and BC and V_0 and V_1 are the initial and final relative velocities of the molecules.

If the collisional probability is averaged over the Maxwell-Boltzmann distribution of velocities, the expression for collinear collisions is found to be

$$P_{1,0} = 2\left(\frac{2\pi}{3}\right)^{\frac{1}{2}} \frac{\theta^1}{\theta} \left(\frac{\theta^1}{T}\right)^{\frac{1}{6}} \exp\left[-\frac{3}{2}\left(\frac{\theta^1}{T}\right)^{\frac{1}{3}} + \frac{\theta}{2T} + \frac{\varepsilon}{kT}\right] \qquad (4)$$

where $\theta = \frac{h\nu}{k}$ and $\theta^1 = \frac{\nu^2 l^2 \mu}{k}$,

and the term ε/kT allows for the additional energy of impact due to the depth (ε) of the attractive well (5).

Eq. (4) is very widely used for comparison with experimental results on relaxation processes, despite the considerable progress which has been made in recent years in the solution of the semi-classical and quantum mechanical problems for the collinear system. This topic has been reviewed in detail by Rapp and Kassal (6); reference should also be made to the work of Heidrich, Wilson and Rapp (7) who used a semi-classical approach to obtain a simple and accurate analytical expression for the collisional transition probabilities between the various energy levels of harmonic oscillator. However,the situation is much less satisfactory for collisions which are not collinear and where anharmonic oscillators and other potential functions are involved.

Inter-molecular forces have commonly been expressed in terms of the Lennard-Jones potential using an r^{-6} attractive potential and an r^{-12} repulsive potential. However, the determination of the appropriate value of l from the Lennard-Jones parameters can pose a porblem, since the value of $P_{1,0}$ given by eq. (4) depends critically on the value of l, but reasonable agreement is obtained by matching the repulsive functions at the turning points for zero kinetic energy and the most probable relative kinetic energy. This problem and the general form of eq. (4) have encouraged experimentalists to look for relationships between log $P_{1,0}$ and $T^{-1/3}$ based on the normally valid assumption that the first term in the exponent predominates over the other two. The best known such formula is due to Millikan and White (8):

$$\log_{10}(\tau P) = 5.0 \times 10^{-4}\, \mu^{\frac{1}{2}}\, \theta^{4/3}\, (T^{-1/3} - 0.015\mu^{\frac{1}{4}}) - 8.00 \ , \qquad (5)$$

where the pressure, P, is in atmospheres and the other quantities have been defined above. Eq. (5) describes the relaxation of most diatomic molecules by both atoms and diatomic molecules to within a factor of two over a wide range of temperatures (Fig. 1). It also applies to some small polyatomic species.

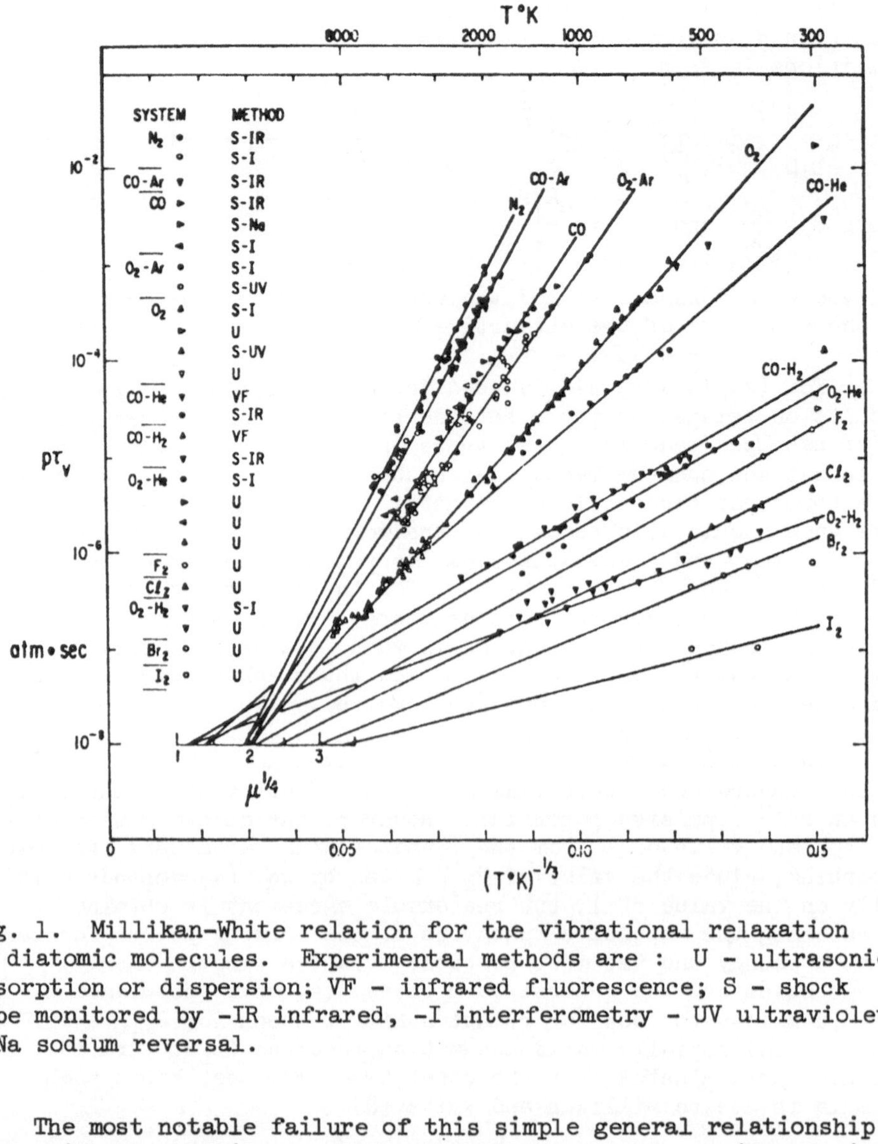

Fig. 1. Millikan-White relation for the vibrational relaxation of diatomic molecules. Experimental methods are : U - ultrasonic absorption or dispersion; VF - infrared fluorescence; S - shock tube monitored by -IR infrared, -I interferometry - UV ultraviolet, - Na sodium reversal.

The most notable failure of this simple general relationship comes with the hydrides. Several factors must be considered with such compounds:

(a) their much larger vibrational amplitude for a given vibrational energy. This is allowed for in eq. (4) but not in eq. (5).

(b) the wider spacing of their rotational levels makes it possible to convert part of the vibrational energy into rotational energy without a large change in the rotational quantum number.

(c) the large dipole moments of the hydrogen halides can cause the term ε/kT in eq. (4) to make a significant contribution.

Until the advent of laser techniques, it had not been practicable to study the relaxation of vibrational energy within the various vibrational degrees of freedom of a polyatomic molecule. Most of the previous measurements were studies of ultrasonic dispersion at room temperature and with a few exceptions (e.g. SO_2) these showed a single relaxation time. This means that there is only one rate determining process in the exchange of translational and vibrational energy and hence that this process is slower than the exchange of energy between the vibrational degrees of freedom.

Lambert and Salter (9) showed that for many polyatomic molecules log $Z_{1,0}$ is directly proportional to the lowest vibration frequency of the molecule, see Fig. 2, providing clear evidence that energy enters the vibrational manifold via the lowest mode. Again the plots for molecules containing hydrogen show faster relaxation than those without hydrogen and molecules containing more than one hydrogen atom form a separate correlation. Clearly factors (a) and (b) discussed for diatomic molecules are important here although it must be remembered that the lowest vibration frequencies of most of the molecules forming correlation I in Fig. 2 do not have large amplitudes as they are not C-H modes. Strongly polar molecules such as H_2O, NH_3 and CH_3CN relax many times faster than predicted by the Lambert-Salter plot. Their behavior corresponds to factor (c) above.

Although the Lambert-Salter plot provides a useful correlation for many polyatomic molecules and is often invoked in the discussion of electronic-to-vibrational energy transfer, there is no adequate theoretical explanation of its validity.

One example of relaxation processes in polyatomic molecules is the CO_2 laser where ν_3 in CO_2 is excited by resonant transfer from vibrationally excited N_2:

$$N_2(v = 1) + CO_2(0,0,0) = N_2(v = 0) + CO_2(0,0,1). \qquad (6)$$

Laser emission occurs to thelevels (1,0,0) and (0.2.0) which form a Fermi diad and from which vibrational relaxation is rapid via the lowest frequency mode, $\nu_2 = 667\ cm^{-1}$. In contrast, recent unpublished experiments by C.B. Moore in which he has used a tunable laser to pump the (1,0,1) and (0,2,1) levels of CO_2 show that these levels are quenched by the transfer of one quantum of ν_3 (the antisymmetric stretching vibration) to another molecule of CO_2 which occurs at one collision in two. In contrast, transfer between (1,0,1) and (0,2,1) by collisions with argon is at least 500 times slower.

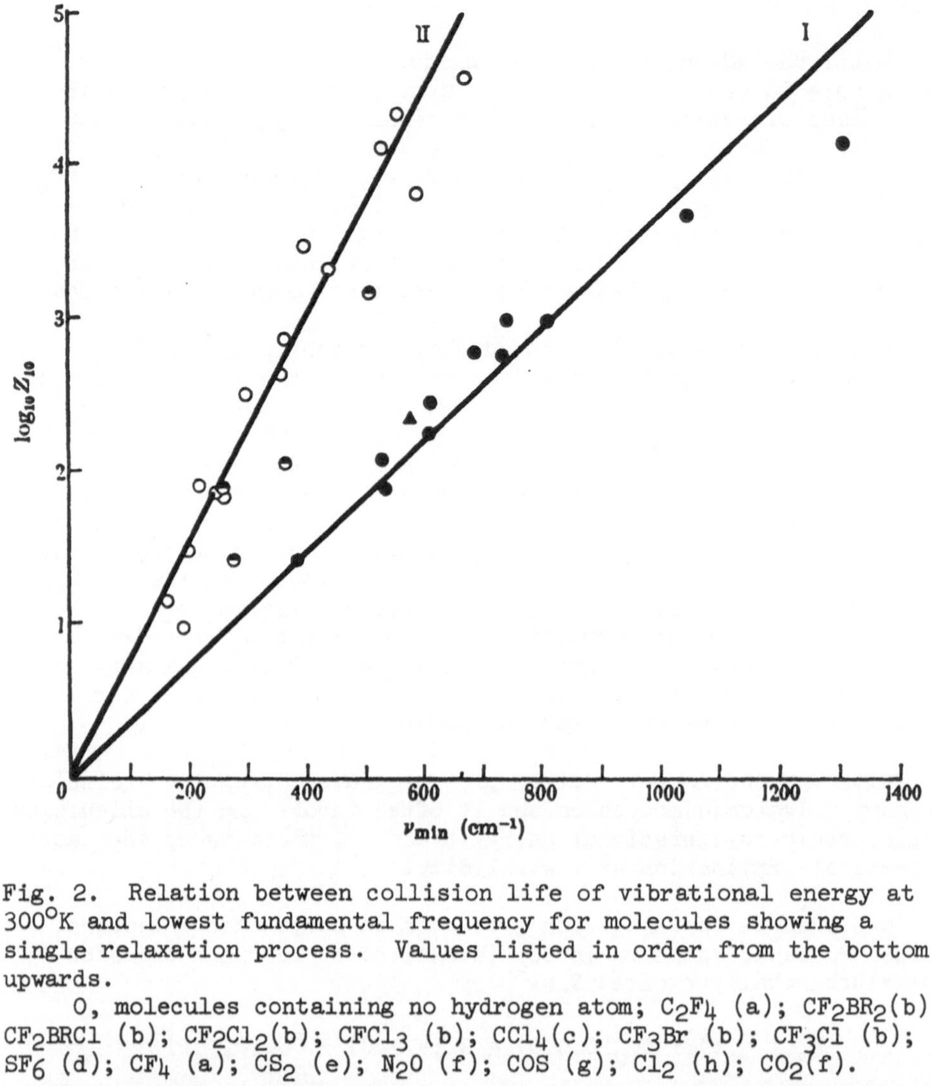

Fig. 2. Relation between collision life of vibrational energy at 300°K and lowest fundamental frequency for molecules showing a single relaxation process. Values listed in order from the bottom upwards.

O, molecules containing no hydrogen atom; C_2F_4 (a); CF_2BR_2(b); CF_2BRCl (b); CF_2Cl_2(b); $CFCl_3$ (b); CCl_4(c); CF_3Br (b); CF_3Cl (b); SF_6 (d); CF_4 (a); CS_2 (e); N_2O (f); COS (g); Cl_2 (h); CO_2(f).

◐, moelcules containing one hydrogen atom; $CHCl_2F$ (b); $CHCl_2$ (c); $CHClF_2$(b); CHF_3 (a)(b).

●, molecules containing two or more hydrogen atoms; CH_2ClF (b); CH_2I (j); CH_2F_3 (a); CH_3Br (b)(j); C_2H_2; CH_2Cl (b); C_2H_4O (e); C_2H_4 (k); cyclo-C_3H_6 (k); CH_3F (j); CH_4 (l)..

▲, deuterated molecule; CD_3Br.

Clearly the advent of tunable lasers is going to provide much information on the redistribution and transfer of vibrational energy in molecules. Studies of the pumping of ν_3 in SF_6 by a CO_2 laser (10) show that vibration-vibration and rotation-translation transfer processes in the pumped levels occur at close to the collision rate, the populations of the excited vibrational levels acquire a Boltzmann distribution characterised by a temperature greater than ambient and these levels decay together with a collisional efficiency of about 1 in 500 in agreement with the Lambert-Salter picture.

The transfer of vibrational energy to and from the higher vibrational levels of polyatomic molecules has considerable importance in the theory of unimolecular reactions where such processes provide the necessary activation. Here the vibrational energies are typically 50 kcal mol^{-1} and the density of vibrational levels is so high that the vibrational manifold is virtually a continuum of levels. Transfer between specific vibrational levels cannot be distinguished and results are normally expressed in terms of the amount of energy removed per gas-kinetic collision. With the more complicated collision partners, values around 5 kcal mol^{-1} are normally found for this quantity although there have been significant differences in the results from various techniques such as 'fall-off' in unimolecular reactions, chemical activation and photochemical activation. However, it now appears that many of the discrepancies can be rationalised by assuming that a fraction of the gas-kinetic collisions result in a major redistribution of energy with the collision partner.

III. VIBRATION-VIBRATION TRANSFER

An example of the highly efficient transfer of vibrational energy involving the antisymmetric stretching mode in CO_2, and transfer of vibrational energy from this level to nearly resonant levels in other molecules is adequately explained by the interaction of the vibrational transition dipole with the corresponding dipole or quadrupole moments of the accepting molecule. However, when the energy discrepancy exceeds about 20 cm^{-1} it is also necessary to invoke some transfer into rotational modes to account for the observed efficiencies and temperature dependences (11).

The 'Lambert-Salter' type of behavior in which the logarithm of the probability of vibrational transfer decreases linearly with the energy discrepancy has also been observed in vibrational energy transfer between diatomic molecules (12). However no detailed theoretical explanation has been attempted.

A particularly interesting problem connected with molecular lasers is the redistribution of vibrational energy, for instance,

$$AB(v = m) + XY(v = n) = AB(v = m - 1) + XY(v = n + 1). \quad (7)$$

For a harmonic oscillator $P^{m,m-1}_{n,n+1} = m(n+1)\,P^{1,0}_{0,1}$, but for a real molecule the anharmonicity will cause the energy discrepancy (ΔE) to vary with m and n. This introduces an additional factor of the order of $\exp(-\Delta E/kT)$ into the transition probability which has an interesting consequence when AB and XY are identical. The addition of a vibrationally cold molecule (e.g. CO) to a reaction system producing highly vibrationally excited molecules of the same species increases the probability of laser action because the ground state CO molecules preferentially accept vibrational energy from the less excited CO molecules because of the smaller energy discrepancy and thereby promote the formation of an inversion in the vibrational distribution.

IV. VIBRATIONAL RELAXATION IN EXCITED ELECTRONIC STATES

The relatively few studies which have been made of vibrational relaxation in excited electronic states have shown this process to be orders of magnitude faster than for the ground state molecule. For instance argon and krypton induce vibrational relaxation of $CO(A^1\Pi, v = 1)$ at virtually every collision (13) whereas vibrational relaxation in the ground state requires $\sim 10^8$ collisions also at room temperature (14). Similarly vibrational relaxation of HD ($B^1\Sigma^+_u$ $v = 3 \rightarrow 2$) by He (15) is 10^6 times faster than the corresponding process for $H_2(X^1\Sigma^+_g$ $v = 1)$ (16). Only a small part of these differences can be attributed to the lower vibration frequencies of excited states. Other important factors may be the lifting of the degeneracy (e.g. in CO $A^1\Pi$) and the mixing of excited states in collisions.

It should be noted that rapid electronic energy transfer between like molecules can give the appearance of rapid vibrational relaxation if it occurs without the transfer of vibrational energy (17).

V. ELECTRONIC-TO-VIBRATIONAL ENERGY TRANSFER

Unlike the systems considered above, where the potential curves of the initial and final states normally have closely similar forms, the interaction of an electronically excited species with a quenching agent will normally differ significantly from the interaction of the corresponding ground state with the quenching agent.

The mutual approach of these surfaces in a collision will favor the conversion of electronic energy into translational energy.

There is good evidence from shock tube studies for the strong coupling of the electronic excitation in atoms to the vibrational energy of the carrier gas (18). However, this does not imply that most of the energy released in the quenching of an atom will appear as vibrational excitation of the quenching agent. For Hg (6^3P) + CO the fraction is about one quarter (19).

The quenching within the spin-orbit multiplets of atoms, where the energies involved can be close to those of a vibrational quantum and the individual multiplet components probably have similar interactions with the quenching agent, should closely resemble the behaviour observed in vibration-translation energy transfer. In fact, there is quite a good correlation between the logarithms of the rate coefficients for the quenching by atoms and small molecules of the multiplets of a number of atoms, for instance, $Se(4^3P_0)$, $Fe(a^5D_3)$ and $Ne(3^3P_{0,1})$ and the energy which has to be converted into translation plus rotation, as is found in the Lambert-Salter plot (20).

Where the excited atom can interact chemically with the quenching agent much higher quenching efficiencies are commonly found. For example $O(^1D)$ is quenched to $O(^3P)$ by N_2, CO and O_2 at about one collision in three and by H_2O, CO_2, O_3, etc. at almost every collision. In contrast the quenching of the higher ^1S state, which is spin allowed, to $O(^1D)$ is about a thousand times slower (21). The efficient quenching of $O(^1D)$ by N_2 and CO can be understood in terms of their bonding interaction to give a highly vibrationally excited ground state N_2O or CO_2 molecule which undergoes a spin-forbidden non-adiabatic transition (predissociation) to yield $O(^3P)$. In contrast $O(^1S)$ does not normally have a bonding interaction with the quenching agent because the corresponding potential surfaces are repelled by some of the five surfaces with the same resultant spin given by $O(^1D)$ plus the quenching agent.

The most extensive data on electronic energy transfer from diatomic molecules comes from the quenching of low-lying metastable $^1\Delta_g$ and $^1\Sigma_g^+$ states of O_2. Except for a few species (I, I_2) which can accept electronic energy, the quenching efficiencies show a linear correlation with the highest fundamental vibration frequency of the quenching agent. Fig. 3 (22) shows data for $O_2(^1\Sigma_g^+)$ where there is good evidence that quenching occurs to the

$^1\Delta_g$ state. Recent studies of the infrared emission by such quenching agents as CO_2, H_2O, and CO show a strong preference for quenching to $O_2(^1\Delta_g, v = 1)$ where the energy gap is 3755 cm^{-1} rather than to v = 0, releasing 5238 cm^{-1}, although the Franck-Condon factors strongly favour $v = 0$. Similar behaviour is observed in the quenching of $O_2(^1\Delta_g)$ by NO (23). The low rate coefficients generally observed for the quenching of these molecules are consistent with weak, short range interactions with the quenching agents.

VI. ELECTRONIC-ELECTRONIC TRANSFER

Almost all the studies in this area have been of electronic energy transfer between atoms. For resonant transfer collision cross-sections as large as 10Å^2 and 500Å^2 are expected for quadrupole and electric dipole transitions, respectively (24). However, these efficiencies should drop sharply as the energy discrepancy, ΔE, increases. Such behaviour is observed for transfer between excited states of alkali metal atoms but scatter due to differences in the interaction potentials makes it difficult to decide whether the cross-sections vary as $(\Delta E)^{-2}$ or show a Lambert-Salter behaviour on ΔE with a stronger dependence on ΔE than is encountered in vibrational relaxation (25). Other studies of atom-atom energy transfer yield somewhat similar results.

The efficient mercury (Hg 6^3P_1) sensitised decomposition of many hydrocarbons is believed to involve electronic energy transfer to yield an unstable triplet state which dissociates. However, the dissociation of H_2 involves an atomic transfer reaction which yields HgH.

Some studies have been made of electronic energy transfer processes involving diatomic molecules; the species studied include N_2 ($A^3\Sigma_u^+$ and $a^1\Pi_g$) and NO ($C^2\Pi$ and $A^2\Sigma^+$). Rates approaching gas-kinetic are often observed. Where vibrational energy can also be transferred, the Franck-Condon principle does not have great importance and the interactions involved are often strong enough for $\sim$ 2000 cm^{-1} of energy to go into translation plus rotation.

It can be seen that many aspects of energy transfer between small molecules need to be investigated further. In particular, rotational energy transfer needs to be studied thoroughly and much can be done with laser techniques now that tunable lasers are available for wide regions of the spectrum.

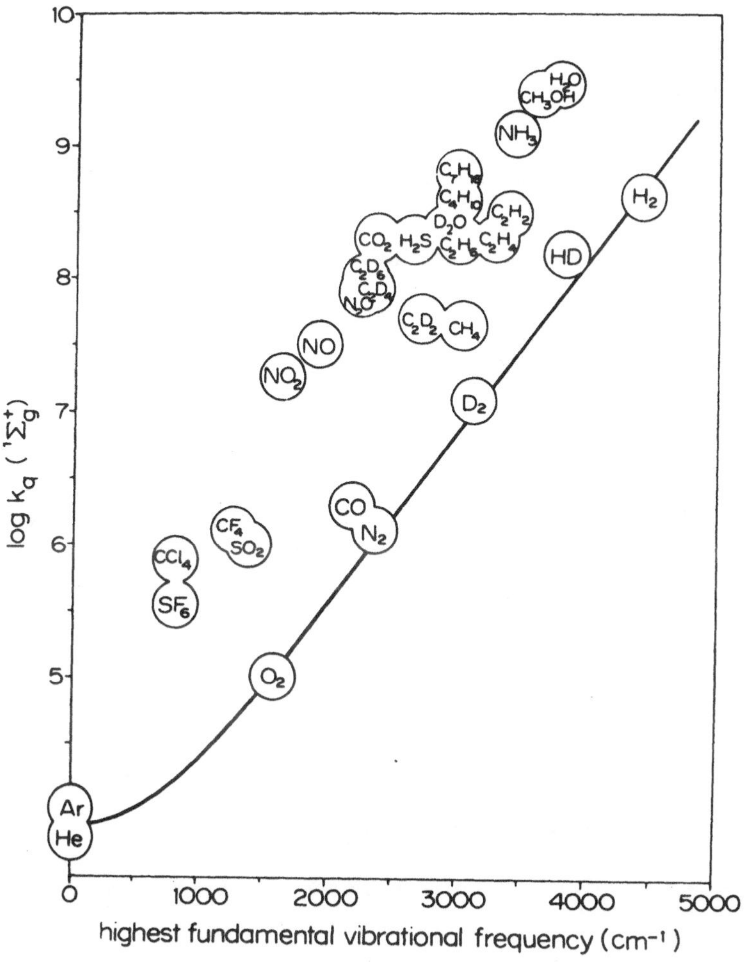

Fig. 3. Dependence of the logarithm of the quenching constant for $O_2(^1\Sigma_g^+)$ in l mol^{-1} s^{-1} units on the highest fundamental vibration frequency of the quenching agent.

REFERENCES

1. L. Landau and E. Teller, Physik. Z. Sowjetunion, 10, 34 (1936).
2. C. Zener, Phys. Rev., 37, 556 (1931).
3. J.M. Jackson and N.F. Mott, Proc.Roy.Soc. A137, 703 (1932).
4. R.N. Schwartz, Z.I. Slawsky and K.F. Herzfeld, J. Chem. Phys. 20, 159 (1952).
5. K.F. Herzfeld, in Thermodynamics and Physics of Matter, Princeton Univ. Press, Section H (1965).
6. D. Rapp and T. Kassal, Chem. Rev., 69, 61 (1969).
7. F.E. Heidrich, K.R. Wilson and D. Rapp, J. Chem. Phys., 54, 3885 (1971).
8. R.C. Millikan and D.R. White, J. Chem. Phys., 39, 3209 (1963).
9. J.D. Lambert and R. Salter, Proc. Roy. Soc., A253, 277 (1959).
10. J.I. Steinfeld, I. Burak, D.G. Sutton and A.V. Nowack, J. Chem. Phys., 52, 5421 (1970).
11. J.C. Stephenson and C.B. Moore, J. Chem. Phys., 56, 1295 (1972).
12. A.B. Callear, in Photochemistry and Reaction Kinetics, (P.G. Ashmore, F.S. Dainton and T.M. Sugden, eds.), Cambridge University Press, p.133 (1967).
13. F.J. Comes and E.H. Fink, Chem. Phys. Letters, 14, 433 (1972).
14. R.C. Millikan, J. Chem. Phys., 43, 1439 (1965).
15. D.L. Akins, E.H. Fink and C.B. Moore, J. Chem. Phys., 52, 1604 (1970).
16. J. Ducuing, C. Joffrin and J.P. Coffinet, Opt. Comm., 2, 245 (1970).
17. L. A. Melton and W. Klemperer, J. Chem. Phys. 55, 1468 (1971).
18. A. G. Gaydon and I. R. Hurle, The Shock Tube in High Temperature Chemistry and Physics, Chapman and Hall, London (1963).
19. G. Karl, P. Kruus and J. C. Polyanyi, J.Chem. Phys. 46, 224 (1967).

20. A. B. Callear and J. D. Lambert, in Comprehensive Chemical Kinetics, Vol. 3 (C.H. Bamford and C.F.H. Tipper, eds.), Elsevier, Amsterdam, p. 182 (1969).

21. R. J. Donovan, D. Husain and L.J. Kirsch, Ann. Rep. Chem. Soc. A, 69, 19 (1972).

22. J. A. Davidson and E. A. Ogryzlo, in Chemiluminescence and Bioluminescence (M.J. Cormier, D. M. Hercules and J. Lee, eds.), Plenum Press, New York, p. 111 (1973).

23. E. A. Ogryzlo and B. A. Thrush, Chem. Phys. Letters 23, 34 (1973); 24, 314 (1974); R.G.O. Thomas and B.A. Thrush, to be published.

24. H.S.W. Massey and E.H.S. Burhop, Electronic and Ionic Impact Phenomena, Oxford Univ. Press (1952).

25. M. Czajkowski, D.A. McGillis and L. Krause, Canad. J. Phys. 44, 741 (1966).

THE CHEMICAL PRODUCTION OF EXCITED SPECIES

B. A. Thrush
Department of Physical Chemistry
University of Cambridge
Cambridge, England

ABSTRACT

The mechanism of formation of electronically excited species in elementary gas reactions is discussed in some detail. It is shown that three-body recombination is normally a more efficient source of electronically excited molecules than two-body association. The formation of electronically excited molecules in atom transfer reactions is considered in terms of the potential surfaces involved and their intersections. A brief comparison is made with the mechanisms of chemiluminescence in organic reactions and biological systems.

I. INTRODUCTION

The energy released in an exothermic chemical process normally exceeds the average thermal energy of the products as given by the Boltzmann distribution law for the reaction temperature. In this sense, many chemical reactions could be regarded as yielding excited species and only strongly exo-energetic processes where the energy distribution in the products shows a strong departure from equipartition or from the Boltzmann law are normally classified as such. The strong interactions between molecules in condensed media give very short relaxation times ($\sim 10^{-12}$s for vibrational energy) and attention is therefore concentrated on gas phase processes.

Apart from the use of flash photolysis which was discussed in a previous chapter (1), these processes have been commonly investigated by using the visible or infrared chemiluminescent emission from the electronically or vibrationally excited species produced. Here attention is concentrated on reactions which yield electronically excited products and can therefore give visible or ultraviolet chemiluminescence. Such reactions occur via a potential surface which is not the lowest energy surface of the system in the product region. Such reactions are few in number, since most elementary reactions proceed by the lowest potential surface, which connects the electronic ground states of the reactants and products. Most of the elementary reactions studied involve the transfer of an atom, for instance between an atom and a diatomic molecule:

$$A + BC = AB + C \quad . \tag{1}$$

Here, much of the energy released in the reaction can appear as vibrational excitation of the newly formed bond, giving infrared chemiluminescence by AB.

The calculation of trajectories for such systems on an empirical or semi-empirical potential surface using Monte Carlo methods enable experimental measurements of the vibrational and rotational excitation of ground state molecules or of reactive scattering in molecular beam systems to be transformed into information about the general form of the potential surface connecting ground state reactants and products. This topic has a considerable literature (2) but such methods have not yet been applied to the formation of electronically excited products either by reaction along a single potential surface, or by the more common situation where the emitting state is formed by a transition between different potential surfaces. Nevertheless, the qualitative findings of such calculations for ground state products can be used to characterize chemiluminescent reactions which are sometimes described as 'direct', 'occurring via a long-lived complex' and so on.

Electronically excited products are commonly formed in combination reactions, where little is known about the production of vibrationally excited ground state species. For two-body atomic combination, the potential surface leading to electronically excited products is simply the appropriate potential curve of the diatomic molecule and the spectral distribution of the emission can be analyzed theoretically to yield information about this potential, whether attractive or repulsive. However, more commonly, and especially with three-body atomic combination, the emitting state is populated by a sequence of fundamental processes, including collision-induced crossings between excited states and vibrational relaxation within excited states, neither of which can be satisfactorily treated theoretically at present. However, the experimental studies yield much information about the rates and mechanisms of the elementary steps occurring.

II. THE FORMATION OF EXCITED SPECIES IN COMBINATION PROCESSES

II.A. Introduction

The formation of electronically excited products by the association of atoms with each other or with a molecule is generally termed an afterglow since such processes are commonly observed by subjecting gases to a pulsed electric discharge or by flowing gases through a discharge. They can also be studied in the outer cones of pre-mixed flames, where an excess concentration of atoms and free radicals is provided by the branched-chain reaction occurring in the inner cone. Since these association processes have a low efficiency in two-body collisions and more commonly require the presence of a third body, the time scale involved is usually milliseconds and a Maxwellian distribution of velocities and translational-rotational equilibrium are maintained in the gas, notwithstanding the presence of about 1% of free atoms in ground or metastable excited states.

Chemiluminescent combination processes can be classed into three groups:

(a) Two-body processes in which the emitting state is formed adiabatically.

(b) Two-body processes in which the emitting state is populated by a non-adiabatic transition between two potential surfaces.

(c) Processes in which the newly formed molecule has been stabilized by a third body before emitting.

Many systems show both two- and three-body chemiluminescence, their relative proportions being governed by the experimental conditions. These two processes can be distinguished because they yield emission from levels of the molecule respectively above and below the relevant dissociation threshold.

II.B. Direct Two-body Combination

For radiative recombination without curve crossing, the probability of a collision pair radiating at a given internuclear distance, r, is given by A_r, the Einstein coefficient for spontaneous emission (3):

$$A_r = \frac{64\,\pi^4\,\nu(r)^3}{3hg}\,|R_e(r)|^2, \qquad \text{(i)}$$

where the frequency of radiation, ν, and the transition moment, R_e, depend on r.

For radiation from a repulsive state to a bound state, as in the radiative recombination of halogen atoms in shock tubes, the collision pairs are assumed to be in equilibrium with the free atoms which have degeneracy g_A, g_B. If g is the degeneracy of the radiating state and σ its symmetry number, the intensity of emission is given by

$$\frac{I}{[A]\,[B]} = I_o = \frac{4\pi g}{g_A g_B \sigma} \int A_r\, r^2 \exp(-\,U_r/kT)\, dr\,. \qquad \text{(ii)}$$

The transitions from a steep potential curve are assumed to originate from the turning points of classical motion for the collision pairs when the energy is U_r. For the evaluation of this integral U_r and A_r are obtained from the intensity of the corresponding absorption spectrum as a function of temperature; for both these procedures it is generally assumed that the eigenfunctions of the repulsive state, well above the dissociation limit, are δ-functions at the turning point and the contributions of the sinusoidal oscillations away from the turning point average to zero (see Herzberg (4), pp 391-4).

The Boltzmann factor causes the emission to lie at larger internuclear distances than the absorption giving a smaller ν^3 factor in the expression for A_r. In addition, these two-body emissions are observed mainly in systems where the atomic transition is forbidden but the molecular transition allowed and the transition moment, R_e, therefore decreases with increasing internuclear distance.

The radiative recombinations of halogen atoms at high temperature give continuous emission in the visible from repulsive states to the bound ground state which arises from the molecular electronic transition $\sigma_u \rightarrow \pi_g$. Spin-orbit splitting in the halogens is quite large and the molecular ground state $(^1\Sigma_g^+)$ correlates with the lower components of the atomic ground states. The emitting states which are best described by Hund's case (c) are $^3\Pi(1_u)$ and $^1\Pi(1_u)$, correlating with two $^2P_{3/2}$ atoms, and $^3\Pi(0_u^+)$, correlating with $^2P_{3/2}$ + $^2P_{1/2}$.

For shock-heated bromine at 1300 to 2600K, emission comes from the repulsive region of the potential curves, 20 - 150 kJ mol^{-1} above the dissociation threshold (5), more recent work suggesting that $^1\Pi(1_u)$ is the dominant emitting state (6). For chlorine, the emission below 480 nm also comes mainly from the $^1\Pi(1_u)$ state but, at longer wavelengths, the $^3\Pi(0_u^+)$ state also contributes (7). These emissions provide information about those regions of the repulsive potential curves of halogens which cannot be reached by light absorption.

The helium afterglow shows emission extending about 50 nm to the long wavelength side of the forbidden atomic transition, He(1s2s^{1}S) $\rightarrow$ (1s^2 ^{1}S). This is due to the allowed molecular transition from the bonding $^1\Sigma_u^+$ state of He_2, arising from He(1s2s^{1}S) + He(1s^2 ^{1}S), to the repulsive $^1\Sigma_g^+$ ground state corresponding to 2 He(1s^2 ^{1}S). Although some of the longer wavelength emission appears to come from bound levels of the $^1\Sigma_u^+$ state this is not necessarily true for the region of the afterglow spectrum close to the atomic transition, although fluctuations in intensity are observed there. Mies and Smith (8) have shown that the eigenfunctions for translational levels immediately above the dissociation limit exhibit maxima similar to those of the highest bound vibrational level. Transitions from these unbound levels to a steep repulsive curve can explain the observed diffuse 'banded' structure.

There is evidence that the emitting state of He_2 has a barrier about 5 kJ mole^{-1} high at r ~ 0.3 nm (9,10) and some two-body emission probably comes from quasi-bound molecules of He_2, which have tunnelled through this barrier (10).

The corresponding molecular continua or faintly-banded emission from the other inert gases come from $^3\Sigma_u^+$ (1_u) states formed from a ground state atom and a metastable $^3P_2$ atom (11). Both two- and three-body emission processes apparently occur (12). The measured rate constants for the two-body processes (13) are consistent with Eq.(2), with U $\simeq$ 0, if the A coefficients lie in the range 10^6 - 10^8 s^{-1}, typical of allowed radiative transitions (14).

II.C. Two-Body Combination with Curve-Crossing (Preassociation)

In systems where two atoms approach each other on one potential surface (Y in Fig. 1) and then make a radiationless transition to the emitting state (Z in Fig. 1) this is termed a non-adiabatic or diabatic process, since the adiabatic curves corresponding to the behaviour at very low velocities do not intersect (see enlarged section, Fig. 1). Their minimum separation from the virtual crossing point is given by the mixing term V_{12}. The probability, P_{12}, of staying on the adiabatic curve depends on the velocity, v, of the system and is usually discussed in terms of the Landau-Zener model which proved a useful if limited model (3,15) as the detailed interactions of potential curves are not normally well known. This model which assumes that V_{12} and v are constant, yields

$$P_{12} = 1 - \exp(-4\pi^2 V_{12}^2/hv\ |F_1 - F_2|) , \qquad \text{(iii)}$$

where F_1 and F_2 are the (constant) slopes of the diabatic curves

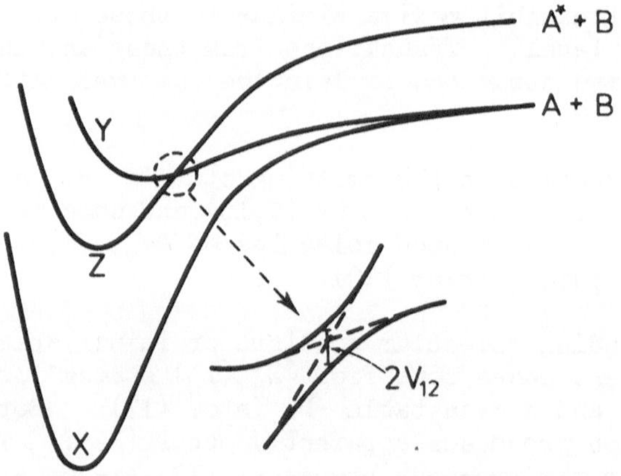

Fig. 1 Potential curves illustrating the two-body (preassociation) afterglow mechanism.

at the crossing point. In a normal collision the system passes twice through the intersection region and the probability of making a non-adiabatic transition becomes (after allowing for an interference term),

$$P = 2P_{12}(1 - P_{12}) , \qquad \text{(iv)}$$

which is a maximum for a particular value of v; for lower or higher velocities the systems behave predominantly adiabatically after two transits of the crossing region. Most of the known systems correspond to small values of V_{12}; if the two interacting states have the same electronic symmetry, V_{12} is normally so large that the non-adiabatic transitions between them have a low probability.

In most examples of this two-body chemiluminescent preassociation (or inverse predissociation) the emitting state, Z, is electronically degenerate (due either to spin or orbital angular momentum) and only one component of the state may be involved depending on the symmetry of the predissociating state, Y, as when a $^2\Pi$ state is predissociated by a $^2\Sigma^+$ state. Furthermore, for states which differ in orbital angular momentum by one unit (e.g. $^2\Pi$ and $^2\Delta$), the interaction increases roughly linearly with the angular momentum of molecular rotation. In fact, most of the preassociation processes which have been investigated are weak ones, prohibited by Kronig's rules due, e.g.: $^5\Sigma_g^+ \rightarrow {}^1\Pi_g$, $^5\Sigma_g^+ \rightarrow {}^3\Pi_g$ in N_2, $^4\Pi \rightarrow {}^2\Pi$ in NO. In these cases, discussion of preassociation or the reverse process, predissociation, involves the interaction of spin and orbital angular momentum in the molecule (4,16). However, the nature of the predissociating state cannot always be deduced from the strength of the predissociation and its dependence on symmetry or quantum number of the rotational levels concerned.

Although preassociation has been discussed in terms of intersection of potential surfaces, predissociation can occur between curves which do not intersect, for instance between the ground, $^2\Pi$, and first excited, $^2\Sigma^+$, states of OH, which correlate with different dissociation products but are both derived from the ^2P_u ground state of the corresponding united atom, F. (17)

The formation of excited molecules by preassociation is normally studied in systems where the combining atoms have a Maxwellian distribution of velocities. The combination process can then be treated either by a statistical mechanical treatment of the equilibrium between excited molecules and free atoms or by regarding the process as resonance scattering and applying the Breit-Wigner formula,

$$A + B \underset{-1}{\overset{1}{\rightleftharpoons}} AB^* \xrightarrow[2]{} AB + h\nu .$$

For a molecular level of angular momentum quantum number J, the energy widths for the non-radiative and radiative scattering processes are given by:

$$\Gamma_J' = \hbar k_{-1} \tag{v}$$

and $$\Gamma_J'' = \hbar k_2 \tag{vi}$$

and $$\Gamma_J = \Gamma_J' + \Gamma_J'' . \tag{vii}$$

Providing that the width of the levels is much smaller than kT, the Breit-Wigner formula can readily be integrated over a Maxwellian distribution of velocities to yield the following expression for the absolute intensity of emission:

$$I_J = I_J^o [A] [B] , \tag{viii}$$

where

$$I_J^o = \frac{\Gamma_J' \Gamma_J''}{\Gamma_J} \hbar^2 \left(\frac{2\pi}{\mu kT}\right)^{3/2} (2J + 1) \exp(- E_J/kT) \tag{ix}$$

Alternatively, the steady-state expression for the intensity of emission from the one level gives:

$$I_J^o = \frac{k_1 k_2}{k_{-1} + k_2} = \frac{K_J k_2}{1 + k_2/k_{-1}} . \tag{x}$$

Applying equilibrium statistical mechanics,

$$K_J = \frac{k_1}{k_{-1}} = \frac{q_{AB}^t}{q_A^t q_B^t} (2J + 1) \exp(- E_J/kT) , \tag{xi}$$

where the q^t are the translational partition functions:

$$q^t = (2\pi mkT)^{3/2} / h^3 . \tag{xii}$$

These expressions for I_J^o are identical and summation over the rotational levels in a vibrational state yields:

$$I^o = k_2 \sum_J \left(\frac{K_J}{1 + k_2/k_{-1}}\right) . \tag{xiii}$$

In other words a level, which lies just above the predissociation threshold, will contribute fully to two-body chemiluminescence providing the rate of predissociation, k_{-1}, is fast compared with radiation, k_2. This is also a criterion for detecting predissociation, that the spectrum should break off in emission in such low-pressure sources as discharges, where these processes compete with each other.

The evaluation of K_J is straightforward, providing that the dissociation energy is known accurately enough to provide reliable values of E_J. Unless the temperature of the system is unusually high, only one or two vibrational levels will be populated in two-body combination. For one vibrational level,

$$\sum_J (2J + 1) \exp(-E_J/kT) = q^r \exp(-E_o/kT) , \qquad \text{(xiv)}$$

where q^r is the rotational partition function and E_o is the energy of the lowest rotational level relative to free atoms. If E_o is negative and not large compared with kT, this expression is close to q^r, providing of course that there is no rotational barrier to preassociation, since the average density of rotational states is independent of energy. Symmetry factors and electronic degeneracies (including those of the combining species) must be incorporated in any calculation.

The most extensively studied example of two-body chemiluminescence is the combination of $N(^4S)$ and $O(^3P)$ to populate level v = 0 of $NO(C^2\Pi)$ which emits to the ground state between 190 and 250 nm (NO δ bands) and to $NO(A^2\Sigma^+$, v = 0) at 1224 nm in the infrared. The fluorescence efficiency of $NO(C^2\Pi$, v = 0) is low confirming that $k_{-1} \gg k_2$, being predissociated probably by the $a^4\Pi$ state which correlates with ground state atoms (18). The predissociation threshold lies between $F_1(5/2)$ and $F_2(7/2)$ and Kley (19) has recently observed that the corresponding low rotational levels of the $A^2\Sigma^+$, v = 0 state are absent from the nitric oxide afterglow at low pressures, when this state is populated only by radiation from $NO(C^2\Pi$, v = 0). Accepting this predissociation limit, the calculated intensity of the two-body emission in the δ-band system agrees with the measured value of $(1.5 \pm 0.4) \times 10^{-17}$ [N] [O] $cm^{-3}\ s^{-1}$, at 300K (20) although the observed temperature coefficient ($T^{-0.35}$) (21) is slightly less than the predicted value of $T^{-0.64}$.

Two-body emission by NO ($b^4\Sigma^-$, v = 5) also appears in the nitric oxide afterglow (19), this shows that the dissociation threshold to $N(^4S) + O(^3P)$ lies between levels v = 4 and v = 5 of the $b^4\Sigma^-$ state and thus establishes the absolute energies of the quartet states of NO.

Although the formation of excited nitrogen molecules by the three-body recombination of ground state nitrogen atoms is well known, only recently two two-body emission processes have been identified. Both involve association of two ground state N atoms along the unobserved but well-established $^5\Sigma_g^+$ surface, followed by inverse predissociation. One populates level v' = 13 of $N_2(B^3\Pi_g)$ plus high rotational levels, J > 32, of v' = 12 (22) and the observed intensities agree adequately with the calculated ones. The second process is more interesting since it populates $N_2(a^1\Pi_g$, v' = 6) which has a comparatively long radiative life ($k_2 = 7 \times 10^3\ s^{-1}$). With such a long radiative life, rotational and vibrational relaxation are important at the pressures used to study afterglows. When these collisional relaxation processes are included, the predicted emission kinetics for $N_2(a^1\Pi_g, v = 6)$ become $I \propto [N]^2/([N_2] + \gamma)$, where γ is governed by the relative rates of predissociation and relaxation. However k_{-1} increases with J and the two-body emission shifts to higher rotational levels as the total pressure is increased, leading to an increase in the effective value of γ in the above expression (23).

At low pressures the absolute intensity of the emission lies within a factor of two of the calculated value.

Apart from emission by $S_2(B^3\Sigma_u^-, v = 10)$ in the recombination of ground state S atoms (24) and a small contribution to the air afterglow emission by two-body combination of O and NO (25), the other examples involve the diatomic hydrides AlH (4) CH (3,26) and OH (27) where one must consider tunnelling through a barrier to predissociation which can arise either from the conservation of angular momentum or by the near-intersection of a bonding curve with a repulsive surface arising from a lower dissociation limit. Unfortunately none of the hydride systems has been thoroughly investigated in the laboratory.

II.D. Three-Body Combination

The bulk of the chemiluminescence associated with atomic recombination comes from bound levels below the dissociation limit, showing that the combining species have been stabilized by collision with a third body. The emitting states are not normally ones which correlate with the atoms from which they have been formed. Most of these processes involve the combination of ground-state atoms, and the excited molecular states, which correlate with ground-state atoms, rarely have allowed transitions to lower electronic states.

It is generally assumed that such states are initially formed in the three-body recombination and are the precursors of the observed emitting states. As this precursor state cannot normally be observed, it is difficult to distinguish the case where the combining species approach along the metastable potential surface and make a collision-induced transition to the radiating state from the case where a steady-state population of the bound levels of the metastable state is established by three-body recombination and collisional removal, and the emitting state is populated from these bound levels. The wide range of levels of the emitting state which are populated and the general lack of clear perturbations in the spectra of these levels show that the crossing between the excited states is predominantly collision-induced and does not rely on the non-adiabatic processes discussed above. Very little is known about the collision-induced mixing of states, which is responsible for these transitions, but there is evidence that the selection rules involving spin conservation are important in transitions between excited states of N_2 (28).

Theoretical calculations show that the observed rate coefficients for three-body recombination, which are typically $\sim 10^{-32}$ cm^6 s^{-1}, represent only a fraction of the collisions which might result in recombination. They also support the view that a considerable proportion of the recombination processes will populate bound excited states.

However, theoretical studies have not been particularly successful in reproducing the considerable dependence of the rate constant on the nature of the third body with its molecular complexity or the negative temperature coefficients which are observed for almost every three-body recombination and are also observed in the associated chemiluminescence. Empirically, these are best expressed in terms of a negative activation energy, i.e. a rate proportional to $\exp(+E/RT)$, where $E \simeq 5$ to 10 kJ mole^{-1} although T^{-n} dependences, where $n \sim 2$, have also been used, particularly for high temperature studies, as in shock tubes.

The presence of these temperature coefficients and the dependence on the nature of the third body provides a useful method of identifying chemiluminescence associated with three-body recombination, since the emitting state is often removed predominantly by collisional quenching, viz.

$$A + B + M \rightarrow AB^* + M \qquad (3)$$

$$AB^* \rightarrow AB + h\nu \qquad (2)$$

$$AB^* + M \rightarrow AB + M, \qquad (4)$$

giving an absolute emission intensity,

$$I_3 = I_3^o [A][B], \qquad (xv)$$

where

$$I_3^o = \frac{k_3 k_2 [M]}{k_2 + k_4 [M]} \qquad \text{(xvi)}$$

and I^o depends on the nature but not the pressure of the third body, M, if $k_2 \ll k_4[M]$.

This general form of pressure dependence will hold even when reaction (3) involves a series of steps providing all the rate-determining processes are collisional ones. For a two-body combination followed by collisional processes, the overall order lies between 1 and 2, i.e. a range one unit lower than for three-body systems.

Even with the nitric oxide afterglow, where population of $C^2\Pi$, v = 0 in a two-body process is highly favoured, the observed afterglow at pressures above 1 Torr is dominated by three-body chemiluminescence, and it is interesting to compare the magnitudes of I^o for the two types of process.

The rate of two-body chemiluminescence is given essentially by the probability of radiation during the duration of a two-body collision, whereas the rate of three-body association is given by the number of stabilising collisions occurring during a two-body collision. If this latter factor is multiplied by the rate of radiation divided by the rate of quenching of the emitting state which often approaches the collision number, this yields the rate of three-body chemiluminescence which appears to be similar to the two-body case with a favourable disposition of energy levels. However the collision cross section for the two-body chemiluminescence is about a factor of ten smaller than for the collisional processes in the three-body system, since the impact parameter in the former corresponds to the bond length of molecule whereas the Van der Waal's radii are appropriate for the latter.

The best known example is the nitrogen afterglow on which there is a vast, often conflicting, literature.

All bound states of N_2 below the dissociation limit to $N(^4S) + N(^4S)$ except $^1\Delta_u$, $^3\Delta_u$ and $^5\Sigma_g^+$ have been identified in the afterglow (Fig. 2). Three-body recombination of ground-state nitrogen atoms leads to efficient population of $N_2(B^3\Pi_g, v' \leq 12)$ and $(B'^3\Sigma_u^-, v' \leq 8)$, which give rise to the familiar yellow afterglow, which extends far into the infrared (29). These states are removed mainly by collisional quenching giving an intensity which is proportional to $[N]^2$. Above 1 Torr, I^o for the $B^3\Pi_g \rightarrow A^3\Sigma_u^+$ N_2

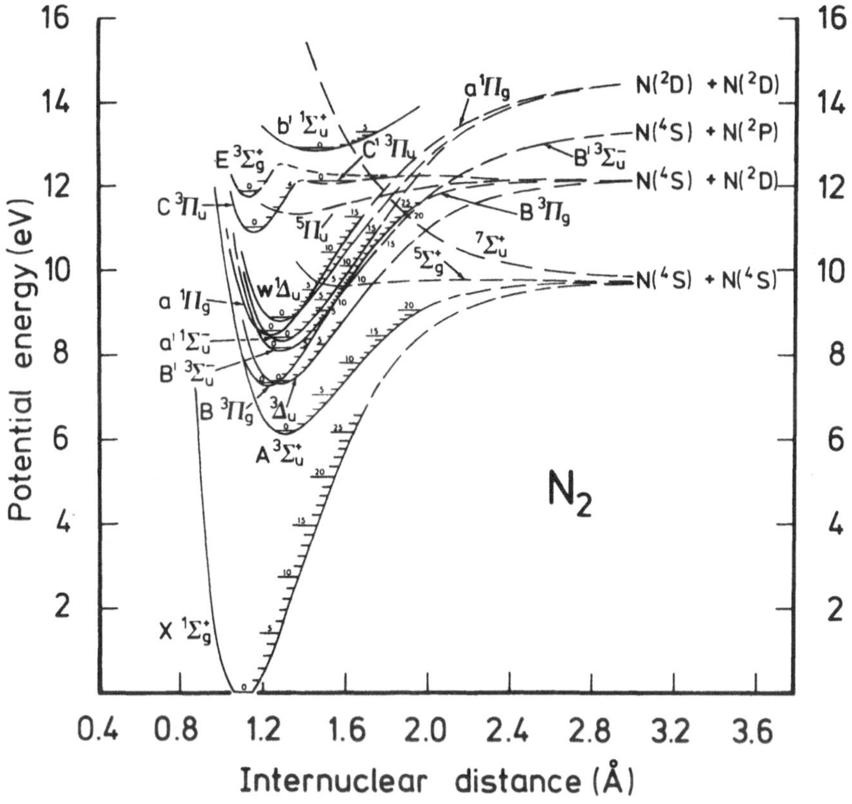

Fig. 2 Potential energy curves of N_2. After Gilmore (1965).

First Positive and $B'^3\Sigma_u^- \rightarrow B^3\Pi_g$ emissions are independent of pressure, but are enhanced when nitrogen is replaced by an inert gas carrier. At lower pressures, $I^o_{B\rightarrow A}$ falls (22) in adequate agreement with data on the quenching of the $B^3\Pi_g$ state by N_2. In N_2 carriers, the vibrational distribution of the $B^3\Pi_g$ state peaks just below the dissociation threshold and then rises again towards lower levels, but changes surprisingly little as the pressure is reduced.

High levels of the $B^3\Pi_g$ state are almost certainly populated by collision-induced crossing from the $A^3\Sigma_u^+$ state, which is populated directly by three-body recombination although the shallow

$^5\Sigma_g^+$ state has also been suggested as precursor. The $B'^3\Sigma_u^-$ state is populated by collision-induced transitions from the $B^3\Pi_g$ state (30), radiation and quenching from it is one source of population of low levels of the B state; another is radiation of First Positive bands and quenching to levels v' = 7 - 9 of the A state, followed by collision-induced transitions to the B state.

This last process prevents observation of emission from v > 6 of the $A^3\Sigma_u^+$ state, whose radiative life is 2.0s. Lower levels have been observed in the nitrogen afterglow with an intensity proportional to [N][M], because the A state is efficiently quenched by ground state nitrogen atoms.

In addition to the two-body process mentioned above, the $a^1\Pi_g$ and $a'^1\Sigma_u^-$ states are populated in the nitrogen afterglow by spin-allowed N-atom induced transitions from triplet states of N_2, particularly $B^3\Pi_g$, giving a rate of population proportional to $[N]^3$ (30).

The three-body chlorine and bromine afterglows both show emission from the $^3\Pi(0_u^+)$ state which correlates with $^2P_{3/2}$ + $^2P_{1/2}$ atoms and the $^3\Pi(1_u)$ state which correlates with two ground state ($^2P_{3/2}$) halogen atoms. The dominant emissions are $^3\Pi(0_u^+) \rightarrow X^1\Sigma_g^+$ for Cl_2 and $^3\Pi(1_u) \rightarrow X^1\Sigma_g^+$ for Br_2 (31,32). In both cases, the intensity of emission from the highest levels is proportional almost to the square of the halogen atom concentration and this dependence decreases almost to first order for the lowest levels. This is explained by three-body recombination into the high levels of the emitting state, followed by a competition between quenching by halogen atoms and vibrational relaxation in which the halogen molecules predominate.

Another interesting deduction from the kinetics of the bromine afterglow is that only about 1% of the recombinations populate the emitting $^3\Pi(1_u)$ state of Br_2 which correlates with ground state atoms. However, this state is shallow with a dissociation energy only one-eighth that of the ground state.

Chemiluminescence associated with recombination processes is observed in premixed laminar hydrogen-air flames (33). In such flames, the branched-chain reaction yields higher concentrations of the chain carriers H and OH than would be present in thermal equilibrium. Processes such as

$$OH + H_2 \rightarrow H_2O + H \qquad (5)$$

rapidly attain equilibrium, but the overall excess of free radicals can only be removed by such three-body processes as

$$H + H + M \rightarrow H_2 + M \qquad (6)$$

and

$$H + OH + M \rightarrow H_2O + M . \qquad (7)$$

In the hydrogen-oxygen flame itself, a blue continuous emission is observed from the chemiluminescent combination of H + OH (34) and analogous continua arise from the combination of alkali metals with OH (35), one being the dominant source of the lilac flame colouration produced by potassium salts. The overall collisional efficiencies are about 10^{-9}.

Chemiluminescence of metal atoms excited by recombination of H and OH is also observed in flames. The efficiencies of excitation of various levels indicate that the atoms are excited by the newly stabilised vibrationally excited molecule rather than by acting as third bodies themselves (36).

The formation of excited triatomic molecules by the chemiluminescent combination of O + NO (air afterglow), O + SO (sulphur dioxide afterglow), O + CO (carbon monoxide flame bands) and H + NO (nitroxyl afterglow) are all three-body processes showing points of interest.

The apparently continuous nature of the air afterglow spectrum and the difficulty of observing a pressure dependence led to a belief that this was a two-body process. It is now known (37) that electronically excited NO_2 molecules are formed in a high proportion of recombinations, O + NO + M, and the intensity and spectral changes in the air afterglow over the pressure range 0.2 mTorr to 1 Torr are wholly consistent with experiments on the visible fluorescence of NO_2, where there are probably two excited electronic states, $^2B_1$ and $^2B_2$, which interact with other states and particularly the ground state, $^2A_1$, in a complex manner. Only at total pressures around 1 mTorr does the two-body chemiluminescence from O + NO predominate.

The O + SO afterglow is a closely similar process, involving largely unresolved emission from several excited states, which are believed to interact with each other and with the ground state (38).

In contrast, the nitroxyl emission from H + NO + M shows resolved rotational structure from a few bands in a single excited electronic state. Interestingly HNO molecules are formed with up to 42 kJ $mole^{-1}$ rotational energy about the near symmetric top axis of the molecule, which lies almost along the N - O bond (39). It has also been shown that a number of third bodies (Ar, H_2, CO_2, N_2O

and SF_6, but not H_2O) give the same ratio of efficiencies for populating the ground and excited states (40).

Unlike the above systems, where the afterglows have negative temperature coefficients similar to those of the three-body recombinations, the carbon monoxide flame band emission, in which $CO(^1\Sigma^+) + O(^3P)$ yield electronically excited $CO_2(^1B_2)$, has the same positive activation energy of 16 kJ mole^{-1} (41) as the overall combination reaction O + CO + M, which is also spin-forbidden. There is no evidence that either of these spin-forbidden processes is affected by other species with unpaired spins and the initial step is probably combination over a barrier caused by an avoided crossing into a $^3B_2$ state, which must lie below the $^1B_2$ state.

III. CHEMILUMINESCENCE IN ATOM TRANSFER PROCESSES

The infrared emission from vibrationally excited products of atom transfer reactions is an important method for investigating reaction dynamics (2). The development of infrared chemical lasers using these reactions has stimulated searches for electronically excited products from such reactions:

$$A + BC \rightarrow AB^* + C \qquad (8)$$

$$A + BC \rightarrow AB + C^* . \qquad (9)$$

Here, the transferred species B is an atom but A and C may contain more than one atom.

Many early studies of such reactions were hampered by the fact that in few atom transfer chemiluminescent systems were the vibrational and electronic energies of the reactants accurately known. In more complex chemical systems such as flames, it is often not certain whether energy transfer or atom transfer is responsible for the observed emission. The present discussion is limited to those atom transfer reactions in which the mechanism of chemiluminescence is well established.

The behaviour of electronically excited species in atom transfer reactions is normally discussed using correlation diagrams in which the reaction channels connecting given reactant and product states are identified and the approximate location of intersections of states of unlike symmetry derived using adiabatic correlation rules (42). The diagram is usually based on the adiabatic approximation in which potential surfaces of states of the same symmetry do not cross but those of different symmetry intersect without mutual interaction. Fig. 3 is a schematic correlation diagram for ground and excited (*) state reactants and products

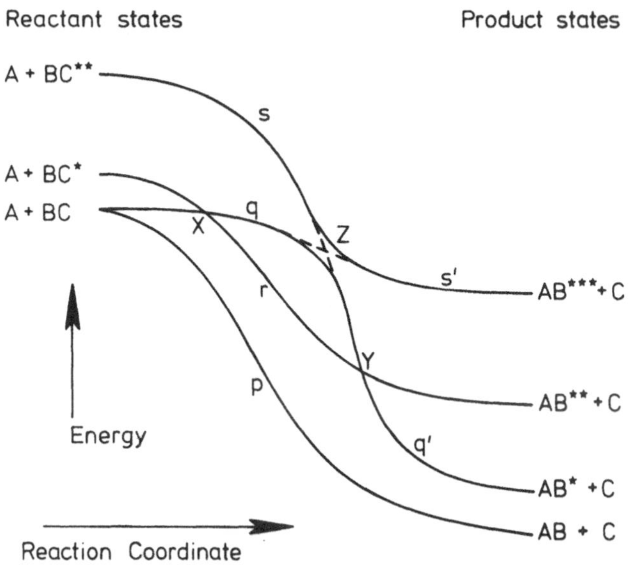

Fig. 3 Schematic correlation diagram for the reaction A + BC → AB + C. ______ Adiabatic Correlation; ----- Diabatic Correlation.

(species A and C have been assumed to have no low-lying excited states). The four adiabatic surfaces are designated p, qq', r and ss'; qq' and ss' have common symmetry and have an 'avoided crossing' at Z, while r has different symmetry. Four principal mechanisms of atom-transfer chemiluminescence can be distinguished (2).

a) An adiabatic process in which the reaction occurs along a single potential surface. The surface p leads to ground state products and surface qq' to the excited state, AB* + C.

b) Due to finite interaction between states of different symmetry, non-adiabatic transitions can occur at the crossing points X or Y so that A + BC yields the products AB** + C, via surface r.

c) For weak coupling between states of the same symmetry and rapid relative motion of the nuclei, the non-crossing rule can break down, as at Z (see Eq. (iii)). In the hypothetical limit, the 'diabatic' surfaces, qs' and sq', intersect without interacting and A + BC yield AB*** + C. This occurs in several metal atom reactions where 'adiabatic' and 'diabatic' behaviour yield ground state and excited state products respectively.

d) If the intermediate ABC* is long-lived, several exit channels may become accessible, each involving one or more non-adiabatic transitions (not shown in Fig. 3). In the limit this would yield a statistical distribution of products from all energetically accessible exit channels.

In all these cases, subsequent collisionless or collision-induced electronic transitions in the excited AB or C products may occur before the observed emitting states are populated.

There is little evidence for efficient processes of type (a) and, in many atom transfer reactions, mechanisms (b), (c) or (d) must be invoked to account for the observed chemiluminescence. The observed yields of electronically excited products are generally much lower than for vibrationally excited products in similar reactions. Because of the high energies of excited electronic states, it is possible that large energy barriers may be partly responsible for the low yields, but little information is yet available on the effect of reagent energy (translational, rotational or vibrational) on yields of electronically excited products.

III.A. Adiabatic Atom Transfer Reactions

The adiabatic process (a) normally yields the vibrationally excited ground state products characteristic of atom transfer reactions. It has also been invoked as an efficient mechanism for the electronic quenching of excited atoms (43),

$$C^* + BA \rightarrow BC + A ; \qquad (10)$$

since the absence of an exothermic adiabatic channel is often associated with a much lower quenching efficiency. Process (10) is essentially the reverse of atom transfer chemiluminescence by mechanism (a) of which there are extremely few well-established examples. One major reason is that, even if an exothermic surface connects reagents to excited products, there must be a lower energy surface leading to ground state products. In fact, if the ground states of BC and AB correlate directly with ground states of their dissociation products B + C and A + B, then an adiabatic surface must exist which connects A + BC to ground state AB + C.

N_3 is an example (the isoelectronic species NCO and NCS are others) of molecules whose ground state ($^2\Pi_g$) does not correlate with the ground states of the separated species ($N(^4S) + N_2(^1\Sigma_g^+)$); chemiluminescence from $N_2(B^3\Pi_g)$ has been observed in the process (44)

$$N(^4S) + N_3(^2\Pi_g) \rightarrow N_2{}^*(B^3\Pi_g) + N_2(X^1\Sigma_g^+) \ . \qquad (11)$$

The reactants correlate with $N_2(A^3\Sigma_u^+) + N_2(X^1\Sigma_g^+)$ and $N_2(B^3\Pi_g) + N_2(X^1\Sigma_g^+)$ via the two available triplet surfaces but not with $2N_2(X^1\Sigma_g^+)$ molecules, which must yield a singlet surface. No estimate of the efficiency of (11) is yet available. A similar situation exists in the formation of electronically excited NO, NCl and NBr in the reactions of O, Cl and Br with N_3.

When exothermic adiabatic channels lead to both ground-state and excited-state products, the relative importance of the two channels determines the efficiency of chemiluminescence. The most important factors are the energy barriers and statistical weights for each channel, but possible non-adiabatic transitions out of each channel must also be considered. These are likely to be more important for the higher energy channel.

A well-studied example is (45)

$$NO(X^2\Pi) + O_3(^1A_1) \rightarrow NO_2 + O_2(X^3\Sigma_g^-) \ , \qquad (12)$$

for which the reagents can combine on two surfaces, of which one leads to ground state NO_2 and the other to electronically excited NO_2; less than 1% of the reactions yield excited O_2 molecules, $a^1\Delta_g$ or $b^1\Sigma_g^+$, (46). The two paths have similar pre-exponential factors and the activation energy of the route yielding $NO_2{}^*$, 17.5 kJ mole^{-1} as compared with 9.6 kJ mole^{-1} for the route to ground state products. Redpath and Menzinger (47) have obtained indirect evidence that the lower multiplet component of $NO(^2\Pi_{1/2})$ only yields ground state NO_2 and the electronically excited NO_2 comes from $NO(^2\Pi_{3/2})$. Recent studies (48,49) employing O_3 containing a single quantum of vibrational energy (the active vibrational mode is not known with certainty), show enhancements of the rate constants of the reactions to yield ground-state and excited-state products by factors of roughly 20 and 6 respectively.

III.B. Non-Adiabatic Atom Transfer Reactions

There is now considerable evidence that non-adiabatic processes can occur with high probability (43), if reaction trajectories pass close to surface intersections. A good example is the quenching of the excited atomic states $C(^1D)$, $N(^2D)$ and $O(^1D)$ by CO and by N_2 at greater than one-tenth of the collision rate, although quenching of the higher states $C(^1S)$ and $O(^1S)$ is several orders of magnitude slower. In each case, the first excited state of the atom, rather than its ground state, correlates directly with the ground state of the intermediate, i.e., CCO, NCO, CO_2, CNN, N_3 and N_2O; the relevant singlet surface (doublet for $^2N(^2D)$) is at least slightly attractive and must cross the triplet surface (quartet for N) leading to ground state species, these conditions clearly favour a non-adiabatic transition via spin-orbit coupling. For N_2O and CO_2 the intermediate state is probably sufficiently long-lived for the system to pass several times through the region of the intersection before redissociating. The weaker quenching of the more highly excited states may be due to a lack of intersections with other surfaces at accessible energies; this probably arises from the antibonding nature of the singlet state formed by (say) $O(^1S)$ plus the quencher due to repulsive interaction with states of similar symmetry formed by $O(^1D)$ and the quencher.

Quenching of $O(^1D)$ by O_2 occurs principally by the chemiluminescent channel (50)

$$O(^1D) + O_2(^3\Sigma_g^-) \rightarrow O_2^*(^1\Sigma_g^+) + O(^3P). \qquad (13)$$

Although reagent and product surfaces of the same symmetry exist, they cannot correlate directly with each other because the reactant (A) and product (C) species are the same. Similar arguments apply to the process

$$N(^4S) + N_2(B^3\Pi_g) \rightarrow N_2^*(a^1\Pi_g, a'^1\Sigma_u^-) + N(^4S) \qquad (14)$$

in the nitrogen afterglow which occurs at about one collision in ten.

This behaviour is paralleled by several chemiluminescent reactions for which no direct surface is available. Process (15) is an example of mechanism (b) (38)

$$SO(^3\Sigma^-) + O_3(^1A_1) \rightarrow SO_2 + O_2(^3\Sigma_g^-) \; . \qquad (15)$$

The single triplet reactant surface correlates with ground state products but chemiluminescence is observed from excited $^3B_1$ and $^1B_1$ states of SO_2, which are formed with lower frequency factors and higher activation energies than the ground state. In this case, crossings between triplet surfaces in the transition state are responsible.

Transfer reactions of alkali and alkaline earth metals yield excited products by non-adiabatic channels. Recent studies have used crossed molecular beams or beams flowing into a low pressure gas rather than the classical diffusion flame method of M. Polanyi (51). Highly excited electronic states are observed, consistent with the large exothermicities of these reactions but implying numerous non-adiabatic processes, possibly favoured by long-lived intermediates (mechanism (d)). Several pseudo-continua have also been observed. With a few exceptions, low photon yields are obtained, suggesting that the chemiluminescence is a minor channel in these systems.

The well known intense Na D-line emission from the sodium/chlorine flame is produced by two independent processes:

$$Cl(^2P) + Na_2(^1\Sigma_g^+) \rightarrow Na^*(3^2P) + NaCl(^1\Sigma^+) \qquad (16)$$

(52) and

$$NaCl^{\dagger} + Na(3^2S) \rightarrow NaCl + Na^*(3^2P) , \qquad (17)$$

in which atom exchange has been shown to be at least an order of magnitude faster than simple vibration-to-electronic energy transfer (53).

Reaction (16) occurs by both mechanisms (a) and (c) (54).

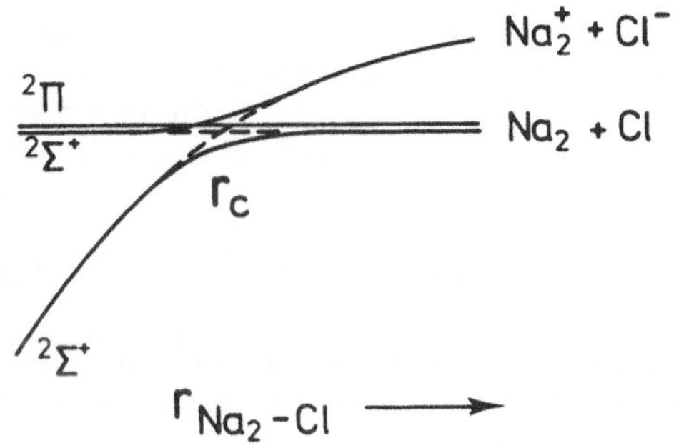

Fig. 4 Schematic reactant potential surfaces for collinear Na_2 + Cl reaction. ______ Adiabatic Correlation; ----- Diabatic Correlation.

As shown in Fig. 4 for a linear collision, the reagent surfaces intersect the $^2\Sigma^+$ ionic surface, correlating with ground state $Na_2^+ + Cl^-$ at an internuclear distance, r_c, of about 1.3 nm. The adiabatic reaction channel, following the $^2\Sigma^+$ surface via an avoided crossing at this point is termed an electron-jump mechanism (55), as the state is largely covalent at $r > r_c$ and largely ionic at $r < r_c$. This comprises the single surface leading to ground-state products, $NaCl(^1\Sigma^+) + Na(3^2S)$. However, the interaction of the $^2\Sigma^+$ surfaces is expected to be weak because the large value of r_c at the crossing point increases the probability of diabatic, rather than adiabatic, behaviour. In this case, Na_2 and Cl approach at $r < r_c$ along the excited $^2\Sigma^+$ surface which, like the unperturbed $^2\Pi$ reactant surface, must correlate with excited products. In detail, this correlation must proceed via an 'electron-jump' at smaller Na_2– Cl separation, in order to yield the ionic product, NaCl, + $Na^*(3^2P)$. The importance of such diabatic behaviour is indicated by the large cross section, 10 - 100 $Å^2$(54), for the chemiluminescent reaction (16). A similar argument applies for reaction via a non-linear intermediate.

In addition to the intense emission from the first excited states of alkali atoms the reactions with the halogens produce very much weaker emission from all the higher states which are accessible energetically. These minor reaction channels must involve non-adiabatic processes and it appears that their probabilities depend only slightly on the orbital symmetry, S, P or D, of the resultant excited atomic state (56).

(17) has been shown to be an exchange reaction by crossing a beam of vibrationally excited KBr molecules with Na atoms and observing $K^*(4^2P)$ resonance emission (53). As with (16), adiabatic correlation rules (i.e. an electron jump to $K.....Na^+ Br^-$) predict formation of ground state products, $K(4^2S)$ + NaBr; the chemiluminescent channel is probably a 'diabatic' avoidance of the electron jump, aided by the high relative velocity of the atoms in the highly vibrationally excited KBr, followed by a non-adiabatic transition into a state correlating with an excited K atom (57). The low probability of vibrational-to-electronic energy transfer without atom transfer, i.e.

$$NaBr\dagger + K(4^2S) \rightarrow NaBr + K(4^2P) \qquad (18)$$

is ascribed to the short lifetime of $(NaK)^+ Br^-$ and the difficulty of reaching the region of the potential surface intersecting the required exit surface.

Like the alkali metal atoms, the metastable states of the inert gas atoms have a single s-electron in the outermost occupied shell. The $^3P_2$ states of xenon, krypton and argon react with halogen molecules to yield predominantly ionic excited electronic states of the diatomic inert gas halides (58). These states have dissociation energies similar to the corresponding alkali metal halides, but emit to yield ground state inert gas and halogen atoms. Non-adiabatic crossing processes must be involved in their formation.

Alkali metal and halogen dimers react with large cross sections $\simeq 100\text{Å}^2$, predominantly to yield ground-state products

$$M_2(X^1\Sigma_g^+) + Y_2(X^1\Sigma_g^+) \rightarrow MY(^1\Sigma^+) + M(^2S) + Y(^2P). \qquad (19)$$

Although weak alkali atom emission from the lowest excited state has also been detected, a more interesting chemiluminescent channel for (19) yields from the first excited (covalent) state of MY (59).

Reaction (19) involves an electron jump to an ionic intermediate

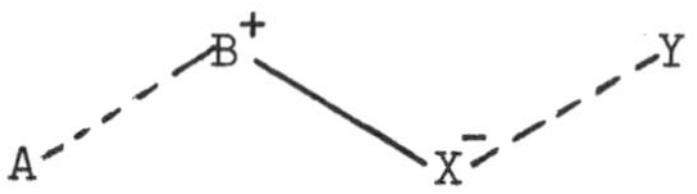

where X is the more electronegative halogen. The major channel involves separation of the end atoms to yield the products A + BX + Y. It has been suggested that the atomic chemiluminescence arises via a further vibration-to-electronic transfer reaction (17) of the separating A and BX species. The molecular chemiluminescence is ascribed to combination of A and Y into the covalent excited state (the atoms are too close to undergo an electron jump to the ionic ground state). The chemiluminescence is thus expected from the less electronegative halide, AY, as observed when using ClF, IBr and ICl (59).

Chemiluminescence has been observed in the reactions of alkaline earth metals with halogens and with oxygen-containing compounds (O_2, O_3, N_2O, NO_2). The mechanisms of these reactions are by no means well established. One example of particular interest is the reaction $Ba(^1S) + N_2O(^1\Sigma^+)$ which appears to give a 99% yield of $BaO\ (a^3\Pi) + N_2\ (^1\Sigma_g^+)$, $BaO\ (A^1\Sigma^+)$ being populated from the triplet state in collisions (60). Although the lowest singlet surface must correlate ground state reactants with ground state products $BaO(X^1\Sigma^+) + N_2(X^1\Sigma_g^+)$, neither ground state BaO nor N_2O correlates with the separated ground state species $Ba(^1S) + O(^3P)$ or $N_2(^1\Sigma_g^+) + O(^3P)$; suggesting that there is a triplet surface below the lowest singlet surface in the region of the transition state.

Although yields of excited states in atom-transfer reactions are usually small, it is often possible to observe almost complete transfer of the available exothermicity of the reaction into vibronic levels of the emitting state. When the thermochemistry of the reaction is unknown and the chemiluminescence provides a useful method for obtaining limits for the bond energy of the emitting species, AB, from the energy of the highest observed level of AB*

$$A + BC \rightarrow AB^* + C\,. \qquad (8)$$

This technique is valid only if the reaction path to yield excited products involves a negligible energy barrier (61).

IV. CHEMILUMINESCENCE IN OTHER CHEMICAL SYSTEMS

The examples given above have been of relatively simple gas phase reactions when the process yielding excited molecules can be identified and its mechanism hopefully understood. Most of this work has been carried out in the last ten years, and many of the familiar examples of chemiluminescent reactions in the gas phase reactions involve systems which are so complex chemically that it is not possible to identify positively the elementary process yielding the emitter. This is, for instance, true of the cool flame emission by formaldehyde in the oxidation of hydrocarbons.

No discussion of the formation of excited species in gas phase reactions would be complete without some reference to the corresponding but very different processes which are important in solutions and in bioluminescence. In solution the formation of an electronically excited aromatic hydrocarbon in the electron jump associated with the neutralisation of the corresponding cation and anion corresponds most closely to the two-body recombination in the gas phase. Similarly the atom transfer reaction (8) is replaced by the four-centre reaction (20)

$$A + BC \rightarrow AB^* + C \qquad (8)$$

$$AC + BD \rightarrow AB^* + CD \qquad (20)$$

where conservation of symmetry enters in a different way.

In process (8), formation of ground-state products is rarely prevented by the correlation rules, the restrictions almost invariably coming from the spin-conservation rule. This arises from Hund's rule of maximum multiplicity; atoms, which have the greatest possibility of orbital degeneracy, can have higher resultant spins than linear and particularly non-linear molecules. In an atom transfer, the plane containing the three atoms whose bonding changes is the only significant symmetry element and the requirement that overall symmetry be maintained in that plane rarely restricts formation of ground-state products.

With a four-centre reaction, (20), where two bonds are broken and two formed, a different type of restriction arises from the requirement that the occupied orbitals in the reactant molecules pass smoothly and without sudden change of symmetry into those of the products.   This is a generalisation of the Woodward-Hoffmann rules of organic chemistry and for instance forbids the direct addition of hydrogen to ethylene to yield ethane (62)

$$
\begin{array}{c}
H \quad H \\
\backslash C \diagup \\
\parallel \\
C \\
\diagup \quad \backslash \\
H \quad H
\end{array}
\quad + \quad
\begin{array}{c}
H \\
\mid \\
H
\end{array}
\quad \rightarrow \quad
\begin{array}{c}
H \\
H \mid H \\
\backslash C \diagup \\
\mid \\
C \\
\diagup \backslash \\
H \mid H \\
H
\end{array}
$$

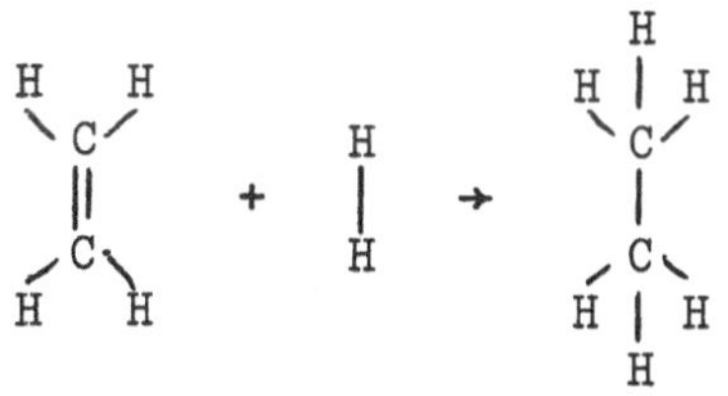

The basic argument is illustrated by the isotope exchange reaction,

$$
\begin{array}{c}
H \\
\mid \\
H
\end{array}
+
\begin{array}{c}
D \\
\mid \\
D
\end{array}
\rightarrow
\begin{array}{c}
H\text{- - -}D \\
\vdots \quad \vdots \\
H\text{- - -}D
\end{array}
\rightarrow
\begin{array}{c}
H\text{——}D \\
+ \\
H\text{——}D
\end{array}
$$

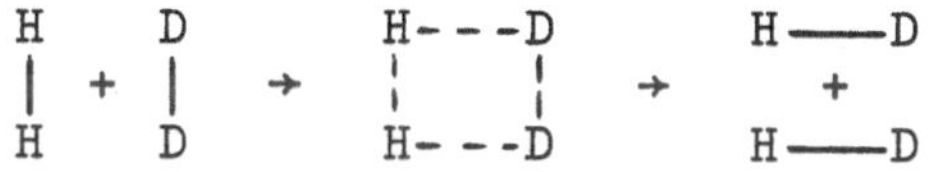

where the occupied molecular orbitals of the transition state are derived from linear combinations of the $\sigma_g 1s$ bonding orbitals on H_2 and D_2.   One of these combinations has no nodal planes and passes over smoothly to the bonding orbitals in H - D, but the other combination has a nodal plane bisecting the incipient H - D bonds and correlates with the $\sigma_u 1s$ antibonding orbital in HD.   Thus the system cannot pass adiabatically from ground state reactants to ground state products in the direct thermal reaction except via a very high barrier.   Such arguments apply also for non-planar transition states (63).

These principles apply also to the best-understood examples of chemiluminescence of organic molecules, which involve the rupture of the dioxetane ring

$$
\begin{array}{c}
O\text{————}O \\
\mid \qquad \mid \\
-C\text{————}C- \\
\mid \qquad \mid
\end{array}
$$

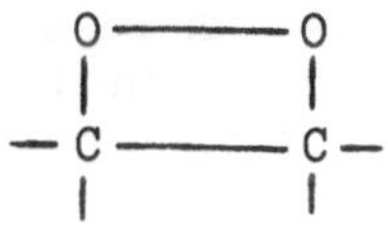

to yield two ketones, one of which is electronically excited (64). Here there is insufficient energy released for the formation of two

electronically excited ketone molecules, which the Woodward-Hoffmann rules would predict, but one excited ketone molecule can be formed with quite high probability by a non-adiabatic process. A mechanism of this type provides the most plausible explanation of such bioluminescent systems as the oxidation of fire-fly luciferin (65).

In the gas phase combustion of hydrocarbons, the electronic emission by CH and OH have been ascribed to the four-centre processes

$$C_2(^1\Sigma_g^+, X^3\Pi_u) + OH(X^2\Pi) \rightarrow CH^*(A^2\Delta, B^2\Sigma^-) + CO(X^1\Sigma^+) \quad (21)$$

$$O_2(X^3\Sigma_g^-) + CH(X^2\Pi) \rightarrow OH^*(A^2\Sigma^+) + CO(X^1\Sigma^+) \text{ ,} \quad (22)$$

for which there is good experimental evidence and plausible rate data giving collisional efficiencies between 0.1 and 0.001 (66,67).

The orbital symmetry rules for four-centre reactions explain the chemiluminescence observed in processes (21) and (22). Here one can separate the orbitals into two groups, the σ orbitals plus those π orbitals lying in the plane of the nuclei in the transition state and those π orbitals lying perpendicular to this plane. For the latter, we have to consider only the three C or O atoms and there is no symmetry restriction in the transfer C -- O -- O or C -- C -- O. For the orbitals lying in the plane of the nuclei, the same correlation problem arises as for the $H_2 + D_2$ system discussed above. For reaction (22), two electrons are in the σ orbitals which correlate with antibonding orbitals of the products, so the direct reaction to ground state products is expected to involve a high energy barrier. Thus a non-adiabatic channel to yield excited OH is favoured. In process (21), where there are adiabatic channels to both ground and excited $(A^2\Delta)$ state CH if the reactant state of C_2 is $X^3\Pi_u$, the orbital correlation diagram predicts a lower energy barrier for the latter, chemiluminescent, channel than for the former. In both examples, the excited state populated is one in which a σ electron has been promoted to a π orbital.

The emphasis on well established chemiluminescent processes may have given the impression that the mechanisms of formation of excited species in elementary chemical reactions are well understood. There are many processes yielding excited species whose mechanisms are not well understood, and even where the overall mechanism has been established, little is known about the detailed form of the potential surfaces involved or of the non-adiabatic processes which are often so important.

The author thanks Dr. M.F. Golde for assistance in preparing this manuscript.

REFERENCES

1. R.G.W. Norrish, this volume.

2. T. Carrington and J.C. Polanyi, in MTP International Review of Science, Vol. 9 (J.C. Polanyi, ed), Butterworths, London, pp. 135-171 (1972).

3. D.R. Bates, Mon. Not. R. Astr. Soc. 111, 303 (1951).

4. G. Herzberg, Spectra of Diatomic Molecules, Van Nostrand, New York (1950).

5. H.B. Palmer, J. Chem. Phys. 26, 648 (1957).

6. R.K. Boyd, G. Burns, T.R. Lawrence and J.H. Lippiatt, J. Chem. Phys. 49, 3804 (1968).

7. H.B. Palmer and R.A. Carabetta, J. Chem. Phys. 49, 2466 (1968).

8. F.H. Mies and A.L. Smith, J. Chem. Phys. 45, 994 (1966).

9. K.W. Chow and A.L. Smith, J. Chem. Phys. 54, 1556 (1971).

10. K.M. Sando, Molec. Phys. 21, 439 (1971).

11. R.C. Michaelson and A.L. Smith, Chem. Phys. Lett., 6, 1 (1970).

12. R. Turner, Phys. Rev., 158, 121 (1967).

13. D.H. Stedman and D.W. Setser, Prog. React. Kin. 6, 193 (1971).

14. A.W. Johnson and J.B. Gerardo, J. Chem. Phys. 59, 1738 (1973).

15. E.E. Nikitin, Theory of Elementary Atomic and Molecular Processes, Moscow (1970).

16. I. Kovacs, Rotational Structure in the Spectra of Diatomic Molecules, Hilger, London (1969).

17. S. Durmaz and J.N. Murrell, Trans. Faraday Soc. 67, 3395 (1971).

18. R.A. Young and R.L. Sharpless, Disc. Faraday Soc. 33, 228 (1962).

19. D. Kley, Habilitationsschrift, Univ. of Bonn (1973).

20. M. Mandelman, T. Carrington and R.A. Young, J. Chem. Phys. 58, 84 (1973).

21. R.W.F. Gross and N. Cohen, J. Chem. Phys. 48, 2582 (1968).

22. K.H. Becker, E.H. Fink, W. Groth, W. Jud and D. Kley, Faraday Disc. Chem. Soc., 53, 35 (1972).

23. M.F. Golde and B.A. Thrush, Rep. Prog. Phys., 36, 1285 (1973).

24. R.W. Fair and B.A. Thrush, Trans. Faraday Soc., 65, 1208 (1969).

25. K.H. Becker, W. Groth and D. Thran, Chem. Phys. Lett. 15, 215 (1972).

26. P.M. Solomon and W. Klemperer, Astrophys. J. 178, 389 (1972).

27. S. Ticktin, G. Spindler and H.I. Schiff, Disc. Faraday Soc. 44, 218 (1967).

28. M.F. Golde and B.A. Thrush, Faraday Disc. Chem. Soc. 53, 52 (1972).

29. E.M. Gartner and B.A. Thrush, Proc. Roy. Soc. in press (1975).

30. M.F. Golde and B.A. Thrush, Proc. R. Soc. A 330, 79, 109, 120 (1972).

31. R.J. Browne and E.A. Ogryzlo, J. Chem. Phys., 52, 5774 (1970).

32. M.A.A. Clyne, J.A. Coxon and A.R. Woon Fat, Faraday Disc. Chem. Soc. 53, 82 (1972).

33. T.M. Sugden, Ann. Rev. Phys. Chem. 13, 369 (1962).

34. P.J. Padley, Trans. Faraday Soc., 56, 449 (1960).

35. C.G. James and T.M. Sugden, Proc. R. Soc. A248, 238 (1958).

36. L.F. Phillips and T.M. Sugden, Trans. Faraday Soc., 57, 2188 (1961).

37. F. Kaufman, in Chemiluminescence and Bioluminescence (M.J. Cormier, D.M. Hercules, J. Lee, eds.), Plenum Press, New York, pp. 83-100 (1973).

38. C.J. Halstead and B.A. Thrush, Proc. R. Soc., A295, 363, 380 (1966).

39. M.J.Y. Clement and D.A. Ramsay, Can. J. Phys., 39, 205 (1961).

40. D.B. Hartley and B.A. Thrush, Proc. R. Soc., A297, 520 (1967).

41. M.A.A. Clyne and B.A. Thrush, Proc. R. Soc., A269, 404 (1962).

42. K.E. Shuler, J. Chem. Phys., 21, 624 (1953).

43. R.J. Donovan and D. Husain, Chem. Rev., 70, 489 (1970).

44. T.C. Clark and M.A.A. Clyne, Trans. Faraday Soc., 66, 877 (1970).

45. P.N. Clough and B.A. Thrush, Trans. Faraday Soc., 63, 915 (1967).

46. M. Gauthier and D.R. Snelling, Chem. Phys. Lett., 20, 178 (1973).

47. A.E. Redpath and M. Menzinger, J. Chem. Phys., 62, 1987 (1975).

48. R.J. Gordon and M.C. Lin, Chem. Phys. Lett., 22, 262 (1973).

49. W. Braun, M.J. Kurylo, A. Kaldor and R.P. Wayne, J. Chem. Phys. 61, 461 (1974).

50. D.R. Snelling, Can. J. Chem., 52, 257 (1974).

51. M. Polanyi, Atomic Reactions, Williams and Norgate, London (1932).

52. W.S. Struve, T. Kitagawa, and D.R. Herschbach, J. Chem. Phys. 54, 2759 (1971).

53. M.C. Moulton and D.R. Herschbach, J. Chem. Phys., 44, 3010 (1966).

54. J.L. Magee, J. Chem. Phys., 7, 652 (1939); 8, 687 (1940).

55. D.R. Herschbach, Adv. Chem. Phys., 10, 319 (1966).

56. D.O. Ham, J. Chem. Phys., 60, 1802 (1974).

57. W.S. Struve, Molec. Phys., 25, 777 (1973).

58. M.F. Golde and B.A. Thrush, Chem. Phys. Letters, 29, 486 (1974).

59. W.S. Struve, J.R. Krenos, D.L. McFadden and D.R. Herschbach, Faraday Disc. Chem. Soc., 55, 314 (1973).

60. C.J. Hsu, W.D. Krugh, and H.B. Palmer, J. Chem. Phys., 60, 5118 (1974).

61. B.A. Thrush, J. Chem. Phys., 58, 5191 (1973).

62. R.B. Woodward and R. Hoffmann, Angewandte Chemie (Int. Ed.) 8, 781 (1969).

63. D.M. Silver and R.M. Stevens, J. Chem. Phys., 59, 3378 (1973).

64. D.R. Kearns, J. Am. Chem. Soc., 91, 6554 (1969).

65. F. McCapra, M. Roth, D. Hysert and K.A. Zaklika, in Chemiluminescence and Bioluminescence, (M.J. Cormier, D.M. Hercules, J. Lee eds.), Plenum Press, New York, pp. 313-323 (1973).

66. R.P. Porter, A.H. Clark, W.E. Kaskan and W.E. Browne, Proc. 11th Symp. Combustion, Combustion Institute, Pittsburgh. p.907 (1967).

67. E.M. Bulewicz, P.J. Padley and R.E. Smith, Proc. R. Soc. A315, 129 (1970).

TWO-PHOTON SPECTROSCOPY IN THE GAS PHASE

E. W. Schlag

Institut für Physikalische Chemie

Technische Universität München

ABSTRACT

In the methods of molecular spectroscopy, molecular information of high precision can be obtained by high resolution measurements in the gas phase. Here one-photon absorption in the ultraviolet has yielded important information. Whereas one-photon absorption selects g→u parity transitions, two-photon processes select g→g transitions. Selection rules also differ in many other ways. As a consequence two-photon spectroscopy opens up the possibility of observing many new transitions in molecular spectroscopy, particularly if they are carried out in the gas phase. We here demonstrate, employing the classic example of benzene, how new transitions are uncovered and a two-photon molecular spectrum in the gas phase assigned using this new technique. It is also shown that the intensity of the laser system suffices to prepare quantum states by two-photon absorption in the collisionless gas phase. This opens up a plethora of new experimental possibilities. Lifetimes are here given as an example.

I. INTRODUCTION

Molecular spectroscopy has yielded a wealth of information about molecules in the last forty years. Spectroscopy in the ultraviolet has yielded information on molecular vibronic states, particularly since the advent of high resolution techniques, such as that carried out in Ottawa. To make full use of high resolution techniques it is essential to study molecules in the gas phase, since otherwise medium effects, in general, strongly perturb molecular absorption lines. Nevertheless many electronic or vibronic levels cannot be detected since the absorption is symmetry- or parity-forbidden. The latter is a particular problem if the excited state is of gerade parity, and hence forbidden. Many unknown molecular states are thus of the wrong parity to be observable in conventional spectroscopy.

These new states of gerade parity are, however, susceptible to the simultaneous absorption of two-photons in a high radiation field. The advent of high intensity lasers makes this in principle an interesting prospect. This should open up an entire new form of spectroscopy, allowing for the measurement of many hitherto unobservable levels. Parallel to the development of one-photon spectroscopy one would, however, expect that this technique can only be fully realized if it is possible to measure spectra in the gas phase. It would furthermore be useful if these spectra were not measurable only point by point, but could be scanned as a function of wavelength.

In principle the latter requirement is satisfied by tunable lasers, but unfortunately, though some of them are tunable, hardly any of them are scannable. It is still extremely difficult to scan a laser with absolute wavelength precision and good resolution over a broader range without detuning the laser in so doing. When the requirement is one of high-resolution scanning, at least over a broad wavelength range, the problem is almost unsolved (1). In short, although it is possible to build lasers of extremely narrow bandwidth, it is almost equally difficult to know where that wavelength is. For molecular spectroscopy, bandwidth narrowing is only of interest, if the absolute wavelength can be specified within this same accuracy. This is almost never the case. In atomic spectroscopy recourse is usually made to well-known lines in the spectrum; in molecular spectroscopy it is the position of these very lines which must be determined.

The second problem hinges on an inherent difficulty in gas phase spectroscopy, the difficulty of low partial pressures. This is doubly a problem in two-photon spectroscopy. This accounts for the fact that hitherto all two-photon spectroscopy has been limited to the condensed phase. In the last few years some two-photon absorption has been reported in the gas phase (2,3,4), but no new assignments could be reported in these cases. Hence the proof of a molecular spectrum assigned with new transitions was still outstanding. We here wish to present the first results, leading to new, hitherto unknown transitions, for the prototype molecular example benzene (5,6).

These are measurements on benzene and benzene-d_6 in the gas phase which demonstrated immediately the power of this new technique by assigning several new frequencies in the first excited electronic state $^1B_{2u}$ which where hitherto unknown. The method behind this new spectral assignment will then be illustrated.

II. THEORY

The fundamental theory for two-photon absorption was developed in 1931 by M. Göppert-Mayer (7) but it wasn't until the advent of lasers that one could observe the phenomenon experimentally. Furthermore, it took the development of tunable high intensity lasers before a spectrum could be reported in the gas phase.

According to Göppert-Mayer the molecular two-photon absorption probability is essentially given by the absolute square of the following transition matrix element

$$M_{fi} \sim \sum_j \frac{\langle f|e_\lambda P|j\rangle\langle j|e_\mu P|i\rangle}{E_f - E_j - \omega_\lambda} + \frac{\langle f|e_\mu P|j\rangle\langle j|e_\lambda P|i\rangle}{E_f - E_j - \omega_\mu} \quad . \qquad (1)$$

Here $|i\rangle$, $|f\rangle$ represent the initial and final molecular states, respectively; e_λ, e_μ the polarization vectors of the two photons absorbed with $\omega_\lambda + \omega_\mu = E_f - E_i$. P is the dipole operator and the sum is taken over all possible intermediate molecular states $|j\rangle$.

The above expression (1) is easily obtained by second order perturbation theory with respect to the interaction of the photons with the molecular system.

On the basis of the Born-Oppenheimer approximation the initial and final states may be described as a product of electronic and vibronic wave functions. For example, $|i\rangle = |S_o, \chi^m_{So}\rangle$, $|f\rangle = |S_1, \chi^n_{S1}\rangle$. Thus the transition probability for a two-photon transition from the m-th vibronic level of the electronic ground state to the n-th level of the first excited singlet state is determined by the product of the probability of finding the molecular system in the initial state $|S_o, \chi^m_{So}\rangle$ (Boltzmann Factor) and the square of the vibronic transition matrix element,

$$M_{S1^n, S_o^m} = \langle \chi^n_{So} | M_{S1\,So} | \chi^m_{So} \rangle \ . \qquad (2)$$

The electronic transition matrix element $M_{S1\,So}$ depends on the nuclear coordinates, $M_{S1\,So} = M_{S1\,So}(Q)$, and may be approximated by

$$M_{S1\ So} = \sum_j \left[\frac{<S_1|e_\lambda P|S_j><S_j|e_\mu P|S_o>}{E_{S1} - E_{Sj} - \omega_\lambda} + \right.$$

$$\left. + \frac{<S_1|e_\mu P|S_j><S_j|e_\lambda P|S_1>}{E_{S1} - E_{Sj} - \omega_\mu} \right]. \tag{2a}$$

The electronic energies E_{S1}, E_{Sj} should be taken at the equilibrium position of the electronic ground state S_o.

The problem in benzene is that two-photon absorption as well as the one-photon absorption is symmetry-forbidden ($M_{S1\ So}$ $(Q_o) = o$). This can be easily seen if one remembers that in two-photon absorption only g-g or u-u transitions are parity allowed and that $|S_1>$ and $|S_o>$ have B_{2u} and A_{1g} symmetry, respectively.

As in the case of the one-photon absorption one has to look for those vibrations which, by disturbing slightly the molecular symmetry, make the $S_o \rightarrow S_1$, two-photon transition allowed. These modes, the inducing modes, have to give a nonzero contribution in the linear approximation of the electronic transition matrix element $M_{S1\ So}$ (Q):

$$M_{S1\ So}(Q) \simeq \sum_i \frac{\partial}{\partial Q_i} M_{S_1 S_o}(Q)|_{Q=Q_o} \times (Q_i - Q_o). \tag{3}$$

Even for such a simple molecule as benzene, a sufficiently reasonable numerical computation of the derivatives $\partial/\partial Q_i$ $M_{S1\ So}(Q)|_{Q=Q_o}$ is still not possible, so that one has to look for some qualitative arguments in evaluating eq. (3). Contrary to the one-photon absorption, where one finds that the inducing mode must be of e_{2g} -symmetry, the formal selection rules are not of very much value in the two-photon case, since a total of nine out of the twenty vibrational modes of benzene can contribute, namely the b_{1u} -modes, v_{12}, v_{13}; b_{2u} -modes, v_{14}, v_{15}; e_{2u} -modes, v_{16}, v_{17}; and e_{1u} -modes, v_{18}, v_{19}, v_{20}. As can be seen from Fig. 1 all of these modes act to couple to the nearest two possible states of E_{1u} and E_{2g} symmetry, either in the first or second step. Thus all these might be possible inducing modes.

However, not every one of these nine modes is assumed to be of equal importance in the two-photon absorption. For example v_{13} and v_{20} will not be very effective inducing modes since they hardly effect the electronic π -system. Also the two out-of-plane

$<S_0/P_1/\Gamma_e>$

S_0	Q_K	Pol.	Γ_e
A_{1g}	-	X,Y	E_{1u}
A_{1g}	-	Z	A_{2u}

$\frac{\partial}{\partial Q_K}<\Gamma_e/P_2/S_1>$

Γ_e	Q_K	Pol.	S_1
E_{1u}	b_{1u}, b_{2u}, e_{1u}	X,Y	B_{2u}
E_{1u}	e_{2u}	Z	B_{2u}
A_{2u}	e_{2u}	X,Y	B_{2u}
A_{2u}	b_{2u}	Z	B_{2u}

$\frac{\partial}{\partial Q_K}<S_0/P_1/\Gamma_e>$

S_0	Q_K	Pol.	Γ_e
A_{1g}	b_{1u}, b_{2u}, e_{1u}	X,Y	E_{2g}
A_{1g}	e_{2u}	Z	E_{2g}
A_{1g}	e_{2u}	X,Y	B_{1g}
A_{1g}	b_{2u}	Z	B_{1g}

$<\Gamma_e/P_2/S_1>$

Γ_e	Q_K	Pol.	S_1
E_{2g}	-	X,Y	B_{2u}
B_{1g}	-	Z	B_{2u}

Fig. 1 Diagram for symmetry selection for Herzberg-Teller coupling (according to eq. 3): two-photon absorption ${}^1B_{2u} \leftarrow {}^1A_{1g}$ inducing vibrations Q_k of benzene. Γe is the symmetry of the intermediate states $|S_j>$.

modes of e_{2u} symmetry are not expected to be the dominant inducing modes since for these modes the derivatives $\partial/\partial Q_i \; M_{S1\;So}(Q)|Q=Q_o$ vanish if the intermediate states $|Sj>$ are of $(\pi\;\pi^*)$ -type. Matrix elements of the dipole operator with $(\sigma\pi)$ -states, however, are usually negligibly small (e.g. A_{2u} and B_{1g} in Fig. 1). Finally, it has been shown by McClain (8) that b_{1u} modes are absent in the two-photon absorption spectrum if the two photons absorbed have identical energies; $\omega_\lambda = \omega_\mu$ in eq. (2a).

Thus, only the two b_{2u} -modes v_{14} and v_{15} and the two e_{1u} -modes v_{18} and v_{19} remain as possible effective inducing modes. It was first thought that the contributions of the b_{2u} -modes should be of minor importance in the two-photon absorption spectrum since, due to an additional selection rule given by Albrecht (9), a cancellation of the most important terms in the derivative of the electronic transition matrix element was expected. This was confirmed by calculations of Honig et al. (10). However, this agreement has to be revised (11) in view of our experimental results (5), where it is demonstrated unequivocally that the v_{14} mode (b_{2u}) is the strongest mode in the spectrum, whereas the v_{19} mode (e_{1u}) is vanishingly small. Albrecht's selection rule is based on the assumption that the electrons do not change their position during the vibrational motion of the nuclei. This assumption, however, is at variance with the idea of the adiabatically coupled motion of electrons and nuclei.

Recently it was shown by Roché et al. (12) for the case of one-photon absorption in benzene that the electron-electron interaction resulting from the joint motion of nuclei and electrons is responsible for the intensity of the $S_o \rightarrow S_1$ transition and that the contributions of the electron-nuclei interaction which has been considered by Albrecht and Honig et al. are almost negligible. On the basis of this nuclear coordinate-dependent electron-electron interaction, Roché et al. calculated that nearly all the vibrational intensity is due to the e_{2g} mode v_6 which is in complete agreement with the experiment. Recent calculations of the two-photon absorption intensities in our laboratory based on the electron-electron interaction used by Roché et al. showed better agreement with the experimental results (5). In detail the calculations demonstrate:

a) that Albrecht's selection rule does not hold and

b) that more than 2/3 of the intensity is induced by the b_{2u} -mode v_{14} if one uses parallel-polarized light.

III. METHODS OF ASSIGNMENT IN TWO-PHOTON SPECTROSCOPY

Vibrations of the symmetry in S_1 proper for this process are known in benzene and hence assignments must be made. Several methods can be considered:

(a) The simplest method of assignment is to consider the effect of change in intensity when one changes the laser light from linear to circular polarization. This will have a profound effect on the intensities which will be expected to be due to the averaging of all space-fixed coordinates with respect to the matrix element discussed above (8,11). This method is however, not exhaustive, but allows some preassignments.

(b) A second method of assignment is in principle simple in its interpretability, but requires a substantial reserve in the available laser power. If the laser intensity is high enough, one would expect two-photon absorption to the red of the electronic origin. This would then be due to absorption originating from thermally populated levels of the ground state making a transition to the origin. This hot band spectrum would require considerable surplus power due to the small number of molecules in a vibrational excited ground state. The Boltzmann factor would be particularly disadvantageous for molecules like benzene, whose lowest vibration of proper symmetry is 398 cm^{-1}. Nevertheless, should it prove possible to see these absorptions it could constitute an unequivocal method of assignment. All these absorptions are from known frequencies in S_o, hence these transitions can be unequivocally assigned. Furthermore the strength and bandshape of these transitions will allow the corresponding transition to be observed in the absorption above the origin. It is generally true for molecules which are not too strongly distorted in S_1, that the ordering of the intensities in the hot band spectrum will directly be reflected in the ordering of the intensities in the normal absorption spectrum, starting from the vibrationless level of S_o. This rule has been substantiated for many systems and is essentially the basis of the mirror image rule in fluorescence spectroscopy. A further assistance can be obtained by observing if the corresponding peaks have the expected similar envelopes.

(c) In some cases the rotational envelope can be employed directly for spectral assignments based on the effect of Coriolis forces on the band (13). This method, though less unequivocal must be employed when the laser intensity is not sufficient for a hot band spectrum in addition to the regular spectrum.

So far these are the only aids to assignment for the gas phase spectrum.

Theoretical considerations (14) once led to the conclusion that v_{19} of e_{1u} symmetry should be the predominant line in the spectrum (see above). Assignment methods (a) and (b) above have now clearly established that instead v_{14} (b_{2u}) is the most powerful inducing mode. Recent theoretical work in our institute (11) has come to the same conclusion, hence confirming this assignment.

The extremely powerful laser system, which was used for the first two-photon hot band spectra of benzene, is shown schematically in Fig. 2. A dye-laser system according to the Hänsch' design (15) is pumped transversely by a 1 MW nitrogen laser. The wavelength is tuned by rotating a diffraction grating (1800 grooves/mm, blazed at 5000 Å) with a stepping motor. The telescope (15 x) avoids damage of the grating and increases the spectral narrowing of the laser light by the expanded illumination of the grating. The bandwidth of our system is 4 cm^{-4}; inserting an in-cavity Fabry-Perot etalon one can achieve a bandwidth of less than 1 cm^{-1}.

For the scan of the complete two-photon excitation spectrum of benzene we have to employ three different dyes:

coumarine	102 ($c = 8 \times 10^{-3}$ mole/ltr)	4700 - 5100 Å,
coumarine	30 ($c = 8 \times 10^{-3}$ mole/ltr)	4950 - 5200 Å,
coumarine	30 + Na-flourescein	5150 - 5500 Å.

A Glan-Thomson prism outside the cavity is used for linear and for circular polarization with a $\lambda/4$ delay plate added. The laser beam is focused by a corrected lens system into the fluorescence cell. The fluorescence coming from the small focus volume is observed with a 56-DUVP photomultiplier. Corning filters in front of the photomultiplier block the exciting blue and green light against the UV. This greatly improves the signal-to-noise ratio relative to one-photon excitation and represents a side benefit of the technique. The PM signal is then observed by a box car and read out on a strip chart recorder.

Fig. 3 shows the two-photon excitation spectrum of benzene (C_6H_6) between 5400 and 4700 Å excitation wavelength measured with linear polarized light.

Figs. 4 and 5 show the result of our measurements in the hot band region of benzene above 5400 Å excitation wavelength. The energy scale is drawn as twice the excitation laser light energy in wavenumbers. It can be seen that there are certain discrete bands well resolved, with intensity behavior differing for linearly and circularly polarized light. However, for comparison, one should recall that the strongest line in this hot spectrum is about 500 times smaller than the strongest line in the excitation spectrum (Figs. 4,6,7).

For the assignment, the excess energies of the bands are calculated by subtracting the 0-0 energy of 38086 cm^{-1} (16) from the transition energy. The resulting value is the ground state frequency of the vibration or of one of its sequence members. A comparison with the known data from IR and Raman results (16) allows us to assign unambiguously the inducing bands. On this basis the hot bands below the $^1B_{2u}$ origin can be assigned unequivocally as seen in Figs. 4 and 5.

The lines in Fig. 4 are listed in Table 1 with respect to their intensity. This intensity is reduced to the square of the exciting laser power, weighted with the Boltzmann factor and normalized to the strongest line v_{14}. In addition we place in Table 1 the intensity ratio of linearly ($\sigma_{\parallel}$) to circularly ($\sigma_{\circlearrowright}$) polarized light as well as the symmetry of the vibration itself. Apart from the unassigned lines X and Y we can read all important details from Table 1:

(1) The strongest inducing mode is the v_{14} mode with b_{2u} summetry.

(2) The b_{2u} vibrations are stronger than e_{1u} (v_{18}) which is stronger than e_{2u} (v_{17}).

(3) ratio $\sigma_{\parallel}/\sigma_{\circlearrowright}$ is much greater than 1 (about 9) for b_{2u} vibrations and smaller than 1 for e_u vibrations. (The lower limit for $\sigma_{\parallel}/\sigma_{\circlearrowright}$ is fixed by the ratio of signal-to-noise).

(4) The v_{1g} mode which was predicted theoretically (10,14) could not be detected with our resolution of the fluorescence signal, and hence its intensity must be at least a factor of 20 below that of the v_{14} mode.

With this knowledge and employing the methods (a) and (b) previously mentioned we can assign the excitation spectrum of benzene at energies above 38086 cm^{-1}, as in Figs. 6 and 7, which show the range above the vibrationless origin $^1B_{2u}$ before the lines reappear as progressions. If we regard the high intensity resolved Figs. 5 and 6, we see two regions which differ in behavior with polarized light. In the region near 39000 cm^{-1} the intensities increase when going from linearly to circularly polarized light. In the region near 39500 cm^{-1} the opposite occurs, i.e. going from linearly to circularly polarized light leads to a decrease in intensities. The absolute intensities in the second region are also several orders of magnitude higher than those in the first region (see Fig. 6).

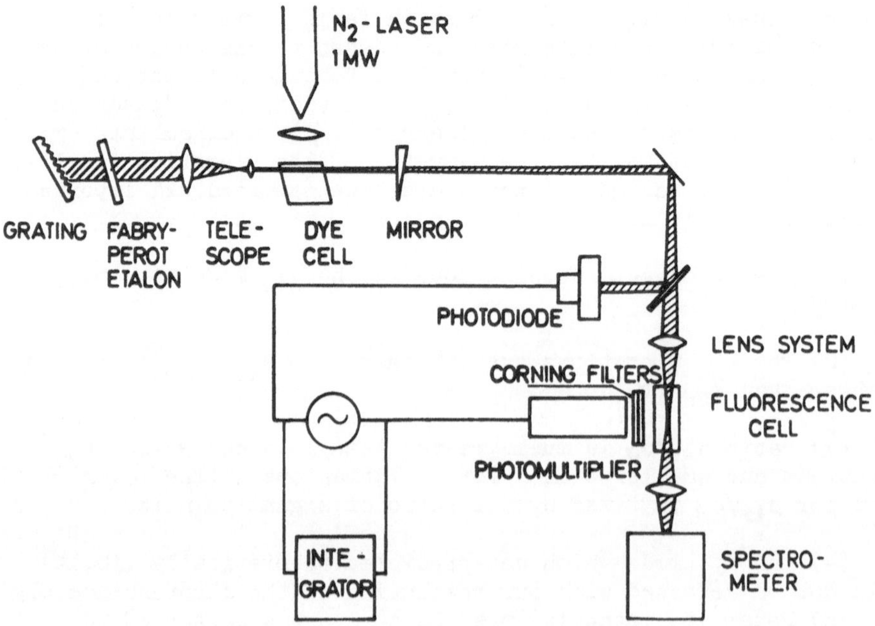

Fig 2 Experimental set-up for measuring two-photon inducing vibrations by a hot band analysis.

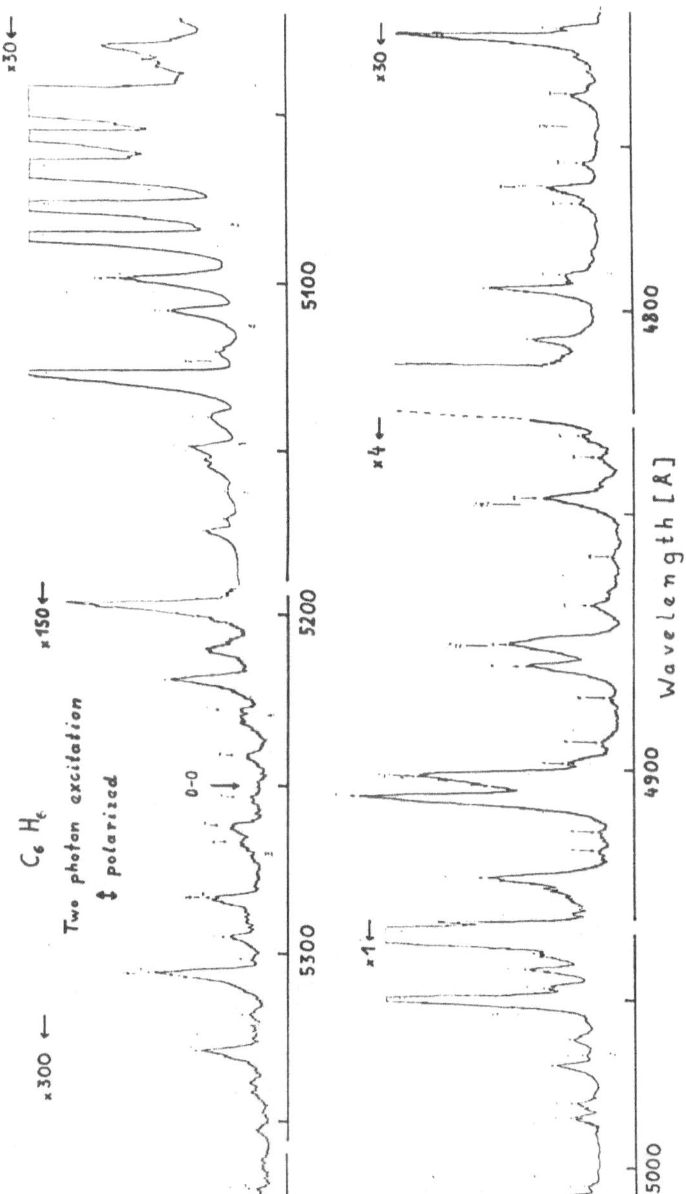

Fig. 3 Survey of the two-photon excitation spectrum of benzene measured with linearly polarized light. The scanning range starts with the hot bands at 5400 Å and ends near the cut-off at 4700 Å.

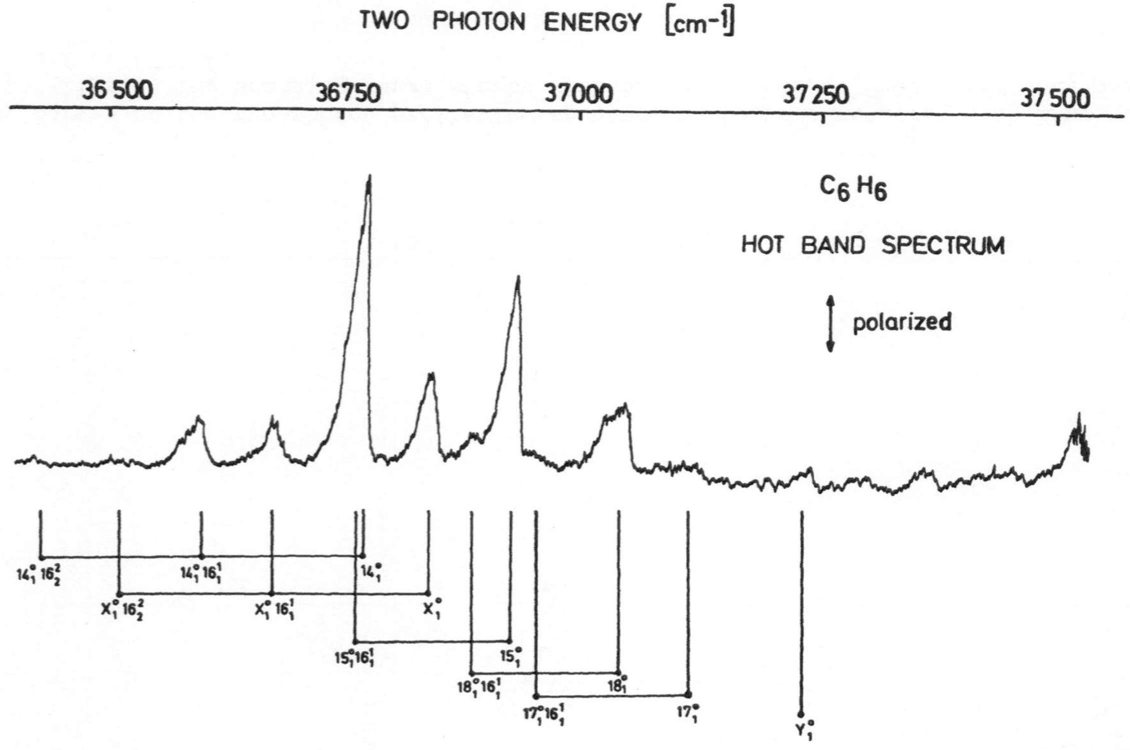

Fig. 4 The two-photon hot band spectrum of benzene for linearly polarized light below the $^1B_{2u}$ origin (38086 cm^{-1}). The vibrations are assigned according to the known groundstate frequencies (16).

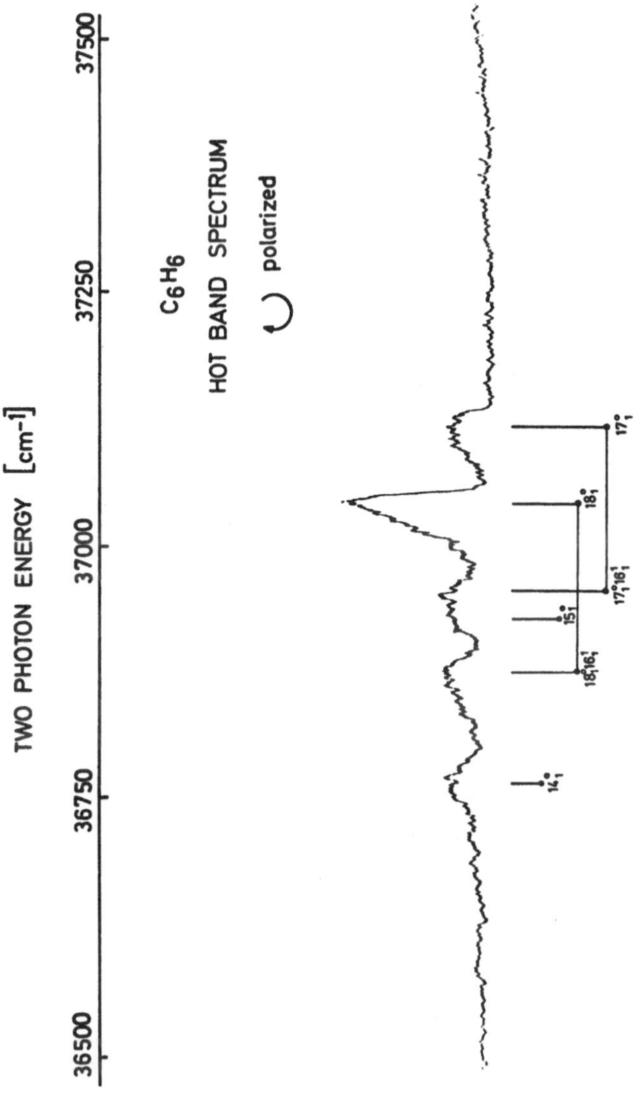

Fig. 5 The two-photon hot band spectrum of benzene for circularly polarized light below the $^1B_{2u}$ origin (38086 cm^{-1}). The vibrations are assigned according to the known groundstate frequencies (16).

Table 1. The two-photon inducing vibrations in the hot band spectra of benzene with normalized intensity and ratio of linearly to circularly polarized light ($\sigma_{\parallel}/\sigma_{\circlearrowright}$). The vibrations are assigned according to the known groundstate frequencies (16).

C_6H_6

Vibration	Vibrational Frequency (cm^{-1})	Boltzmann Factors	Normalized Intensity	$\sigma_{\parallel}/\sigma_{\circlearrowright}$	Vibrational Symmetry
14	1309	$1.75 \cdot 10^{-3}$	1000	9	b_{2u}
15	1146	3.77	224	>6	b_{2u}
X	1245	2.47	198	>3	(b_{2u})?
18	1037	6.58	45	0.5	e_{1u}
17	967	9.40	8	0.5	e_{2u}
Y	(847)	(16.7)	5		

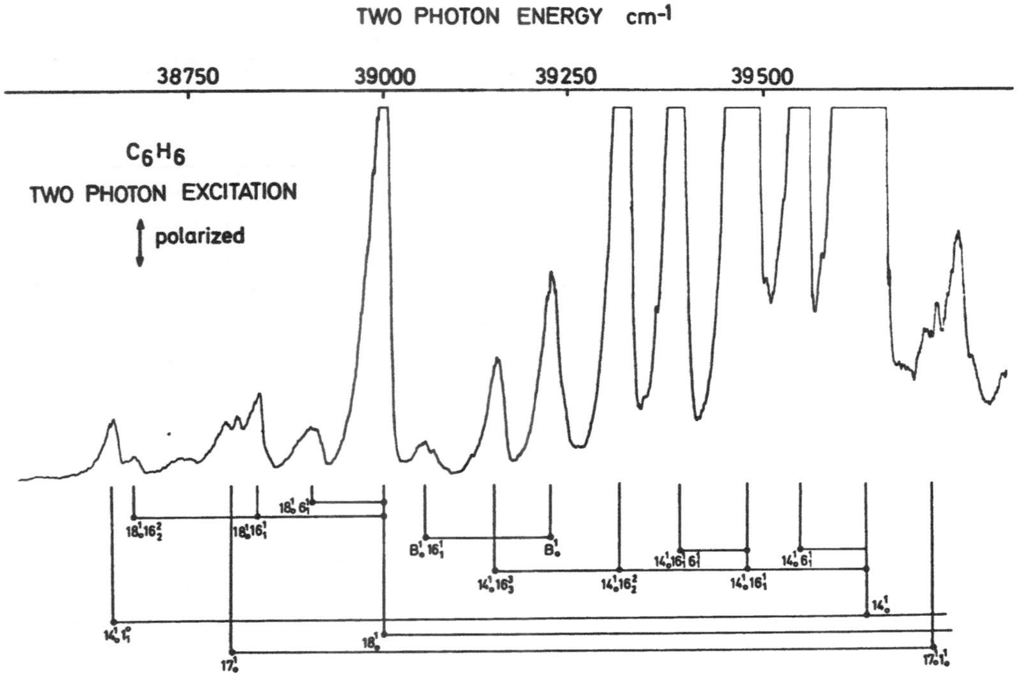

Fig. 6 The two-photon excitation spectrum of benzene for linearly polarized light above the $^1B_{2u}$ origin. The vibrations are assigned according to the results of the hot band analysis (see also Table 1).

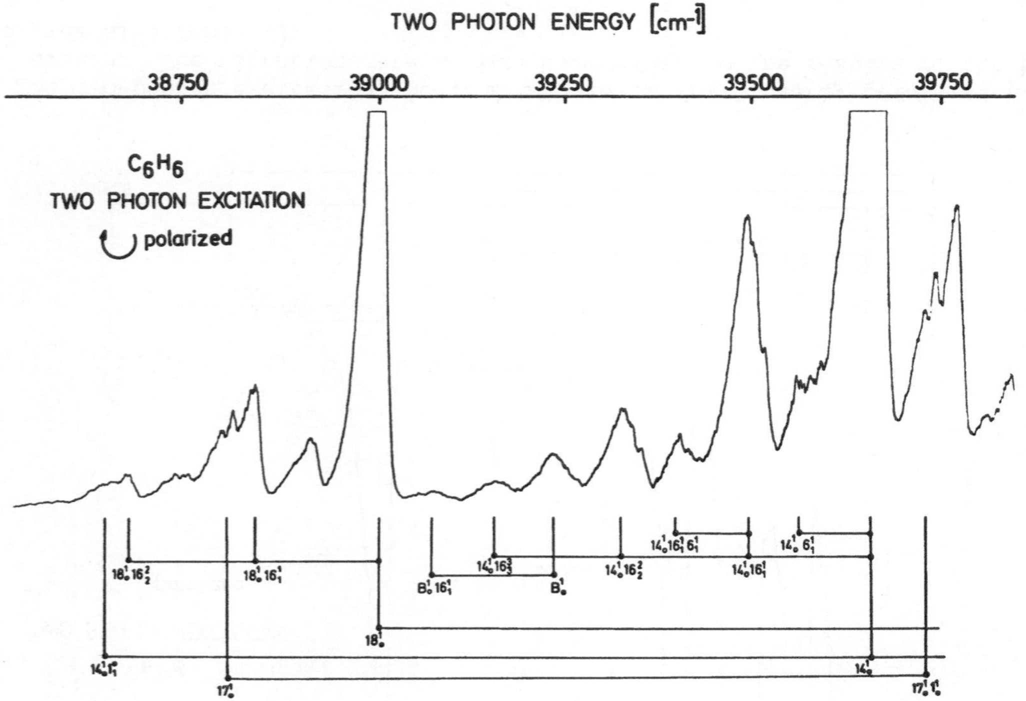

Fig. 7 The two-photon excitation spectrum of benzene for circularly polarized light above the $^1B_{2u}$ origin. The vibrations are assigned according to the results of the hot band analysis (see also Table 1).

Together with Table 1 from the hot band measurements we are now able to assign this spectrum. After subtracting the sequences with the known frequency shift of 161 cm^{-1} for the v_{16} and 86 cm^{-1} for v_6 mode, we name the heads of the sequences and progressions according to the found relations (Table 1).

In this way we obtain the hitherto unknown frequencies of the same ungerade parity vibrations in the excited $^1B_{2u}$ state: 14^1 : 1564 cm^{-1}; 18^1 : 917 cm^{-1}; 17^1 : 722 cm^{-1}. Finally it should also be remarked that v_{14} is accompanied by an unusually large inverse frequency shift from the ground state frequency of about 255 cm^{-1}. It can be seen here that this forms a reasonable basis for assignments in this new spectroscopy. In the meantime, by improving signal-averaging techniques and converting the laser to continuous scanning, it has been possible to observe hundreds of bands of this spectrum in our laboratory.

IV. LIFETIMES

A further interesting challenge might be to ask questions regarding possible differences in the photophysical and photochemical behavior of these newly discovered states. Again the difficulty is that this demands an additional surplus of laser power. In particular it is necessary to perform all these experiments at pressures below 1 Torr. At these particle densities two-photon absorption becomes quite problematical. Nevertheless such low pressures are required to ensure that the molecule will not suffer a collision prior to emission. Only in this way can the original vibronic state be isolated on the time scale of the experiment and the lifetime observed. In spite of the loss in intensities, some measurements were possible and are shown in Fig. 8. These then constitute the first lifetimes of isolated two-photon prepared states. The full triangles (▲) are the nonradiative lifetimes of two-photon excited states of the v_{14} and v_{18} vibration. For comparison the hollow circles (o) are the nonradiative lifetimes of one-photon excited states from Spears and Rice (17). Though we did not measure the quantum yield for the fluorescence, we use the fact that the two-photon excited and isolated states must be depopulated via a one-photon process. This means that the two lowest observed states 18^1 and 14^1 should approximately have a radiative lifetime similar to the origin whose radiative lifetime is $\tau_{rad} = 455$ nsec (6). The nonradiative lifetime is therefore $\tau_{nr} = k_{nr}^{-1} = (k_{obs} - k_{rad})^{-1}$. In analogy to the one-photon excitation, the behavior of the progression slope of the v_{18} mode and the steeper decrease of sequence members of the v_{14} mode agrees with the results from the one-photon case. One can here clearly observe the trend towards a cut-off of all

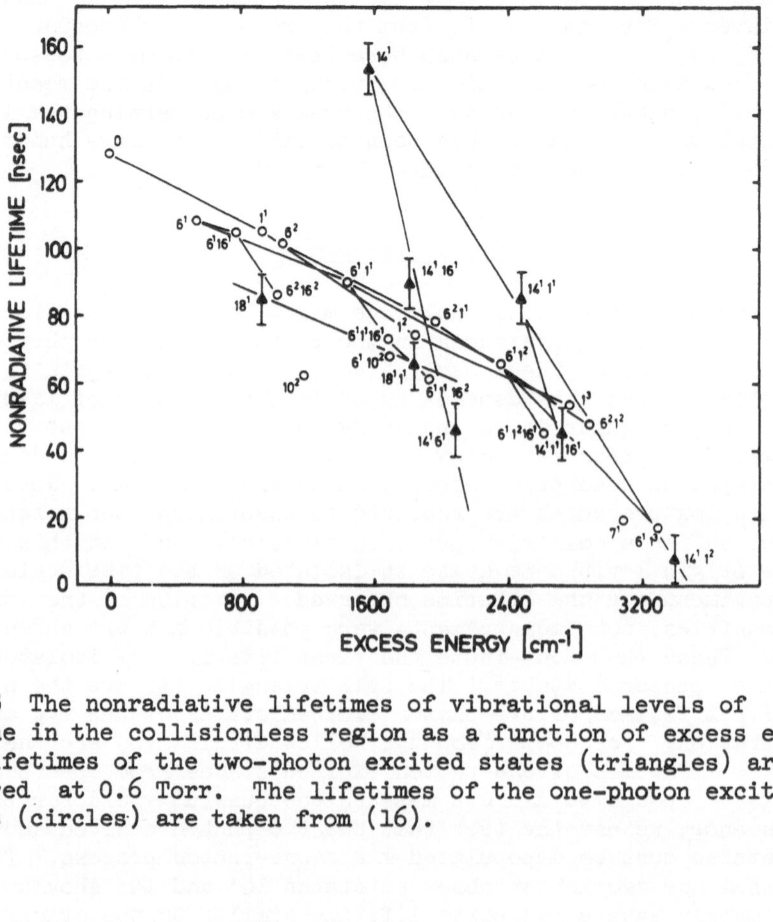

Fig. 8 The nonradiative lifetimes of vibrational levels of benzene in the collisionless region as a function of excess energy. The lifetimes of the two-photon excited states (triangles) are measured at 0.6 Torr. The lifetimes of the one-photon excited states (circles) are taken from (16).

fluorescence approximately 3200 cm^{-1} above the vibrationless S_1. Surprisingly the lifetime of the mode v_{14} is longer than that of the vibrationless origin. An explanation is given by us in terms of the large inverse frequency shift of about 255 cm^{-1} discussed above which enlarges the energy gap between S_1 and T_1 and therefore reduces the nonradiative rate to the triplet states with respect to the "normal" one-photon case (6).

Such effects clearly give us new information about hitherto unknown states. The importance of such new states in new photochemical and photosynthetic processes is apparent. The two-photon preparation of isolated quantum states should have a rich future as a novel tool in the understanding of molecular systems.

REFERENCES

1. See, however, U. Boesl, Diplomarbeit Technische, Universität München, West Germany (1973).

2. R. M. Hochstrasser, J. E. Wessel, and H. N. Sung, J. Chem. Phys. 60, 317 (1974).

3. R. G. Bray, R. M. Hochstrasser, and J. E. Wessel, Chem. Phys. Letters 27, 167 (1974).

4. L. Wunsch, H. J. Neusser, and E. W. Schlag, "Radiationless Processes." General Discussion Meeting, Schliersee, West Germany (1974).

5. L. Wunsch, H. J. Neusser, and E. W. Schlag, Chem. Phys. Letters 31, 433 (1975).

6. L. Wunsch, H. J. Neusser, and E. W. Schlag, Chem Phys. Letters 32, 210 (1975).

7. M. Göppert-Mayer, Ann. Physik 9, 273 (1931).

8. D. M. Friedrich and W. M. McClain, to be published.

9. A. C. Albrecht, J. Chem. Phys. 33, 156, 169 (1960).

10. B. Honig, J. Jortner, and A. Szöke, J. Chem. Phys. 46, 2714 (1967).

11. F. Metz, Chem. Phys. Letters, in press.

12. M. Roché and H. H. Jaffé, J. Chem. Phys. 60, 1193 (1974).

13. J. H. Callomon, T. M. Dunn, and I. M. Mills, Phil. Trans. Roy. Soc. A 259, 499 (1966).

14. R. M. Hochstrasser and J. E. Wessel, Chem. Phys. Letters 24, 1 (1974).

15. T. W. Hänsch, Appl. Opt. 11, 895 (1972).

16. C. S. Parmenter, Advan. Chem. Phys. 22, 365 (1972).

17. K. G. Spears and S. A. Rice, J. Chem. Phys. 55, 5561 (1971).

LIFETIME SPECTROSCOPY

E. W. Schlag

Institut für Physikalische Chemie

Technische Universität München

ABSTRACT

The direct study of lifetimes of various molecular vibronic states within an electronic manifold is shown to have produced much new data about the dynamics in the photoexcited states in the short decade of its existence. It is demonstrated that it is now possible, with the use of modern high resolution tunable lasers, to obtain lifetimes with a resolution comparable to high resolution absorptivity measurements. This directly yields dynamic information about the photoexcited state. In this way one has the possibility of obtaining highly precise spectroscopic information with respect to "lateral" transitions, such as Franck-Condon factors, vibrational induction effects, etc.

I. INTRODUCTION

In order to study the dynamics of photoexcited states we must consider first of all, in general, the problems involved in the study of time-dependent processes. Measurements of time-dependent processes have certain inherent problems. Most patently, the measurement of a time-dependent process demands the presence of a suitable clock for timing the process. Nevertheless many dynamic processes are studied without the obvious presence of such a clock. For example, unimolecular rates are studied, if with a clock at all, then with one having a time scale of minutes, whereas the elementary processes are now known to be 11 - 12 orders of magnitude faster. The relevance of employing such a slow clock in the study of fast processes is an important issue in the feasibility of the study of elementary processes. In order to study an elementary molecular process, which in turn has an internal time scale t(molec.), it is usually best to have a clock, external or internal, performing a measurement t(meas.) such that

$$t(\text{molec.}) \sim t(\text{meas.}), \qquad (1)$$

without here going into the detailed considerations of the interference between these terms as treated in the quantum theory of measurement.

It is by no means customary to employ an external clock for the measurement. Quite often internal clocks are employed which are considered to be more or less well known. A typical example here is to measure the time for energy randomization in a molecule via a beam experiment by considering the angular anisotropy of the scattered fragments. In this case the internal clock becomes the period of rotation of the molecule. This must be known in order to measure the time for randomization. Another example is the case of unimolecular reactions in which the half-life of the reactive state is measured by the amount of product that is formed relative to stabilized reactants. In this case the time between collisions t(coll.) is the internal clock against which the reaction half-life is measured. Hence the accuracy in the measurability of this reactive half-life is dependent on the accuracy to which t(coll.) is known, say from accurate measurements of the collision cross sections, preferably for the particular vibronic states under study. Such information is often known only approximately, if at all. It is also subject to a broad distribution of values in a thermal ensemble. Hence one can say that in many cases the use of internal clocks, though facile, usually compares one unknown quantity with another, also often not well known. Quenching studies in photochemistry here present a similar example.

Clearly, in order to obtain data about dynamic processes which are reliable and not subject to severe qualifying statements based on our assumed understanding of the clock process, we need to employ external clocks. Only such clocks can be readily calibrated and hence allow an absolute determination of the half-life of a dynamic process. We have been of the opinion that absolute measurements are worth the extra effort in order to get away from the qualifying statements necessary in all relative measurements. The circle of possibilities of such measurements, though still limited, is clearly increasing.

A typical example for an external clock would be a well defined pulse impressed on the system under study. The measured response of the system would then reflect solely the transfer function of the system, and not be in addition mitigated by a poorly determined excitation function, as is often the case with internal clocks; the latter however usually lend themselves to simpler instrumentation.

In 1965 we started a program to apply these principles to the study of the photoexcited state. This meant finding a reliable external clock on the time scale of nanoseconds, since t(molec.) for individual excited states is clearly also on this time scale. The method of choice then was to construct a modulated light source with a modulation frequency of 30 MHz. This could readily be done by constructing a time-dependent transmission grating employing a sound wave in a water bath (Debye-Sears Effect). A sample could thus be irradiated while impressing a rapid external clock. The sample emits fluorescence after a given time characteristic of the lifetimes present for the excited state. At this time, however, we decided that in order to obtain meaningful information about the photoexcited state it was not sufficient to obtain one single lifetime for the excited electronic state but rather that it might be important to perform absolute measurements on the lifetimes of as many different quantum states within the available electronic manifolds as could be observed.

For this, however, it became necessary to study the lifetime of the excited vibronic state on a time scale before the next collision, since the collisions would destroy the integrity of the state. Hence the pressure had to be lowered below 1 Torr in order to assure that the fluorescence emission under study could be observed prior to the destructive collision. The concomitant loss in signal strength due to the low sample density was severe. In order to illuminate a given vibronic state, state selection was done employing a high pressure arc lamp and a monochromator, with a band pass of 30 Å, leading to further loss in signal. The signal was electronically monitored by measurement of the phase delay of the fluorescing light relative to the incoming, modulated light (1,2). At that time such measurements were made difficult by the lack of commercial availability of broad band amplification and detection equipment. In general then three conditions had to be fulfilled:

(1) The process must be timeable with an external clock on a nanosecond time scale, t(molec.) $\sim$ t(meas.);

(2) Individual vibronic states must be studied, with a concomitant loss of signal due to (a) wavelength selection with a monochromator, and (b) pressures below 1 Torr to prevent thermalization of the sample, due to the requirement t(meas.) $<$ t(coll.).

The results on β-naphthylamine, the first molecule to be studied by this method, are given in Fig. 1.

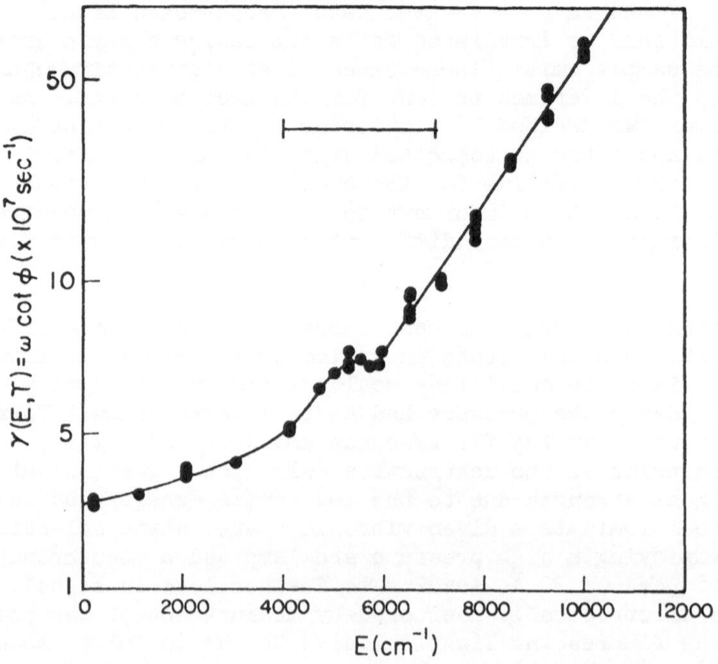

Fig. 1 Measured phase angles expressed as γ(E, T) for the measurement at 140° C for β-naphthylamine.

Here, the measurement was performed by measuring the phase delay relative to fluorescence. Although this is a radiative process, it must be emphasized that this fluorescence process nevertheless measures the total lifetime of the vibronic state due to all processes affecting this state. In general these processes can be radiative or nonradiative, i.e. this observation of fluorescence delay is ipso facto also a means of measuring radiationless processes. In fact, the measurement of lifetimes has become one of the predominant methods for the measurement of radiationless processes, particularly in the study of individual quantum states.

The radiationless processes involved can be of many types, but in general we distinguish:

(1) Internal conversion (IC), i.e. crossing to another electronic manifold under conservation of spin;

(2) Intersystems crossing (ISC), i.e. crossing, but not conserving spin, e.g. formation of triplets;

(3) Chemical reaction, i.e. forming new species.

In this discussion we exclude inelastic processes due to collisions. This separation is not quite strict, however, since the role of collisions in the induction of radiationless processes is not yet completely understood.

Even if we can take the view that in addition to fluorescence we only have one additional competing radiationless process, typically ISC, then the single measurement of the lifetime τ still cannot give information about both processes, since

$$\tau^{-1} = k_{rad} + k_{non\text{-}rad}. \qquad (2)$$

Hence another measurement, usually the quantum yield, is required:

$$\phi = \frac{k(rad.)}{k(rad.) + k(non\text{-}rad.)} \qquad (3)$$

or

$$\phi = k(rad.)\cdot\tau. \qquad (4)$$

Only the measurement of both quantities gives us the values of both elementary constants. However, in this way it is possible to obtain absolute values for these quantities on the time scale of nanoseconds by comparison with a laboratory clock. For β-naphthylamine an estimate of ϕ can be made based on quenching data by Neporent et al. (3). If it is further assumed that k(rad.) is the same for all vibronic states, the results in Fig. 2 are

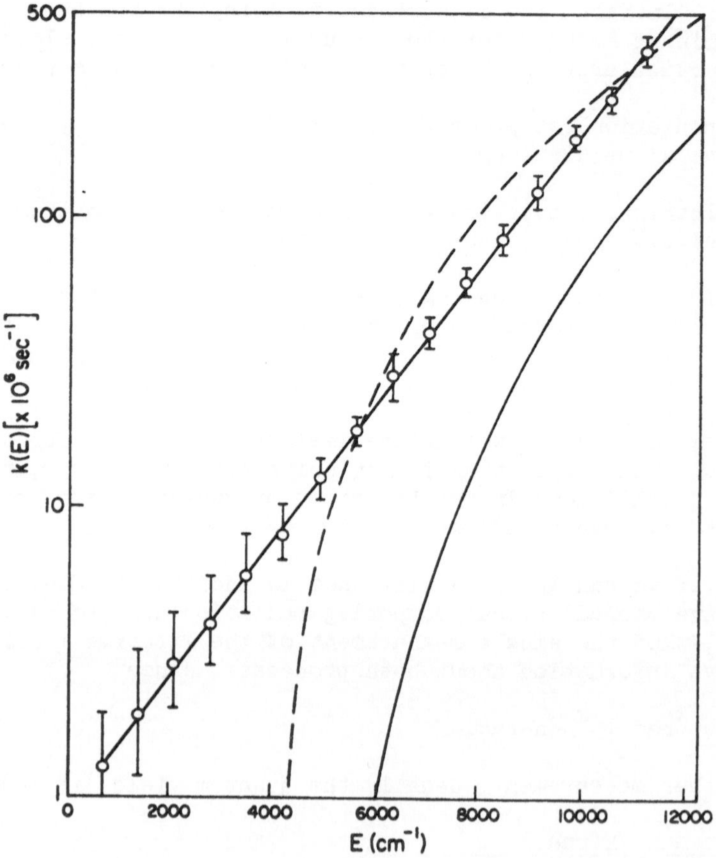

Fig. 2 Experimental elementary rate constant k(E) (inverse lifetime for radiationless transition); ---- and ——: theoretical curves.

obtained. Here the rather surprising result was obtained that the non-radiative lifetime in the excited state is a very strong function of the vibrational (and rotational) energy above the first electronic origin, and not just determined by the nature of the electronic state involved. The second electronic state is not reflected in the lifetimes observed. It must be emphasized here, however, that the variable in Fig. 2 is the excess energy beyond the 0 - 0 absorption origin and not the excitation energy of the sample. Even if one shifts the energy scale of excitation by the 0 - 0 absorption energy, this still does not correspond to an excess energy scale. A particular quantum state in excitation can be produced from a thermally excited ground state, hence the excitation energy in general is no measure of the excess energy. This can be seen clearly for the case of naphthalene in Fig. 3 in which Fig. 3a shows strong variations (4). These variations disappear upon re-analysis of the individual bands and after plotting on an excess energy scale. It must be emphasized that this correction is also necessary for large molecules like β-naphthylamine in which we deconvoluted the hot-bands out of the observed spectral lifetime response in order to obtain Fig. 2.

Interestingly enough, although it follows from the above considerations that lifetimes and quantum yields must be known pairwise for all vibronic states of interest, such information is rarely available. Most measurements at good or moderate resolution are lifetime measurements. It is usually assumed that the radiative component is nearly constant so that all strong effects can be attributed to the effect of radiationless processes. It is generally believed that quantum yields, perhaps because of their strong and honored tradition, are well understood and easy to measure. Unfortunately, neither is true. It is extremely difficult to measure accurate quantum yields of individual vibronic states. It can probably be maintained that to date no accurate values have been published with reasonably good resolution. Some of the major discrepancies between laboratories result not from differing lifetimes, but from differing quantum yields. This may also be the case for recent results on aniline. The fact that quantum yields demand an accurate and hence wavelength-independent measurement of flux makes these measurements problematical. Most measurements are based on results relative to a "known" standard. This standard is usually only known for the thermalized state of the system, and then not too accurately. The accumulation of errors, including spectral response shifts in the photocathodes of the detectors produce major effects. The ideal situation would be to measure quantum yields and lifetimes in the same experiment simultaneously. This is possible with present techniques. Presently data are often obtained from different set-ups, under different resolutions even in different laboratories.

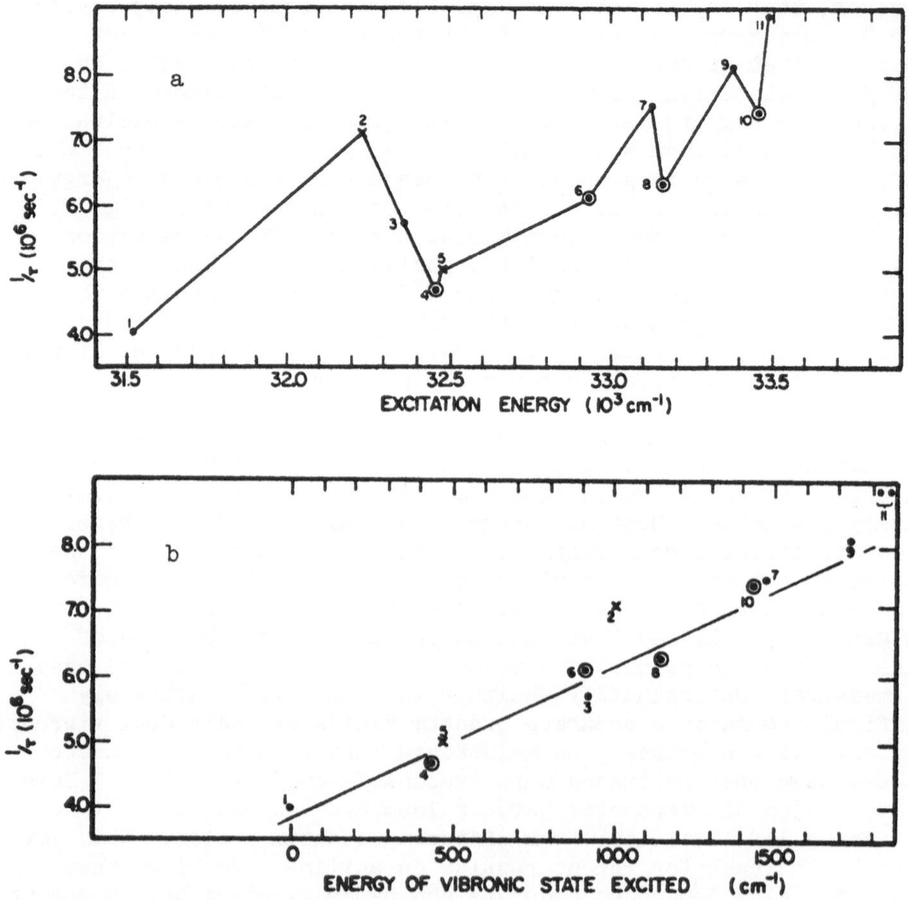

Fig. 3 Fluorescence decay rates of naphthalene vapor of 0.07 Torr, 25°C and 3 Å excitation bandwidth according to (4): as function of excitation energy (a), as function of excess energy (b). Encircled points refer to origin transitions, crossed points denote transitions with uncertain assignments.

In the short tradition of radiationless processes it has become standard to refer to individual quantum states and single vibronic levels (SVL). It should be emphasized that these are unfortunate misnomers. As in all spectroscopy one can perform measurements at good or poor resolution. The better the resolution, the more states that can be studied; but these are hardly ever single states, except for the here-excluded case of diatomic molecules. Most measurements to date have been in the 1-10 Å resolution range. For the case of naphthalene about 20,000 rovibronic states are contained within this bandwidth. Nevertheless, in absorption spectroscopy it has proven very useful, as demonstrated by many workers, Herzberg and Ramsay among others (see discussion elsewhere in this book), to perform high resolution measurements even though one doesn't study single level absorptivities. Hence similarly it might be better to refer to lifetime measurements of vibronic states as a type of lifetime spectroscopy. For this purpose high resolution is usually defined as a resolution of 250,000. The upper limit is usually considered to be around 600,000 at which point the Doppler width of the molecules has been reached. At this point the relative random motion of the molecules produces a Doppler shift negating all further efforts to narrow the spectrometer slits. Experiments below the Doppler limit are possible and have been developed, but so far this region has been considered as a subject apart from "high resolution" spectroscopy. For a molecule like benzene even within the Doppler width a number of rotational states are still imbedded. Hence, for ultimate rotational resolution experiments within the Doppler width are necessary. An obvious way to eliminate the relative motion of molecules is to employ a molecular beam. Such experiments are presently being done. A more typical method for working within the Doppler profile is to cause the sample to abosrb light from opposite directions (5). Now at irradiation levels which lead to a substantial population of the excited level, non-linear effects are produced. These non-linear effects can be used to produce a dip in the absorption profile, hence marking molecules with no relative kinetic energy. A more sophisticated technique is to mix two modulated signals via this non-linear effect and observe the sum frequency in the fluorescing signal (6). This leads to a considerable enhancement in sensitivity above the dip method. Two-photon experiments provide one of the most elegant methods for studies within the Doppler profile. In this case the two-photon absorption is produced again by opposing beams focused not only on the same sample, but absorbed by the same molecule (7). This leads to a cancellation of the Doppler shift in one direction by subtracting the same shift from the absorption in the other direction. The import of these experiments for radiationless transitions is apparent.

It must be emphasized that not only does the Doppler width represent a barrier to experimentation within the customary spectroscopic framework, but so does the time scale of the clock employed, via the Heisenberg relation. It might also be mentioned that the limits have also only been approached in most recent laser experiments. Conventional light sources limit measurements essentially. It can again be seen that this is a continuous range of spectroscopic techniques, and that the concept of single levels is probably never fulfilled for molecules with more than two atoms, at least until the Doppler limit. Even NO_2 is already extremely difficult here.

II. NEW TECHNIQUES

The original technique of measuring the phase lag of a fluorescing signal relative to the modulation of the exciting light has been discussed above. The three conditions cited above, however, make the signal strength so weak that it becomes impossible to improve the resolution.

A second technique which was developed is that of single-photon timing. Here the sample is modulated or pulsed, and the low level of emitted photons resolved into individual pulses. In the limit in which less than one photon per pulse or period arrives one can time-correlate these photons to the incident signal which produced them. The histogram of many such photons produces again a decay curve or a phase lag. Hence one can have either single-photon counting or single-photon phase fluoremetry. The limit of this techique requires counting up to a day or two for each wavelength setting, to achieve the best resolution of rarely better than 1 Å (8). This technique has gained much popularity in recent years and commercial instruments have become available.

We wish here to introduce a third technique, that of laser-flash induced fluorescence (LIF). Technically dye-lasers are capable of spectral outputs down to about 4000 Å, and can be doubled with low efficiency down to 2300 Å. In order to justify the extra effort of laser techniques it was decided that this was only meaningful if data could be obtained substantially better than the 1 Å resolution available up to now.

The design conditions then are: (1) resolution in the picometer range; (2) arbitrary selectability of any wavelength to an absolute accuracy equivalent to the resolution, and (3) sufficient power after frequency doubling to assure sufficient signal strength.

The latter condition is particularly important since the flux will be too high for single-photon techniques, though perhaps too noisy to measure a lifetime with each shot. This could be problematical since conventional signal averaging techniques do not function on so short a time scale. This, though, may soon change. Condition (2) is important since a unique program must be developed to pick an absolute wavelength out of the range of several hundred Å available to the dye-laser with picometer resolution. For this purpose a calibration procedure had to be developed. Condition (1) is easily achieved with an intracavity etalon. Condition (2) was solved in a satisfactory manner (9), and condition (3) is demonstrated by the output of a single shot, shown in Fig. 4. It can be seen that a single shot, under high resolution, is adequate to measure a lifetime at the natural vapor pressure of 60 mTorr of naphthalene. Hence it can be considered demonstrated that laser-flash induced fluorescence allows a further extension of the measurement capabilities in excited states. With respect to measurements made by single-photon techniques, the new LIF technique has the advantage of:

(1) two orders of magnitude increase in resolution (close to the Doppler limit), and

(2) one shot - one lifetime, hence a simplicity in evaluation.

This technique then moves dynamic measurements into the range of high resolution spectroscopy.

A further refinement, easy to implement with LIF experiments, is to incorporate a quantum yield measurement with the lifetime measurement. In principle, by measuring all the parameters of a single laser shot, both lifetime and quantum yield are determined in a single shot, on the same apparatus. Such a setup is shown in Fig. 5. The light source is a flashlamp-pumped dye-laser, which works with rhodamine 6G and emits in the spectral region between 5700 and 6250 Å. It produces laser light pulses, which have a length of about 250 nsec and a peak power of about 15 kW. What is remarkable is the small spectral bandwidth of 0.1 Å. For excitation of organic molecules into S_1, especially of naphthalene, which has been investigated in some of our experiments, UV between 2800 Å and 3150 Å is required. Therefore a setup of two lenses and a KDP-crystal is used to double the frequency of the laser light as to be seen in Fig. 5. The bandwidth of the thus generated UV light pulse is equal to or less than half the bandwidth of the original light. Thus we have an exciting UV light source at our disposal with a power of about 500 W and a bandwidth of 0.05 Å. As a frequency control of the laser light we use a high resolution 3 m spectrometer (Jobin Yvon THRP). Limited by the photographic resolution we are able to determine the absolute wavelength to an accuracy of 0.04 Å. This corresponds to an accuracy of 0.02 Å for the wavelength of the exciting UV light.

Fig. 4 Fluorescence decay curve of an $S_o - S_1$ transition in naphthalene obtained with one laser shot on the screen of an oscillograph.

To measure lifetimes the experimental setup in Fig. 5 has been complemented with a Pockels cell system, in order to obtain rectangular pulses. As the fluorescence decay time of naphthalene has a time scale of some 200 nsec, the exciting light pulse must have a short enough fall time, so as to not falsify the original fluorescence decay curve. The Pockels cell delivers a square pulse with a fall time of 2 nsec. The entire Pockels cell system consists of two polarizers and a KDP crystal, which changes the polarization when a high voltage is applied. This pulse-slicing system has been built up between the dye-laser and the frequency-doubling system in the setup of Fig. 5. With this apparatus we are now in a position to measure both the lifetime τ and quantum yield ϕ of one vibronic state synchronously in one single laser shot.

In the following we will discuss the separate measurement of quantum yields and lifetimes, as already completed on this apparatus. The quantum yields were measured (10) and compared, where possible, to known values. So here, in Fig. 6 it is shown that the quantum yield of naphthalene at high pressures (thermalized system) is $\phi = 0.18$ in agreement with known values (16). These measurements were made both in the vibrationless ground state and in the neighborhood of the b_{1g} vibration. If now the pressure is lowered below 1 Torr (into the low pressure region), it is seen that the quantum yield is no longer a constant, but rather a strong function of wavelength. This is caused by the fact that various quantum states under isolated molecule conditions have grossly different quantum yields. In particular, if in conventional 1 Å experiments a $\phi = 0.18$ is observed, then within this bandwidth the value on a wavelength scale of picometers can vary from 0.12 - 0.40. This shows that medium resolution measurements constitute a gross averaging over strongly varying yields. This result is indeed alarming and leads one to reconsider much of the data taken at lower resolution. It is particularly interesting to note that this variation takes place over a very narrow range of frequencies, which tends to force one to the conclusion that this is an effect produced by the rotational quantum states of the system. This is again astonishing, as previous measurements of quantum yields in the neighborhood of a vibronic band were interpreted to demonstrate the absence of rotational effects on vibronic level lifetimes (11); however, again these are taken at lower resolution. Although we found this variation of ϕ to be strong, this by itself does not permit one to say whether the effect has a radiative or radiationless origin. In cases of diatomic molecules where such effects are seen for individually resolved levels such a strong variation has been attributed to the onset of predissociative processes, i.e. a radiationless process. This was deemed reasonable as rotational

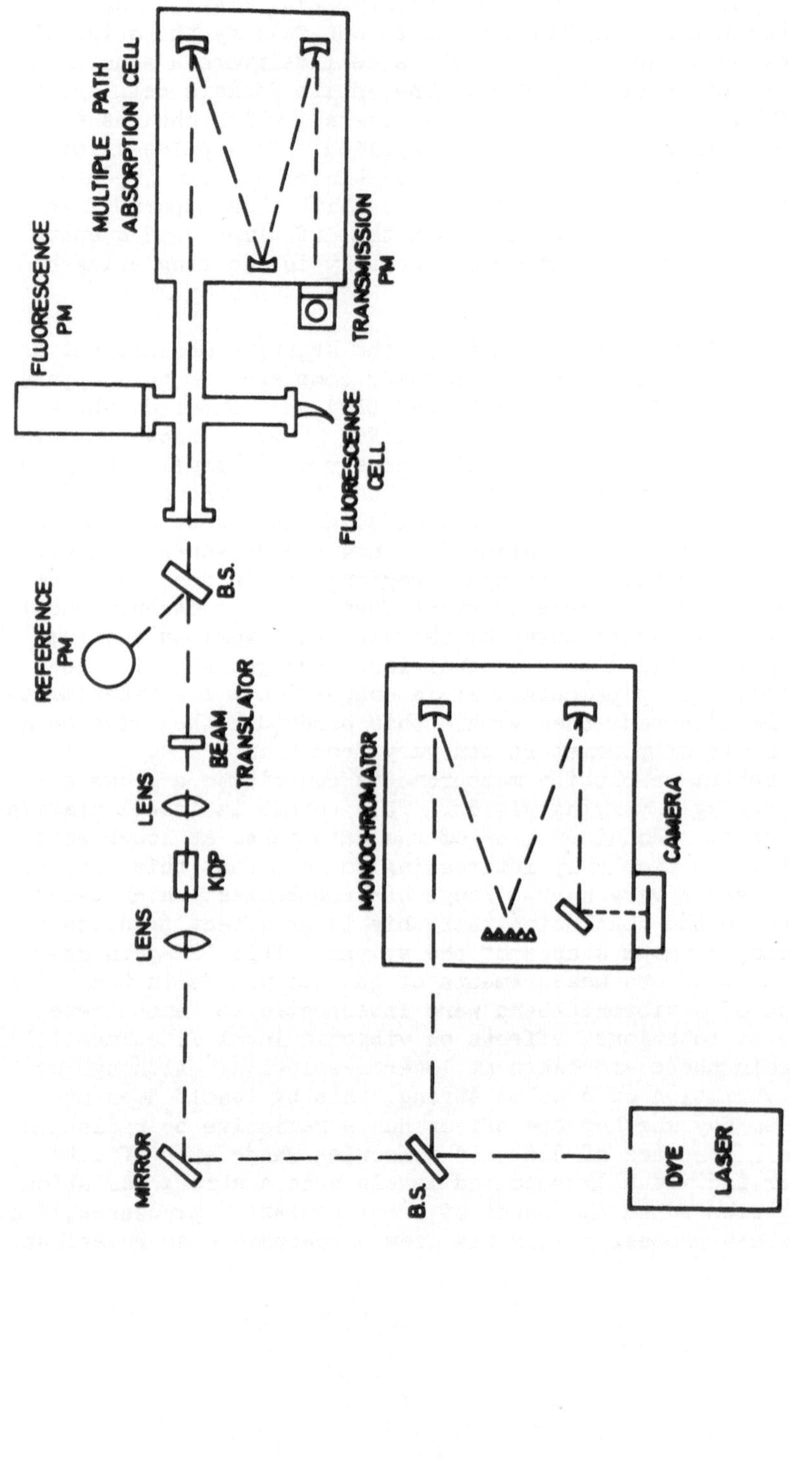

Fig. 5 Experimental setup for measurements of absolute quantum yield ϕ of molecules in the collision-free gas phase.

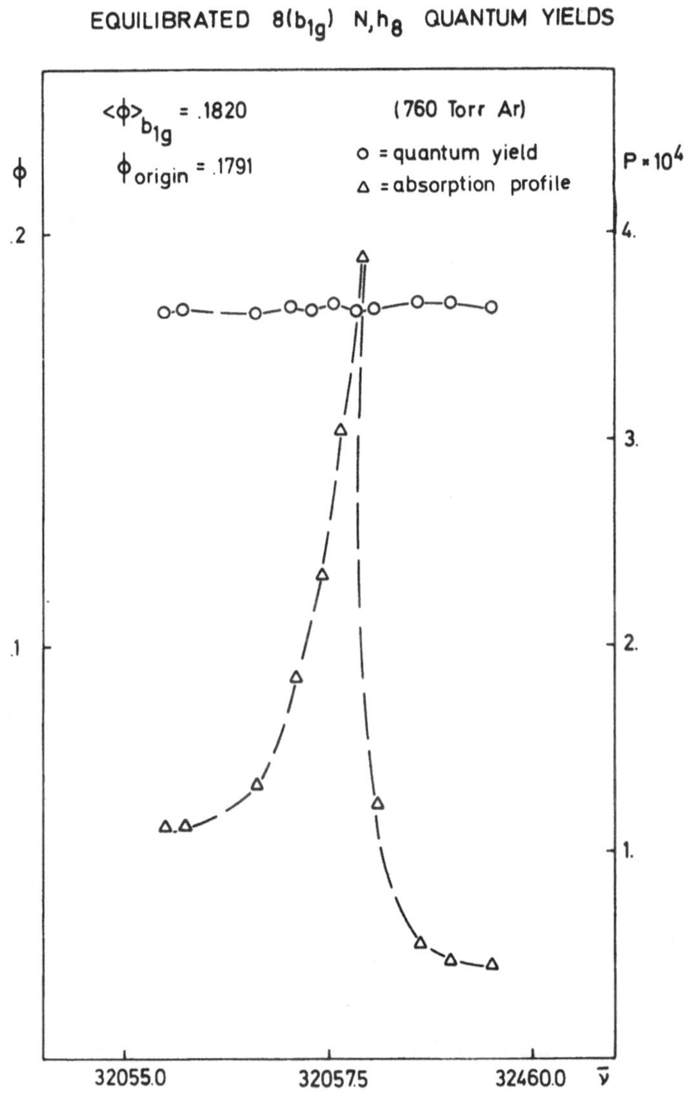

Fig. 6 Quantum yield of naphthalene, measured at high pressure around the vibrationless origin.

effects are generally neglected in any consideration of radiative processes. We are here in the unique situation of being able to measure both components, since a high resolution measurement of both lifetime and quantum yield was possible. The result is that rotational effects can also have a strong effect on the radiative rate constant. This can be seen in Fig. 7. This result was astonishing and shows that rotational effects cannot be ignored in radiative transitions. Further experiments must be carried out before definitive statements can be made about the origin of this novel effect.

High-resolution lifetimes by LIF have been performed first on naphthalene and perdeuteronaphthalene (12). The Jablonski diagram for naphthalene is given in Fig. 8. In this diagram the three possible spin levels of the triplet are shown separately. First order spin-orbit coupling would only populate the z-component, whereas second order spin-orbit coupling would also populate x and y components by vibronically mixing in higher states in a Herzberg-Teller process. In particular, vibrations of b_{2g} and b_{3g} symmetry could act as promoting modes in the process. It is known from low temperature solid measurements (13) that the z-component is not populated (less than 2%). This precludes the first-order spin-orbit effect and forces us to consider a Herzberg-Teller mechanism. The increased importance of these second-order terms is now also understood theoretically (14). Unfortunately only b_{2g} modes are known for naphthalene, the b_{3g} modes not being identified in the absorption spectrum.

This absorption spectrum of naphthalene is dominated by bands where b_{1g} vibrations and by progressions in which $b_{1g} + a_g$ vibrations are excited. Also some totally symmetric a_g vibrations can be excited by absorption. Another feature of the absorption spectrum is the very typical sequence pattern, where the same vibronic quanta are excited in the ground S_o and excited S_1 state. Mainly three vibrations are causing these sequences, an unknown vibration with a bandshift of 6 cm^{-1}, a b_{1u} vibration with a bandshift of 10 cm^{-1} and a b_{2g} vibration with a bandshift of 55 cm^{-1}. When using one-photon absorption as a preparation method, only these vibronic symmetries can be investigated in a lifetime measurement. By two-photon absorption vibrations with other symmetries can also be excited.

In order to obtain accurate lifetimes for each band and to observe the dependence of the lifetime on small changes of the excitation energy, a lifetime spectrum was plotted by smooth, accurate tuning of the laser wavelength over a range of some wave-numbers around the maximum of the absorption. In Fig. 9 the lifetime spectrum around the origin of S_1 is displayed. At the bottom the spectral position of the A^0, A^{001}, A^{01} and A^1

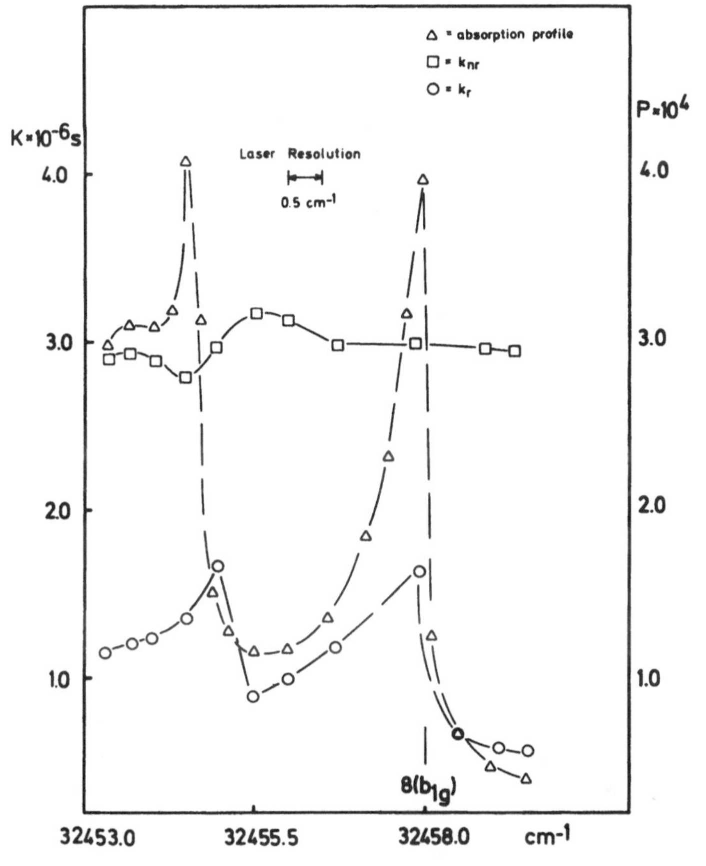

Fig. 7 Radiative and nonradiative rate constants as a function of excitation energy for the $8(b_{1g})$ band in naphthalene at 0.07 Torr.

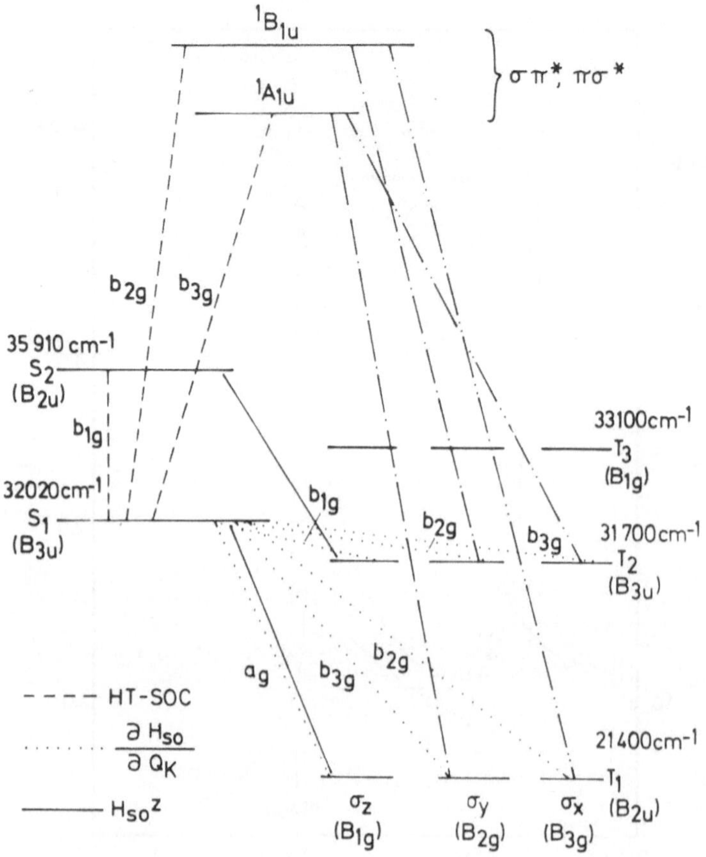

Fig. 8 Mechanisms for radiationless transitions among excited states of naphthalene shown in a Jablonski diagram.

absorption bands are marked which belong to the origin, 6 cm^{-1}, 10 cm^{-1} and 55 cm^{-1} sequences, respectively. The very expanded energy scale showing a range of only some 10 cm^{-1} should be emphasized, as well as the narrow laser bandwidth on this scale. Due to this high resolution the lifetimes of vibronic levels excited by the principal bands and their sequence bands are definitely not distorted by one another. In Fig. 9 not only can the selected vibronic bands be observed, but also some band structure in the lifetime. So a very sharp edge to the blue and a slower decay towards the red can be seen at the origin. Also the lifetime peak and the absorption peak coincide within the resolution of the experiment. This need not necessarily be so. Thus for the first time a rotational structure in the lifetime spectrum has been observed (although not fully resolved).

Under such a high resolution one observes that the lifetimes can be easily classified into those producing exponential or non-exponential decay. At all investigated levels in S_1 the lifetime spectrum was investigated in the neighborhood of each level. Thus it was possible to examine the quality of the exponential decay of these lifetimes near and at the peaks. For most absorption peaks and their red wings good exponential decays could be observed. Thus we have the assurance of observing good lifetimes with real structure and not the effect of two or more decay rates with varying weights producing such structure.

Observing the plot of lifetimes vs. excitation energy in Fig. 9, one observes that the smooth decrease of lifetime with energy observed in Fig. 1 and 2 is no longer present, also in contrast to Fig 3b. This is particularly puzzling as both spectra refer to naphthalene. It is a clear demonstration of the fact that increased resolution often reveals structure not observed under low resolution. This need not always be so since when a molecule is large enough, and hence the levels dense enough, one would expect them to act as a quasi-continuum, and hence a smooth function would be expected at any resolution. β-naphtylamine may be such an example, particularly at higher energies. The molecule is then said to act as its own heat bath, an assumption similar to that employed in the theory of unimolecular reactions. Naphthalene clearly is not such a case, the states being separate and non-communicating. Hence we observe considerable variations. It can be seen that the b_{2g} sequence does have a profound effect on the rate, as expected on symmetry grounds. This sequence occurs, however, at several places in the spectrum, always being added to a different base frequency. It can be seen that this profoundly affects the ability of the b_{2g} to act as a promoting mode. Hence it is clear that the effect as a promoting mode is a strong function of the place in the spectrum. Furthermore, it should be noted that in one case the base frequency is the $8(b_{1g})$ mode and in another case the $7(b_{1g})$ mode. Nevertheless the behavior of the b_{2g}

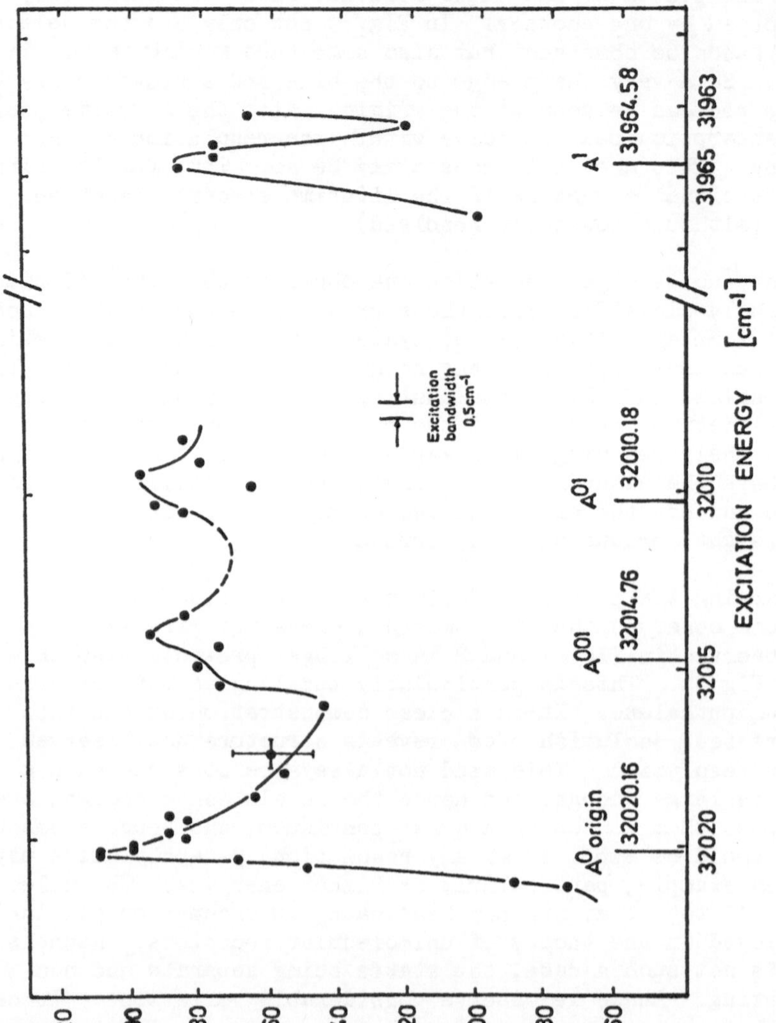

Fig. 9 Measured lifetime spectrum around the electronic origin of S_1 in naphthalene-h_8. The positions of the bands are taken from (15).

mode riding on top of these is totally different. Since the symmetries here are the same, this must be taken to be direct evidence for the fact that the symmetry alone cannot determine the intensity of the Herzberg-Teller induced spin-orbit coupled process but that in addition Franck-Condon effects must play an important role in these radiationless processes, in fact nearly as important a role as the symmetry considerations themselves.

The deuterium isotope effect has also been observed (12). Here the isotope effect is measured as a function of excess energy, but presented in reduced form of decay rates k_H/k_D (not at constant excess energy, but at constant initial quantum state). This is shown in Fig. 10. In general it is expected that the deuterated molecule has a lower rate constant since the level spacing in the deuterated molecule is smaller, and hence the quantum state discrepancy in ISC larger. Hence $k_H/k_D > 1$ will be termed a "normal" isotope effect. It is seen that the effect at the origin is normal, then crossing unity, and hence becoming inverted. At high energies the isotope effect again tends to become normal. Furthermore, it is seen that this is predominantly a function of energy content, and not a function of symmetry. The symmetry of a vibration hence cannot be deemed responsible for non-normal behavior. The smooth curve obtained seems to connect all points with equal parity.

This reduced representation tends to accentuate effects not often seen by comparing the excess energy plots themselves. Although both systems are normalized to the same initial quantum state in the singlet S_1, both isotopes must end up in different final states in the triplet, having different Franck-Condon factors and different transition symmetries. This alone should produce a strong fluctuation in Fig. 10, even on this reduced representation. The fact that this is not observed can only be taken to mean that the detailed vibrational information is not essential, only the total energy being an important variable. This would be true for a situation in which final states communicate on the time scale of the ISC process, even though initial states (Fig. 9) clearly do not communicate. Detailed quantum yields would have to be taken to check this interpretation. A similar plot of the isotopic lifetimes of benzene reveals considerable scatter, hence these considerations cannot be applied to benzene, although benzene has not been measured at high resolution to date.

III. CONCLUSION AND FUTURE

The introduction of high resolution laser techniques has made new experiments of increased precision and resolution possible. In many polyatomic molecules this increased resolution has been

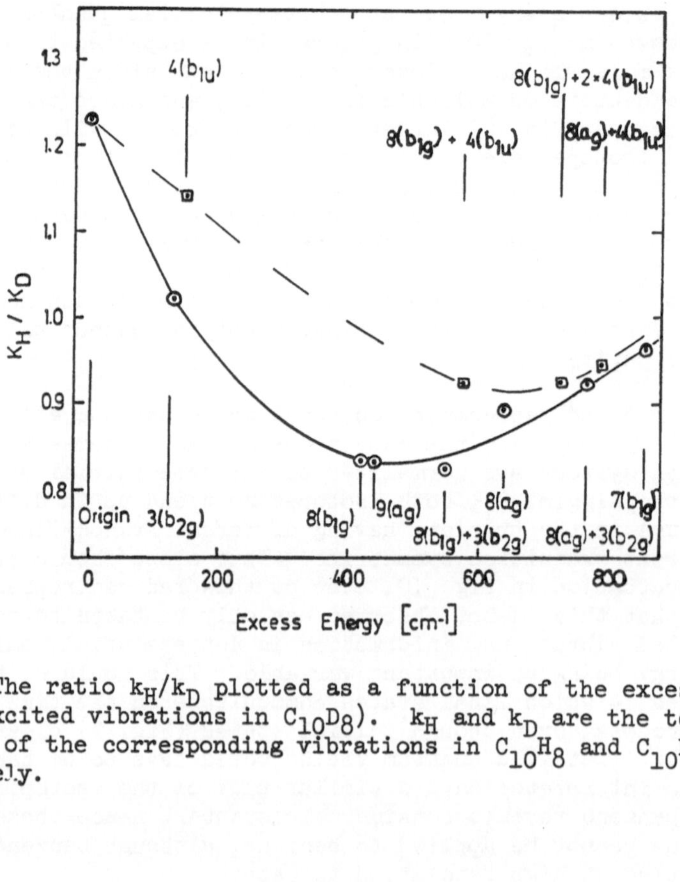

Fig. 10 The ratio k_H/k_D plotted as a function of the excess energy (of the excited vibrations in $C_{10}D_8$). k_H and k_D are the total decay rates of the corresponding vibrations in $C_{10}H_8$ and $C_{10}D_8$, respectively.

demonstrated to be essential if even reasonable rate constants are desired. Similar conclusions apply to lifetime measurements and quantum yields.

In this talk we have only discussed the method of studying the rate constants, radiative and radiationless, originating from defined quantum levels in the electronic manifold: a sort of vibronic lifetime or kinetic yield spectroscopy. Clearly much more work is required for the understanding of radiationless and radiative processes. In particular chemical reactions should become an interesting field. One of the prime difficulties with present techniques arises when several competing radiationless processes occur. In that case only their sum is determined by present techniques. It would be necessary to monitor not only the disappearance from quantum levels, but also the appearance of new species in their respective quantum levels. Again this would have to be done at low pressures on a nanosecond time scale. Such experiments are still very difficult to perform.

REFERENCES

1. E. W. Schlag and H. von Weyssenhoff, J. Chem. Phys. 51, 2508 (1969).

2. E. W. Schlag, S. Schneider, and D. W. Chandler, Chem. Phys. Letters 11, 412 (1971).

3. B. S. Neporent, Zh. Fiz. Khim. 21, 1111 (1974).

4. J. C. Hsieh, U. Laor, and P. K. Ludwig, Chem. Phys. Letters 10, 412 (1971).

5. T. W. Hänsch, I. S. Shahin, and A. L. Schawlow, Phys. Rev. Letters 27, 707 (1971).

6. M. S. Sorem and A. L. Schawlow, Opt. Commun. 5, 148 (1972).

7. F. Biraben, B. Cagnac, and G. Grynberg, Phys. Rev. Letters 32, 643 (1974); M. D. Levenson and N. Bloembergen, Phys. Rev. Letters 32, 646 (1974).

8. A. E. W. Knight, B. K. Selinger, and I. G. Ross, Australia J. Chem. 26, 1159 (1973).

9. U. Boesl, Diplomarbeit, TU München (1973).

10. W. E. Howard and E. W. Schlag, to be published.

11. C. S. Parmenter and M. D. Schuh, Chem. Phys. Letters 13, 120 (1972).

12. U. Boesl, H. J. Neusser, and E. W. Schlag, Chem. Phys. Letters 31, 1 (1975); U. Boesl, H. J. Neusser, and E. W. Schlag, Chem. Phys. Letters 31, 7 (1975).

13. H. Sixl and M. Schwoerer, Chem. Phys. Letters 6, 21 (1970).

14. F. Metz, to be published.

15. D. P. Craig, J. M. Hollas, M. F. Reddies, and S. C. Wait, Phil. Trans. Roy. Soc. London A 253, 543 (1961).

16. M. Stockburger, H. Gattermann, and W. Klusmann, Radiationless Processes, General Discussion Meeting, Schliersee, West Germany (1974).

THE STUDY OF ELECTRONIC SPECTRA IN CRYSTALLINE SOLID SOLUTIONS

D. S. McClure

Department of Chemistry, Princeton University

Princeton, New Jersey

ABSTRACT

Crystalline solid solutions are useful for obtaining electronic spectra of large molecules because the molecular environment is uniform for all solute molecules, thus forming one condition for obtaining narrow lines. At temperatures of 2°K linewidths under 1 cm^{-1} are often observed. In addition, single oriented crystals give polarization information. The spectra of pyridine, pyrazine, naphthalene, benzoic acid dimer and stilbene in suitable crystalline matrices are presented and analyzed as examples of this method.

I. INTRODUCTION

The complexity of the spectra of vapors of large molecules is a barrier to their analysis. Because of the high temperature needed in order to get enough vapor pressure, many rotational levels of the initial state are populated, and high resolving powers are needed to separate the lines. Even with a molecule of the moderate size of benzene, however, the Doppler width causes adjacent lines to merge, and no improvement of the spectrograph can help. This problem has been partially circumvented by fitting calculated contours to observed rotational contours, and thus extracting the rotational contants (1, 2).

For a molecule having low frequency internal vibrational or torsional modes, the complexity of the electronic spectrum in the vapor phase may be very great. In a molecule such as stilbene (diphenyl ethylene), the vapor spectrum is nearly useless. Even for a more rigid molecule of this size, such as naphthalene, the vapor spectrum has not been analyzed.

The development of laser methods may solve these problems by eliminating the effect of Doppler broadening and achieving much higher resolving powers. These methods will be extraordinarily complex, however.

As we increase the size of the molecule further, we reach another obstacle toward getting its vapor absorption spectrum; its vapor pressure is too low, or it is thermally unstable at the required temperatures.

Solution spectra are usually too broad even for a vibrational analysis. Shpolsky found that certain combinations of solvent and solute when frozen gave rather sharp spectra (3). An example is naphthalene in n-pentane (Fig. 1). Here the solution was cooled to $4^{o}K$ in a 0.5 mm cell. The line widths here are about 10 cm^{-1}.

A much better scheme is to use solid crystalline solutions. These provide a means of achieving sharp and polarized spectra so that vibrational analyses may be performed on them. They do not, of course, give any information on the moments of inertia. On the other hand, the larger the molecule, the less significant is the overall moment of inertia, compared to the internal bond length, angle and force constant changes within the molecule. These internal changes can, in principle, be deduced from a vibrational analysis coupled with a Franck-Condon analysis of the transition from the ground state.

Some of the results achieved by the use of the crystalline solid solution method will be reviewed in this article.

II. NAPHTHALENE IN DURENE

A crystalline solid solution between two organic substances can be made quite often when the size, shape and polarity of the two molecules are similar. A prime requirement, however, is that the host crystal shall be transparent in the spectral region of the guest molecule. The first such system to be exploited in a detailed way in spectroscopy was durene (1,2,4,5 tetramethylbenzene) as host, with naphthalene as guest (4). The smaller size of the conjugated system of the host is responsible for its being transparent in the region of the first and part of the second transitions of naphthalene. The spectrum of this system at $2^{o}K$ is shown in Fig. 2. The durene crystal, Fig. 3, which is monoclinic, has been cut so as to reveal the bc face. The naphthalene molecules align themselves

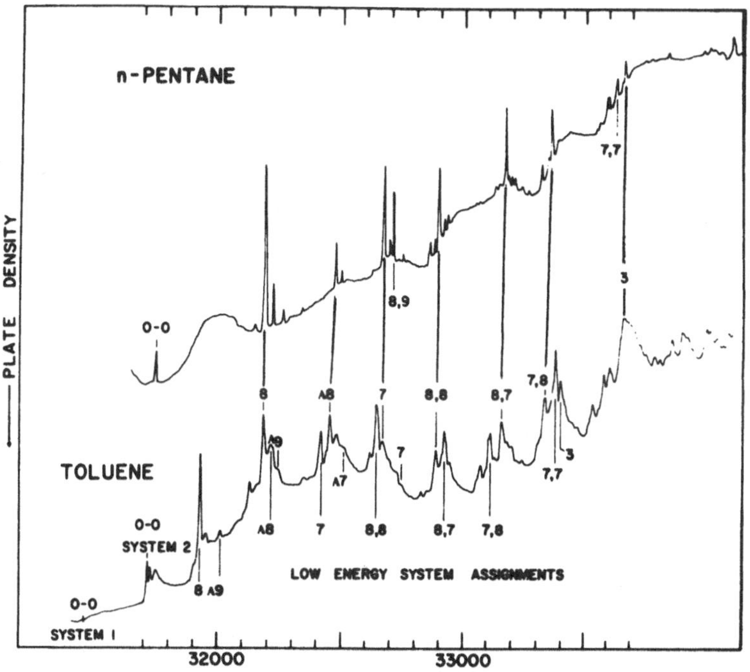

Fig. 1. Absorption spectrum of naphthalene in polycrystalline n-pentane and in polycrystalline toluene. Toluene has two prominent sites (T = 2°K).

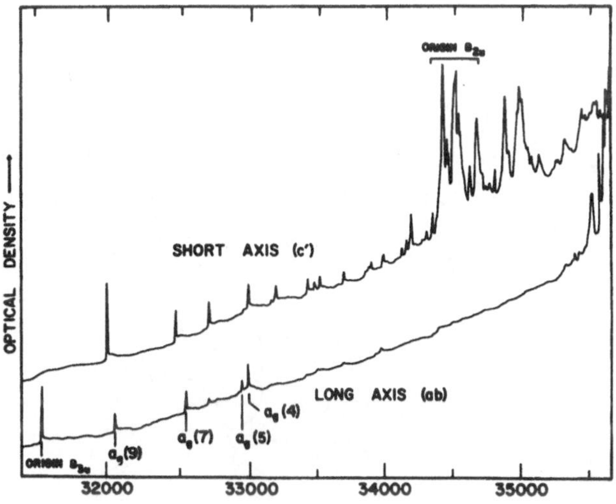

Fig. 2. Absorption spectrum of naphthalene in single crystal durene, b c' face at 2°K. Spectrum on short axis is vibronically allowed in B_{3u} region, fully allowed in B_{2u} region. Spectrum on long axis is formally, but weakly allowed.

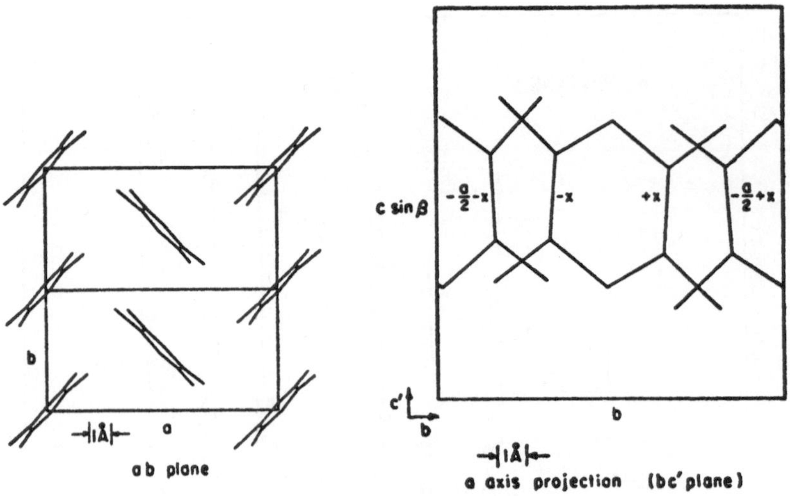

Fig. 3. Crystal structure of durene.

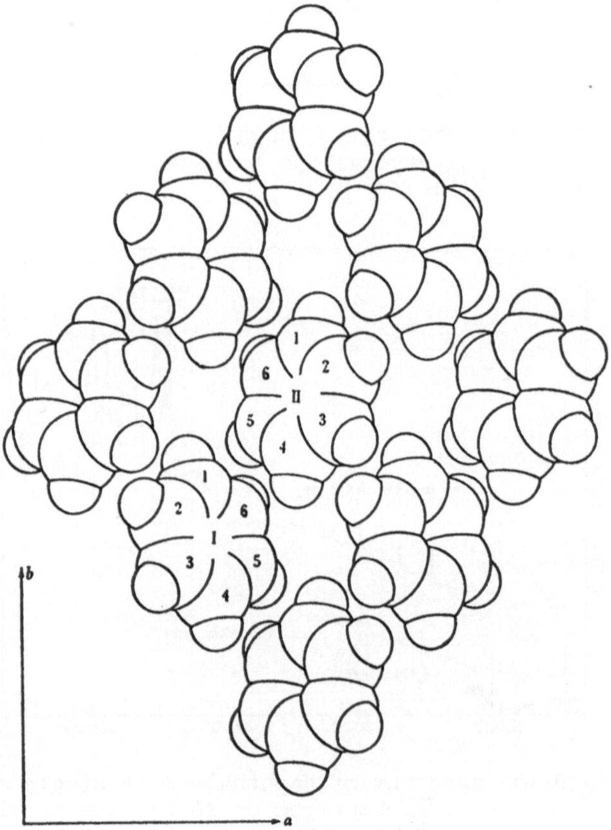

Fig. 4. Diagram of the crystal structure of benzene viewed down the c-axis.

so that only the short axis transitions absorb when light is polarized along the C'-axis, and only the long axis transitions absorb when light is polarized along the b-axis. This system clearly revealed that naphthalene has an allowed long-axis transition, but that the vibronically induced short-axis transition is actually stronger, and the confusion over the assignment of the first excited singlet level was eliminated. The transition to the second excited singlet state is seen in Fig. 2 to be entirely short-axis polarized.

This early work settled two important state assignments. Some of the same information could have been obtained from the rotational contours of the absorption bands. In the case of the first transition, this analysis was begun after the mixed crystal work was completed, and it was found that the allowed bands have single heads, while the vibronically induced bands have double heads, spaced 2.4-3.0 cm^{-1} apart. Still, no moments of inertia have been derived for this molecule. In the case of the second transition, the bands are apparently too broad even to deduce meaningful rotational contours.

The reason for the breadth of the bands in the second singlet transition of naphthalene is the vibronic interference between overtones of the lower state, $^1B_{3u}$, and the upper $^1B_{2u}$ state. This phenomenon has been investigated by Wessel, using mixed crystals of naphthalene in durene and naphthalene in p-xylene (6). The latter matrix is just as good as durene though harder to handle, since its melting point is below room temperature. Nevertheless, single crystals can be grown and oriented, and the differently polarized transitions of the embedded naphthalene molecules can be separated.

In the early work, it was assumed that naphthalene could only be oriented with its long axis in the same direction as the long axis of the replaced durene molecule. This is certainly a reasonable idea, but it needed to be tested. The EPR work of Hutchison and Mangum did confirm the orientation (7). This was done by observing the proton hyperfine structure in the EPR signal.

III. ORIENTATION OF MOLECULES IN SOLID SOLUTION

The work of Hutchison and Mangum settled the question of the orientation of naphthalene in durene. They could not see any deviations from the durene-like orientation. It should be noted, however, that this work was done at 77°K. Both host and guest molecules belong to the same point group so the force system acting on a naphthalene in a durene site has the same symmetry as that acting on a durene. This circumstance favors the host-orientation for the guest, but does not guarantee it. The same symmetry would be present if the naphthalene were rotated by 90° so as to have its long axis along the short axis of the host molecule site. The forces, however, should, and do, favor the normal orientation.

The symmetry argument does not eliminate the possibility of multiple minima; the same point symmetry could be achieved for several orientations around the C'-axis of durene, since this is no symmetry axis. The fact that no other orientations occur simply means that the force field is actually smooth and only one minimum exists.

Hochstrasser and Small, investigating phenanthrene in biphenyl observed a doubling of most of the absorption, fluorescence and phosphorescence lines in spectra taken at 4°K (8). They made a very complete spectral study of this system and followed it with further experimental and theoretical work on the details of the orientation of phenanthrene in biphenyl. The results are quite definite that there are two orientations of phenanthrene in biphenyl differing by rotations of 5-10° around the long axis from the biphenyl position. The minima differ in energy by only a few cm^{-1} and are separated by a barrier which is ineffective above about 20°K.

They also observed a tripling of the lines of pyrene in biphenyl, a doubling of anthracene in biphenyl and a broadening of the lines of naphthalene in biphenyl. These results show that a single unique position of a guest molecule in a host crystal cannot always be assumed.

Hochstrasser and Small also found that the energy of excitation of phenanthrene depends on which site it is in, and that this site dependence is not the same for the lowest singlet and lowest triplet. The fluorescence doublets are split by 17 ± 2 cm^{-1}, while the phosphorescence doublets are split by -10 cm^{-1}. The ground state energy difference was found to be about 4 cm^{-1} by observing the rate of interchange of molecules between the two sites in the triplet state, and the barrier in the triplet state was found to be 84 cm^{-1}. Thus the energy differences were found to be 4, 14 and -13 cm^{-1} in ground, triplet and singlet states respectively. Thus, a rather minor difference in the local environment causes appreciable shifts in the spectrum, and in the present case these shifts are opposite in direction for singlet and triplet states.

An electron spin resonance study of phenanthrene in biphenyl was made by Brandon, Gerkin and Hutchison (9). No evidence for the double minimum in the potential was found, but the work was done at 77°K, and at this temperature the interconversion of molecules between the two minima is probably occurring rapidly. They did find that both phenanthrene and naphthalene in biphenyl are oriented in the same way, within a degree or two, but, remarkably that they are both rotated around their long axes by 6° away from the biphenyl planes. This statement, of course, applies to the triplet states of the embedded molecules, and not necessarily to the ground state.

The fact that there can be a different population of two crystal potential minima in different electronic states is related to the Franck-Condon factor for a transition in the crystalline matrix. The Franck-Condon factors relate to the crystal phonon modes excited in the transition rather than to molecular modes. The multiple minima in the guest-host potential makes the calculation of these factors fairly complex, because one has to know the barriers in each state and whether or not the wavefunction is confined to one or not. This subject, however, would bring us into a complicated problem of solid state physics and away from our initial goal of elucidating molecular spectra. Interesting though it may be, the problems of multiple minima and lattice modes will be avoided as far as possible in the rest of this article, but they remind us that spectral simplification compared to the vapor is not entirely achieved, and warn us that the matrix does affect the spectrum.

IV. SPECTRUM ANALYSIS

Some crystalline matrix spectra are free of the major complications of multiple sites and apparently provide an easy way to observe the absorption and emission spectra of large molecules. The three main benefits are the elimination of overlapping structure by the use of temperatures near $0^{o}K$, the narrow line width and the polarization. These have been illustrated already for naphthalene in durene in Fig. 2. This is an example for which the solid state was essential for the solution of the problem of spectral assignment. In this section, we want to discuss other such examples.

IV.A. Pyrazine and Pyridine

The history of the analysis of the lowest $n \rightarrow \pi$ spectra of these molecules shows that some of the problems were not solved by use of the vapor spectra: mixed crystal spectra were used to establish vibrational assignments in pyrazine. The comparison of vapor and mixed crystal spectra for both molecules is instructive. The most successful matrix for these molecules is benzene, though cyclohexane, p-dioxane, durene, p-xylene and acetonitrile have been used by various investigators. Energy level diagrams for benzene, pyrazine and pyridine are shown on the next page in Fig. 5.

Using single crystal benzene whose structure is shown in Fig. 4, Narva was able to distinguish lines polarized in the three different symmetry directions of pyrazine(10). To separate the X and Y directions (see Fig. 5) required careful intensity measurements

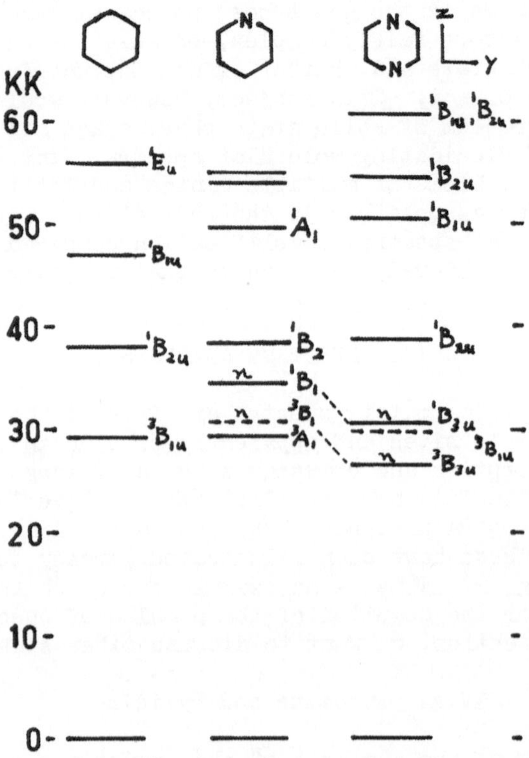

Fig. 5. Energy level diagram for benzene, pyridine and pyrazine. The n-π* levels are marked by n.

since in the ab or bc plane of benzene, the polarization ratios for these axes differ by only a factor of 2. An overall view of the spectrum corresponding to the first singlet-singlet transition, $^1A_g \rightarrow {}^1B_{3u}$, is shown in Fig. 6; the polarization ratios appear correctly in Fig. 7 where lines due to the important 10a and 6a vibrational modes are shown. Fig. 6 does show the reversed polarization of the Z (or N-N polarized) lines. Polarization results like these are much easier to get from crystals than from rotational contours.

Similar polarization results were obtained by Brownrigg in the case of pyridine in benzene single crystals (11).

In the vapor, one can assign vibrational levels of an upper state by making transitions from specific thermally excited ground levels, or by means of a tunable laser, a particular upper state can be excited, and the ground state vibrations into which it decays can be noted and used for making an assignment of the upper state mode. These methods, in conjunction with isotope substitution make it possible to do a fairly good job of assigning excited state modes.

In the solid state, relaxation out of initially excited vibrational levels is so fast that emission is observed only from the zero vibrational level, and since low temperatures must be used, no "hot-bands" are possible. The vibrational assignments must be made on the basis of polarization, isotope shift and by analogy to the ground state.

In the case of pyrazine in benzene the three possible allowed vibrational symmetries, a_g, b_{1g} and b_{2g} could be distinguished for each spectral line. The lines were so sharp in many cases that the satellites due to ^{13}C and ^{15}N in natural abundance could be measured, and used with the Teller-Redlich product rule to test assignments. These satellites are usually swamped by rotational structure from the stronger main component in vapor spectra. Deuterium substitution was also used.

Because we did not make accurate intensity measurements in pyridine, the ^{13}C lines were not as useful in this case. They are present, but an accurately known intensity ratio to the main line is good verification of their identity. For pyrazine one calculates a ratio of 0.044 for the ^{13}C line relative to the ^{12}C line, and a value of 0.047 is observed in the case of the origin. Similarly $^{13}C_2$ and ^{15}N lines can be verified. Isotopic substitution produces both a zero point energy shift and a vibrational frequency shift, both of which are useful.

The most important vibrational assignments in pyrazine are related to mode 10a and its overtones and combinations with 6a.

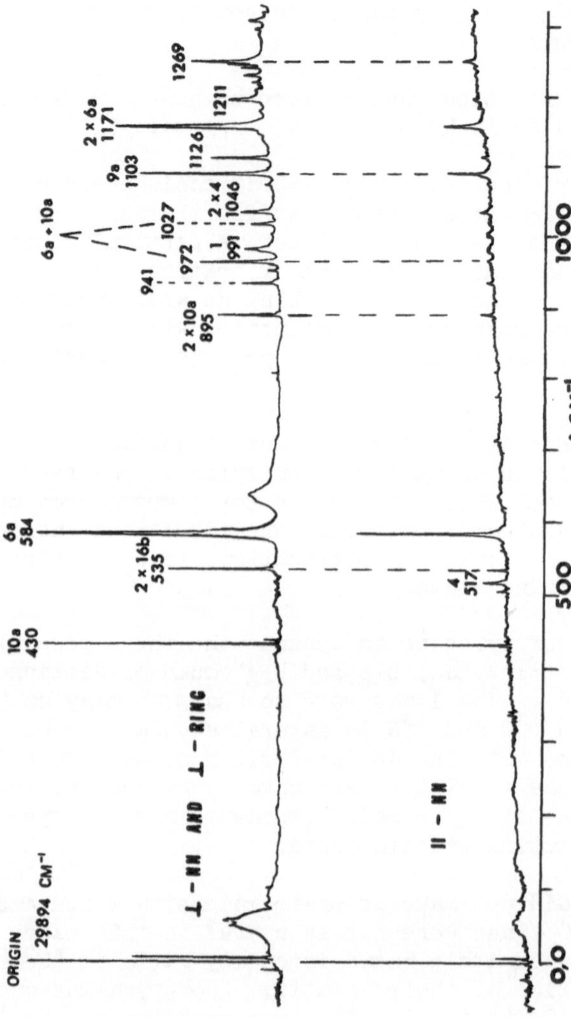

Fig. 6. Pyrazine H_4 in a benzene crystal, 2°K.

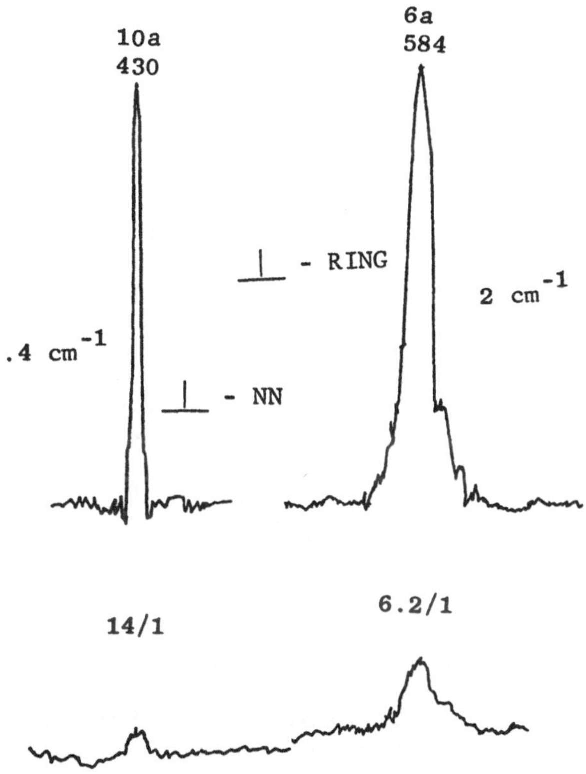

Fig. 7. Line widths and polarization ratios for two lines of pyrazine in benzene (2°K).

The atomic motions in these modes are shown in Fig. 8. The 10a mode appears because of vibronic coupling between the $^1B_{3u}$ $n\pi$ state at 29883 cm^{-1} and the $^1B_{2u}$ $\pi\pi$ state, 8000 cm^{-1} higher. It is identified first by its polarization, perpendicular to the N-N axis, in the molecular plane. Innes showed that this mode drops from a value of 918 cm^{-1} in the ground state to 383 in the $^1B_{3u}$ ($n\pi$) state (12). The assignment was claimed for many years to be mode 5 until the early indications of the solid state work prompted a reexamination of it. The vapor data now agrees with the recent single crystal data. The deuterium isotope effect was not sufficient to distinguish these possibilities. The 10a mode has a node along the N-N line, and therefore should have no ^{15}N isotope shift. The crystal work shows that the shift is no more than 0.1 cm^{-1}, whereas the alternative assignment as ν_5 requires a shift of 2 cm^{-1}. Furthermore, it has the correct ^{13}C shift (observed 1.0, calc. 1.08 cm^{-1}).

One of the remarkable observations about the out-of-plane modes of pyrazine and pyridine is their solvent sensitivity. The 10a mode of pyrazine shifts from 383 (vapor) to 430 (benzene) to 486 (pure crystal). In cyclohexane and durene, the frequency is nearly the same as in the vapor. The solvent effect in pyrazine is shown in Fig. 9 (13, 14).

The prominent out-of-plane mode in pyridine, 16b, is only 58.8 cm^{-1} in the vapor, but rises to 151 in benzene.

The solvent effect is one way to identify these modes in the solid state and relate them to vapor values. More significant is the fact that the overtone can now be picked out of other possible lines by its solvent effect and identified. This is not a simple matter otherwise because the overtone is far from being harmonic, and it was not possible to make this assignment with assurance in the vapor. In fact, the wrong assignment was made.

Using the single crystal data, the ^{15}N, ^{13}C and ^2H isotope shifts could also be used to verify the assignment. A final verification was to use the Franck-Condon Principle: the overtone belongs to A_g and is therefore allowed, but since the fundamental is now totally symmetric, there is no displacement and the intensity can only be induced by the frequency shift:

$$\frac{I_{02}}{I_{00}} = \frac{1}{2}\left[\frac{\omega'' - \omega'}{\omega'' + \omega'}\right]^2 .$$

Having accurate intensity measurements, the value of I_{02}/I_{00} = 0.061 from this formula could be compared with observation, giving 0.045 for H_4 and 0.06 for D_4 pyrazine, (for the mode 10a).

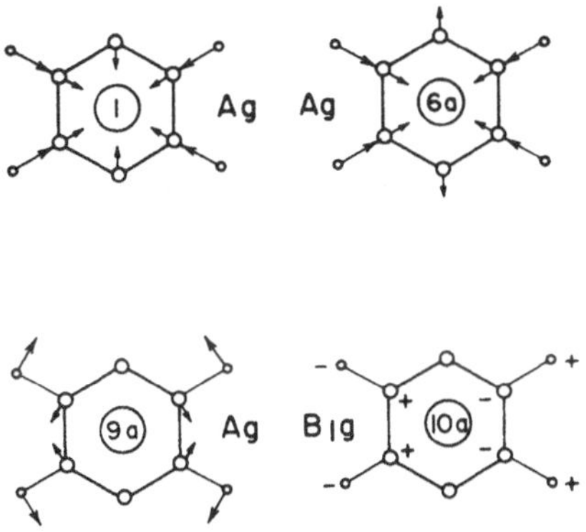

Fig. 8. Vibrational modes important in pyrazine.

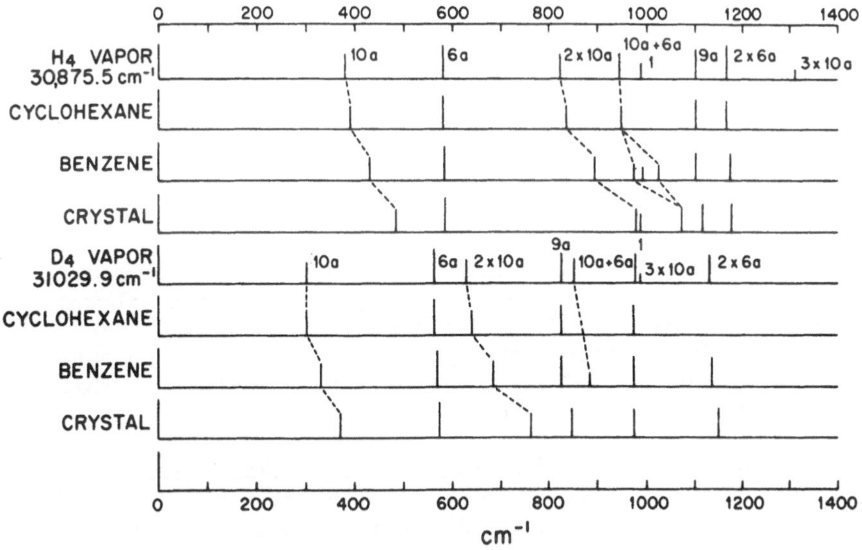

Fig. 9. Solvent effect on out-of-plane mode 10a and in-plane modes of pyrazine in $^1B_{3u}$ state.

The anharmonic progression of 10a can be identified in the vapor and in crystals up to three or four members. It is a positive anharmonicity, and can be fitted by adding a quartic term to a harmonic potential. The quartic term decreases as the solvent effect increases, and it has nearly disappeared in the pure crystal.

Another significant anharmonic effect is found in the combinations of 10a with 6a. These always have a negative anharmonic term of 20 cm^{-1} or more. Again, to establish these combinations, many indications are necessary. The solvent effect moves these combinations in the same way as it moves the fundamental 10a. The polarizations of lines assigned to 10a + n 6a are all in the Y direction. The deuterium isotope shifts also confirm these assignments.

The theory of the 6a - 10a coupling is that there is a quartic term, V_{AABB}, in the potential which mixes these modes during out-of-plane displacements; here the symmetry is reduced to C_2, and then both a_g and b_{1g} modes belong to the A_g class. This quartic term is added to the vibrational Hamiltonian to produce a coupled Hamiltonian: $T_A + V_A + T_B + V_B + V_{AABB} Q_A^2 Q_B^2$ for the A and B modes. This can be solved using products of the separate oscillators' functions and the results are fitted to experiment. Using one line to get the parameter of the coupling, several other lines can be predicted in this highly anharmonic set, and the values agree well with experiment.

The result of this study was to find a set of potential parameters and also to investigate their origin. Since the $^1B_{2u}$ ($\pi\pi$) state lies only 8000 cm^{-1} above the $^1B_{3u}$ ($n\pi$) state, the out-of-plane b_{1g} mode can be expected to couple the two states, producing the observed intensity in the odd quantum number levels of mode 10a. Using the carefully measured intensity of mode 10a relative to the origin, the electronic portion of the vibronic coupling was evaluated. This electronic matrix element was shown to be big enough to explain the quartic oscillator component of the 10a mode and also the reduction of the force constant of this mode in the $^1B_{3u}$ state, compared to the ground state.

For mode 4, however, for which there is also a large quartic oscillator component, the vibronic coupling with the distant $^1B_{1u}$ state is not enough to explain the quartic factor and the reduction of force constant.

The in-state modification of the potential function of pyrazine in its $n\pi$ state seems to be caused by the presence of the extra electron placed in the π-orbitals. Each carbon atom in the excited state may have about one eighth of an extra electron in its π orbital which will tend to repel electrons in the CH, CC and CN bonds. We believe in fact that the carbon orbitals become partially rehybridized in the out-of-plane configuration departing from sp^2 toward sp^3, and that this effect reduces the out-of-plane force constants.

In the ground state, the carbons are analogous to N in the NH_3^+ ion which has sp^2 hybridization and one π-electron and is planar. Neutral NH_3 has one more "π-electron" which forces the molecule into a tetrahedral hybrid bonding configuration; the carbons in the excited state of pyrazine have a small fraction of an extra electron which applies a force in the same direction but cannot cause a double minimum as in NH_3.

The reduced frequency of the 10a + 6a combinations is in agreement with the rehybridization picture. The 6a mode reduces the 120° C-C-N angle toward the smaller tetrahedral angle in one phase of this motion, and if 10a is simultaneously moving the H atom out-of-plane, the force required for the two motions will be reduced.

The excited n π state of pyridine should show some of the same peculiarities as were observed for pyrazine. Ramsay has shown, in fact, that pyridine becomes slightly non-planar, but that the distortion is along the mode 16b and not 10a(15). Since 16b belongs to b_1 of C_{2v} it must have its vibronic coupling with the second of the π states, rather than with the lowest one shown in Fig. 4. However, Ramsay's data shows only even numbers of quanta of 16b, and no vibronic coupling needs to be invoked to explain their appearance.

We began to study pyridine in single crystal benzene hoping to resolve a problem with the spectrum of pyridine vapor which had been noted by Sponer long ago (16). This was the occurrence of an upper state vibration of only 139 cm^{-1}. There was no evidence for such a frequency in the crystal, and we began to experiment to see if this line represented another electronic state having a different solvent shift from the n π state. For example, the vapor under increasing high pressure of N_2 showed that the origin and the 139 band both shifted together. The problem was solved by Ramsay who showed that 139 is the second quantum level of 16b, and that this mode is in a strongly quartic potential.

Later we identified 16b in the crystal as a weak line having the opposite polarization from the rest of the spectral lines, but appearing as a single quantum level at 151 cm^{-1}. An appreciably stronger line appearing at 306 cm^{-1} was identified as the second quantum level; almost all of the quartic anharmonicity has been removed by the crystal, and the frequency has been more than doubled. Furthermore, single quanta can now be seen. The spectrum of pyridine is shown in Fig. 10.

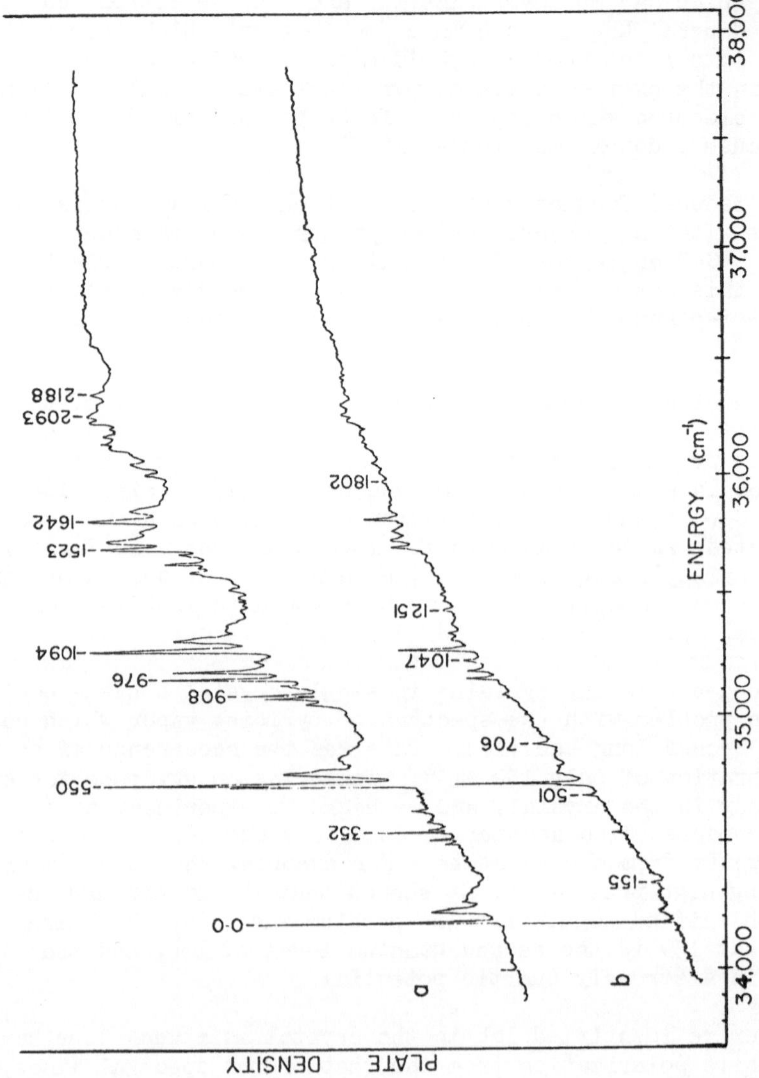

Fig. 10. Pyridine h-5 in benzene. Polarized singlet absorption, 4.2°K.

In the ground state 16b has a frequency of 406 cm^{-1} and the Franck-Condon principle explains its high two quantum intensity in the vapor in terms of eq. 1. In the crystal the intensity is much lower, but the frequency drop is lower, and eq. 1 still explains the intensity satisfactorily.

The single quantum intensity is caused by vibronic coupling to the B_2 state near 50,000 cm^{-1}. Very little intensity is provided by this distant state, and none is seen in the vapor, probably because the wavefunction for 16b is much more extended, and overlaps poorly with that of the B_2 state. The more compressed wavefunction in the crystal has a better overlap.

We have studied the analogous, but larger molecules of phenazine and acridine. These are anthracene-like heterocycles with nitrogens at the 9, 10 and 9 positions, respectively. Out-of-plane vibrations can be observed in the $n\pi$ spectra of phenazine in a biphenyl crystal. These modes appear fairly strongly because the $\pi\pi$ states from which intensity is derived are very close to the $n\pi$ state. The evidence is, however, in favor of only small reductions in the frequency of these modes in the excited state. It is not clear yet if these modes have quartic oscillator components. It does appear that the effect of the excitation is even more diluted in these molecules than in pyrazine and pyridine.

IV.B. Trans-Stilbene

Stilbene, or trans-diphenylethylene is a molecule whose ordinary solution spectrum is a nearly featureless band 5000 cm^{-1} broad centered at 34,000 cm^{-1}. Its vapor spectrum is uninformative. The trans → cis photochemical transformation is of great interest as an example of this important type of isomerization route in a large molecule. Therefore, it was important to learn what one could about the excited states of stilbene.

The molecule, dibenzyl, or diphenyl ethane, forms a convenient host crystal for stilbene. In fact, stilbene is always present in small amounts in dibenzyl as it is an oxidation product. The host crystal orientation permits isolation of the long-axis spectrum of stilbene, and it is found that the first singlet-singlet transition is completely polarized in the direction of the long axis of dibenzyl(17). At 4.2°K, the spectral lines are only a few cm^{-1} wide and only one prominent site is occupied. The resulting spectrum, Fig. 11, can be analyzed in great detail. With the help of perdeuterated stilbene and the ground state vibrational assignments, the vibrational assignments of the excited state could be found. Furthermore, an analysis of the Franck-Condon factors and a Pariser-Parr-Pople calculation made possible a rather complete study of changes in the force constants and bond lengths in the excited state.

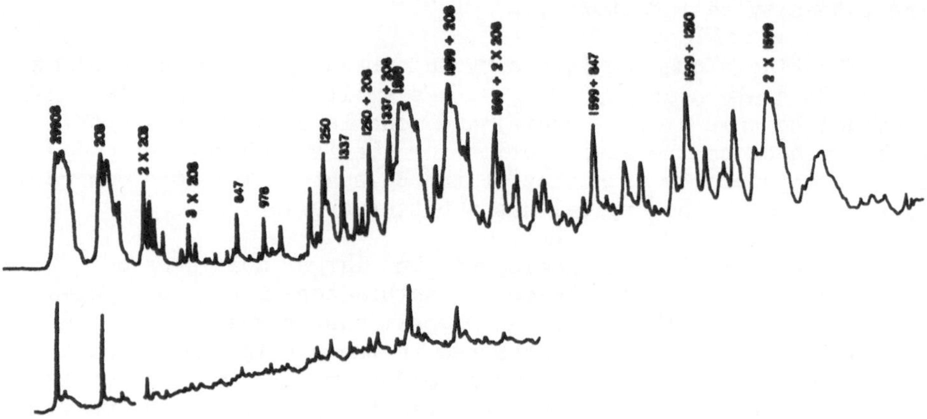

Fig. 11. Absorption spectrum of stilbene in dibenzyl at 20°K.

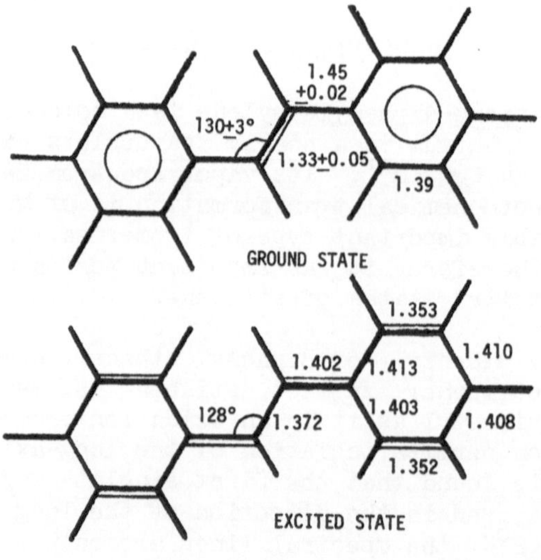

Fig. 12. The geometry of stilbene in ground and excited state is deduced from a Franck-Condon analysis. Bond lengths in Å (18).

The results of this work, done by Chen-Hanson Ting, are shown in Fig. 12 (18).

IV.C. Benzoic Acid Dimer

This is an example of a molecule which cannot be obtained in the gas phase. The hydrogen-bond strength is so small that most of the vapor is monomer. Extremely sharp, well resolved dimer spectra can, however, be achieved in several matrices, particularly single crystal benzene (19).

Figs. 13, 14, 15 and 16 show the absorption, fluorescence, triplet-singlet excitation, and phosphorescence of benzoic acid in benzene at 2°K. From these spectra, and with some help from calculations and solution spectra, a detailed understanding of the hydrogen bonds in the ground, triplet and lowest singlet states has been derived.

The benzoic acid dimer must replace two benzene molecules. Fig. 5 shows that two translationally equivalent molecules along the b-axis are spaced about the right distance apart and have the coplanar orientation needed to incorporate the benzoic acid dimer. The absorption spectra are polarized in accord with this orientation. Fig. 14 shows only one prominent vibration, oppositely polarized with respect to the rest of the spectra, the mode 6b, which induces a long axis, or b-polarized line.

One of the important features of these spectra is the presence in fluorescence of certain odd-parity modes. These are not expected to occur in a $^1A_g \rightarrow {}^1B_u$ transition in a molecule having the D_{2h} symmetry expected for benzoic acid dimer. No such modes appear in absorption or phosphorescence. These facts seem to eliminate the possibility that the cause of the appearance of the odd modes is that the molecule is twisted in the benzene crystal. There may be such a twist, which would remove the center of symmetry, but then all the spectra should show some evidence of it. The most intense odd modes are the ones which involve the carboxyl group the most. In a dimer, the monomer halves contribute partial vibrations which can be combined into odd and even normal modes of the whole molecule. Matrices in which carboxyl group atoms are moving in odd parity relative to one another are strong, while the corresponding odd ring modes are not present. The identification of these modes is quite certain, as the infrared and Raman spectra have been carefully studied.

The interpretation of the occurrence of these odd modes which we believe to be correct is that the hydrogens move toward the excited half of the dimer and trap the excitation there. The emission, therefore, comes from one half of the dimer, and thus the odd and even dimer modes have the same symmetry with respect to the emitting moiety.

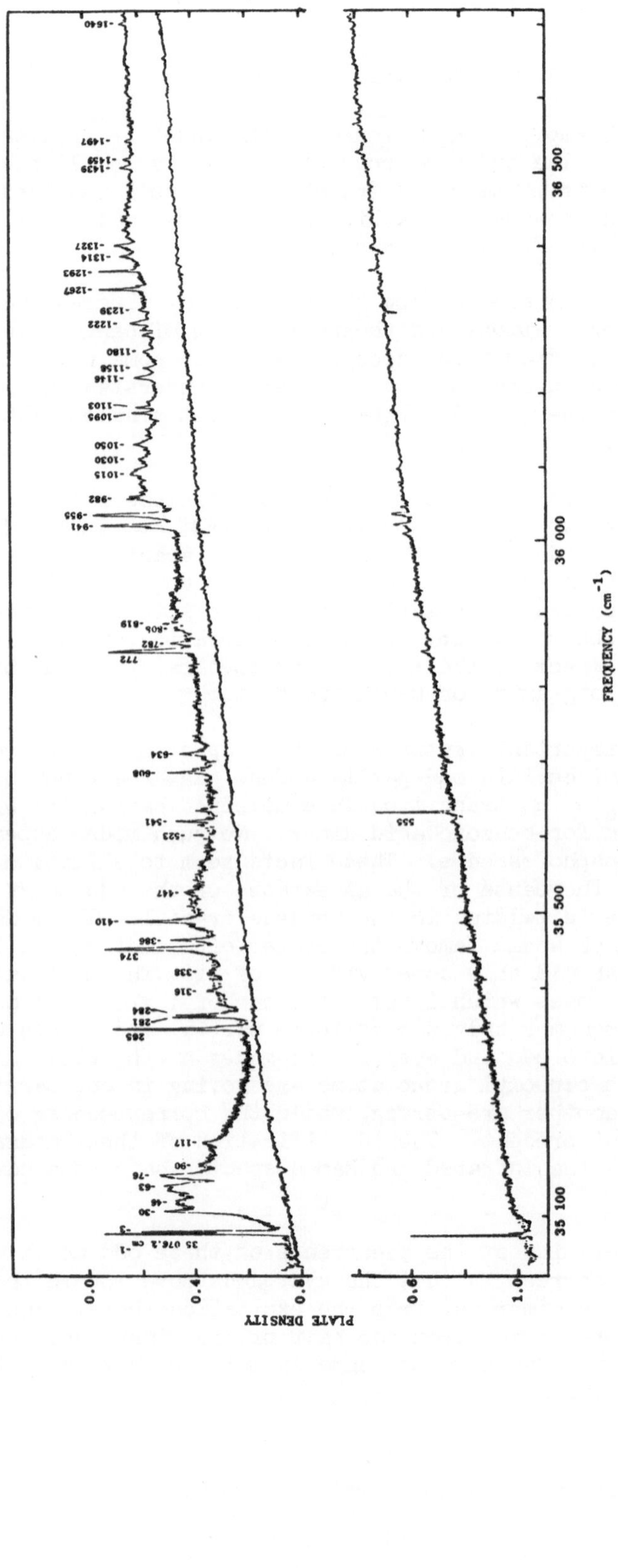

Fig. 13. Absorption spectrum of benzoic acid dimer in benzene single crystal. Top trace a or c-axis, lower trace b-axis. Origin at 35076.2 and 35079.2 cm^{-1}. Note the oppositely polarized line at 585 due to mode 6b.

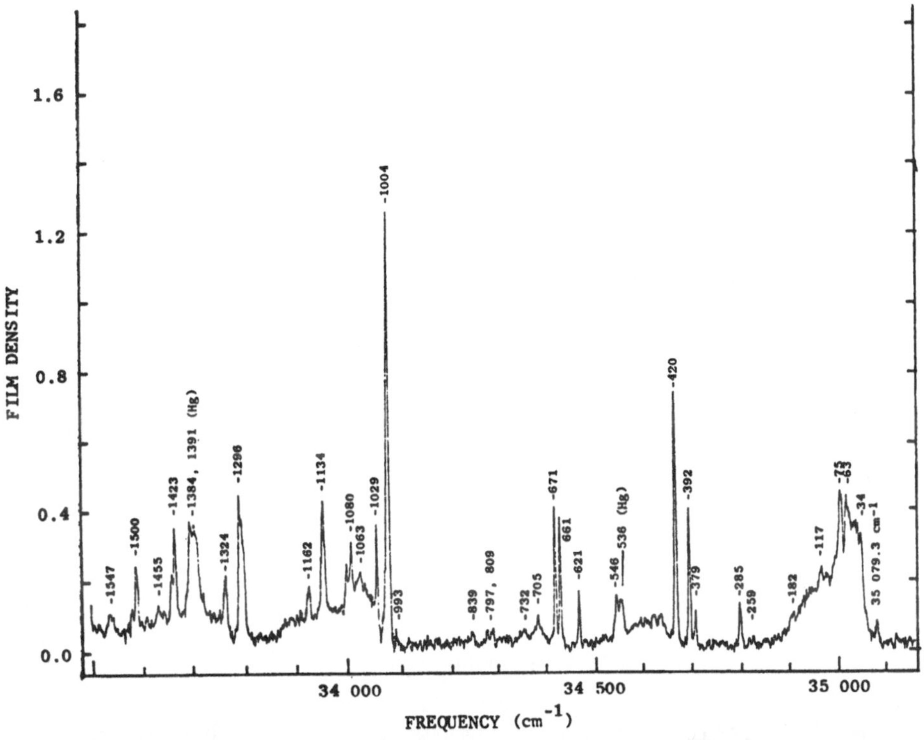

Fig. 14. Fluorescence of benzoic acid dimer in single benzene crystal. The origin at 35079.3 cm^{-1} is reabsorbed. The line 392 is the odd parity 6a type mode while 420 is the even parity type 6a mode.

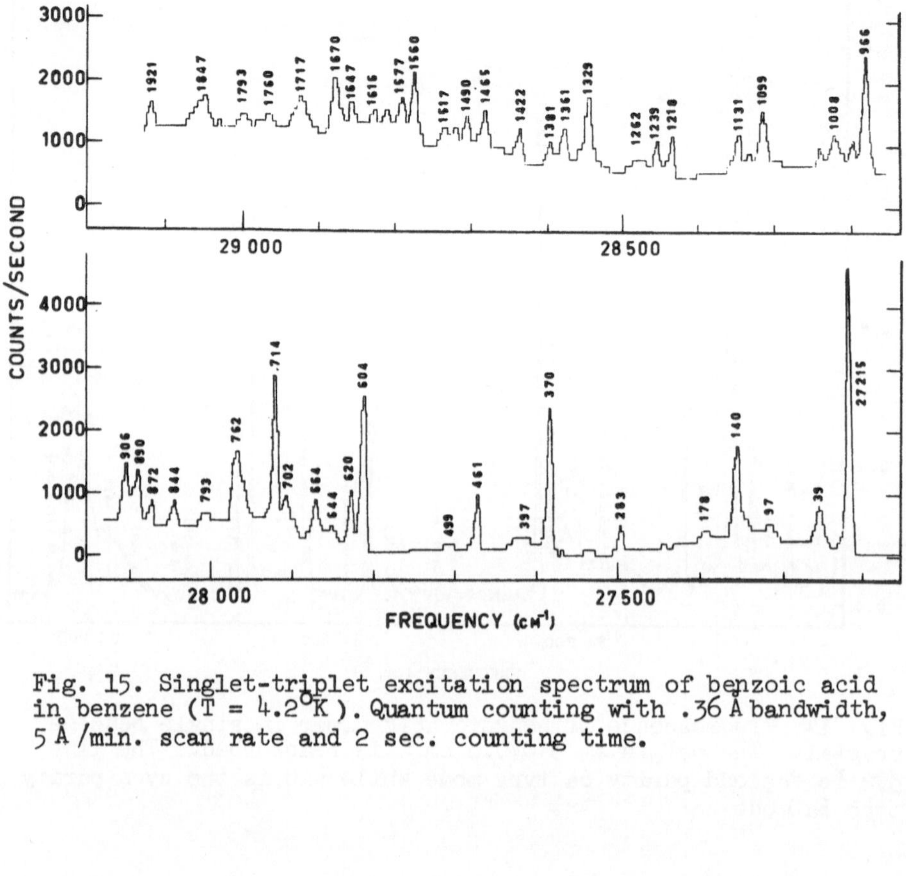

Fig. 15. Singlet-triplet excitation spectrum of benzoic acid in benzene ($T = 4.2^\circ K$). Quantum counting with .36 Å bandwidth, 5 Å /min. scan rate and 2 sec. counting time.

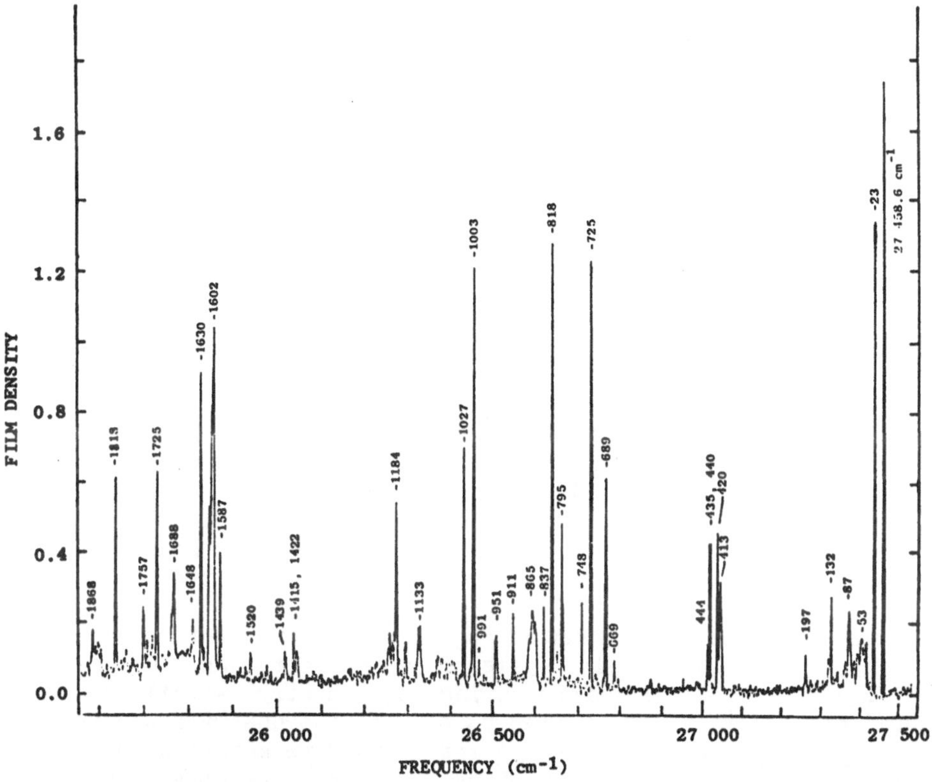

Fig. 16. Phosphorescence of benzoic acid dimer in benzene single crystal. The doubling of the origin is thought to be due to an extra site ($T = 2^{o}K$).

The significance of this fact is that the attraction of the carbonyl oxygen of the carboxyl group which the hydrogen shift demonstrates, correlates with the reduced acidity of benzoic acid in its excited state. Studies of solution spectra show that the -OH group is a weaker acid and the -C=O group is a better H-bonding group in the first singlet state, compared to the ground state.

IV.D. Vibronic Interference

The crystalline solid solution method appears to be one way to get the necessary data to understand internal conversion, or the transfer of energy from one electronic state to another. Naphthalene is the first system where this was studied in detail. The experimental evidence for the coupling of two states in naphthalene is shown in Fig.17. Here one sees the region of the spectrum where the vibrational overtones of the $^1B_{3u}$ state overlap the beginning of the $^1B_{2u}$ state. The spectrum is short-axis polarized so that the $^1B_{3u}$ spectrum is made of an odd number of b_{1g} modes with added a_g modes, resulting in vibronic states belonging to $b_{1g} \times B_{3u} = B_{2u}$. Thus, the two sets of states have the same symmetry and perturb each other. The complicated spectrum produced in the region of overlap is called the vibronic interference spectrum.

The solvent shift of the B_{2u} state relative to the B_{3u} state is greater in p-xylene than in durene so that the overlap of the two states occurs earlier in xylene than in durene. Therefore, one can observe both the unperturbed levels in the durene solution spectrum and the same levels perturbed in the xylene solution spectrum. Fig. 17 shows this region of the two spectra.

The short-axis spectrum of Fig. 2 shows several forbidden origins, i.e., vibrational levels of b_{1g} symmetry superimposed on the $^1B_{3u}$ state, and therefore having B_{2u} symmetry overall. Symmetric vibrations add to these origins to produce the observed spectrum. Since there are three active b_{1g} modes and five active a_g modes, there are a great many combinations having B_{2u} symmetry in the region 2500 cm^{-1} above the true origin, in the vicinity of the origin of the $^1B_{2u}$ state. These combination levels are the ones being intensified by interaction with $^1B_{2u}$. The average density of levels of the correct symmetry is about 100 per cm^{-1}, but less than one per cm^{-1} has much intensity. This fact makes the overlap region appear as a set of discrete lines each on the order of one cm^{-1} wide.

John Wessel found it possible to calculate the spectrum of naphthalene in xylene from the spectrum in durene(6). The overtone levels in durene between about 2250 and 2650 cm^{-1} from the $^1B_{3u}$

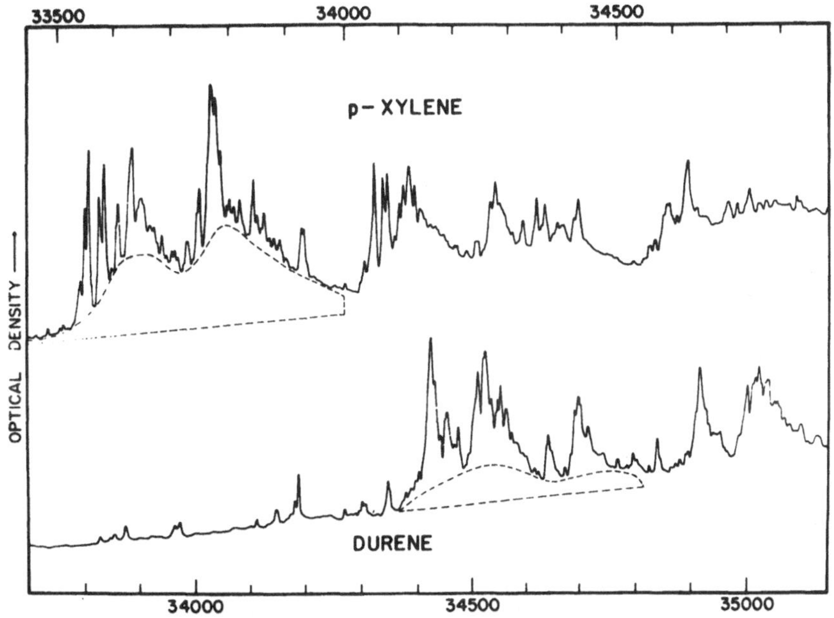

Fig. 17. Spectrum of naphthalene in p-xylene and in durene in the interference region. Short-axis polarized, 2°K. Xylene shifted 270 cm^{-1} to blue so $^1B_{3u}$ origins coincide.

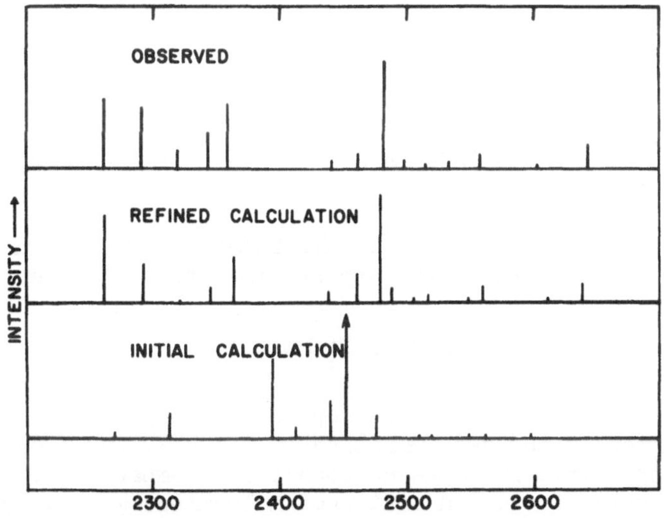

Fig. 18. Top: Simplified representation of observed spectrum in xylene in Fig. 17 between 33,552 and 33,931. Bottom: Initial calculation using $E(B_{2u})$ = 2450. Middle: Final calculation using $E(B_{2u})$ = 2482. Abscissa is frequency from xylene origin.

origin are considered to be unperturbed levels, and are entered as diagonal elements 2 to N in the matrix below. A perturbing electronic level of B_{2u} symmetry is added as diagonal element 1, and off-diagonal elements $(1|h'|N)$ and $(N|h'|1)$ are produced. h' is the vibrational-electronic coupling operator. The perturbation matrix thus has the form

$$\begin{pmatrix} E(B_{2u})_{11} & (1|h'|2) & (1|h'|3) & --- \\ (2|h'|1) & E(B_{3u} \times b_{1g})_{22} & & \\ (3|h'|1) & & E(B_{3u} \times b_{2g})_{33} & \\ \vdots & & & \ddots \end{pmatrix} .$$

The off-diagonal terms are taken to be proportional to the intensity observed in the durene solution. Specifically it is assumed that $I(n) = (1|h'|n)^2/\Delta E^2$ where ΔE is the energy from the level n to the B_{2u} origin in the durene matrix.

The diagonalization of this matrix is performed several times, allowing the position of the perturbing electronic level to vary until a good fit is obtained. The results of one such process are shown in Fig. 18. The initial choice of B_{2u} is 2450, and was moved to 2480 to obtain the "refined fit".

These calculations led to detailed assignments of the interference structure, and therefore to a good understanding of what is going on to produce such a spectrum. Recently Langhoff and Robinson have used Wessel's data and obtained an even better analysis of it (20).

Another clear example of vibronic interference is found in the acridine spectrum where the $n\pi^*$ spectrum runs into the $\pi\pi$ spectrum (10).

V. CONCLUSIONS

The earliest work in spectroscopy of large polyatomic molecules was directed toward the determination of the symmetry types of the excited states. This work was necessary for comparison to the newly developing theory of excited states of molecules. The theoretical work of Goeppert-Mayer and Sklar on benzene (21) was compared with the experimental work of Radle and Beck (22) and Sponer (23). It was not possible then to distinguish the B_{1u} from the B_{2u} assignment, but the theory was quite definitely in favor of the latter. The next more complex ring system was naphthalene whose lowest two excited state assignments were determined by the mixed crystal method (4). The Pariser-Parr-Pople theory was in agreement with these results, and made many more predictions which were later verified by experiment(24). Assignments of the aromatic molecules can be considered fairly secure, due to the semi-empirical work of Platt (25) and the theoretical work of Pariser, Parr and Pople and the experimental work.

Work on their heterocyclic analogues has recently progressed with the solution of some spectroscopic puzzles for pyridine and pyrazine, and the analysis of the spectra of acridine and phenazine. The new features here are the positions and assignments of the $n \rightarrow \pi$ type transitions, and the effects of the hetero atom on the strength and position of the $\pi \rightarrow \pi$ type transitions. The experimental work on heterocyclic molecules has not been stimulated by the appearance of a good theoretical description of the excited states of these molecules, as was the case for the aromatic hydrocarbons. The theory is much more difficult for heterocyclics because both σ- and π-electrons have to be considered. Work by Clementi (26) and Whitten (27) on pyrazine has been important. Recent work by Wadt and Goddard has been illuminating (28). They have made a real attempt to correlate theory and experiment, and to describe accurately the electronic changes upon $n \rightarrow \pi$ type excitation.

Most of what a spectroscopist learns from a spectrum is almost never computed by a theorist. So far, theories have concentrated on calculating energy levels for a fixed nuclear configuration, and transition probabilities to these levels. The agreement between experiment and theory is supposed to show how good the theory is. This is a necessary but not sufficient condition. The spectroscopist measures the vibrational energy levels of an excited state. These levels probe the wavefunction much more sensitively than does the total energy. A theoretical study of potential functions of key excited molecules should ultimately be made.

We are very far from knowing very much about the potential function for any large polyatomic molecule. No one has yet made a normal coordinate study of the first excited state of benzene, even though most of the vibrations in this state have been assigned, and are known for several isotopic species. Our study of pyrazine has shown some of the complexities which can arise in excited states. There will always be more interaction between nearby electronic states leading to quartic anharmonicity and negative harmonic contributions than in the ground state. Valence angles and bond lengths of the ground state structure will change as a result of rehybridization and redistribution of electrons. For large molecules, however, these changes are small and there is a good chance to find them out provided the large problem of handling the many-atom normal coordinate analysis is solved, and the perturbations due to non-harmonic effects can be included.

The new direction of polyatomic molecule spectroscopy should be to probe these potential functions. However, spectroscopic methods are limited by the Franck-Condon principle to a region of configuration space very near to the initial state of the transition. Some of the most interesting parts of a potential function are those which lead to isomers or to photochemical products. There are no straightforward methods to get at these regions, and combinations of spectroscopy, photochemistry and theory are necessary.

REFERENCES

1. J.H. Callomon, T.M. Dunn and I.M. Mills, Proc. Roy. Soc. A 259, 499 (1966).

2. S.N. Thakur and K.K. Innes, J. Mol. Spect. 52, 130 (1974).

3. E.V. Shpolskii, Sov. Phys. Usp. 6, 411 (1963); J.L. Richards and S.A. Rice, J. Chem. Phys. 54, 2014 (1971).

4. D.S. McClure, J.Chem. Phys. 22, 1668 (1954); ibid. 24, 1 (1956).

5. D.P. Craig, J.M. Hollas, M.F. Redies and S.C. Wait, Proc. Chem. Soc., 361 (1959).

6. John Wessel, Ph.D. Thesis, University of Chicago (1970).

7. C.A. Hutchison and B.W. Mangum, J. Chem. Phys. 34, 908 (1961).

8. R. Hochstrasser and J. Small, J. Chem. Phys. 45, 2270 (1966).

9. R.W. Brandon, R.E. Gerkin and C.A. Hutchison, J. Chem. Phys. 41, 3717 (1964).

10. D.L. Narva and D.S. McClure, Chemical Physics, to be published; also, D. Narva, Ph.D. Thesis, Princeton University (1975).

11. J. P. Brownrigg, Ph.D. Thesis, University of Chicago (1974).

12. K.K. Innes, J.D. Simmons and S.G. Tilford, J. Mol. Spect. 11, 247 (1963).

13. E.J. Zalewski, D.S. McClure and D.L. Narva, J. Chem. Phys. 61, 2964 (1974).

14. I. Suzuka, N. Mikami and M. Ito, J. Mol. Spect. 52, 21 (1974).

15. J. P. Jesson, H.W. Kroto and D.A. Ramsay, J. Chem. Phys. 56, 6257 (1972).

16. H. Sponer and H. Stuckeln, J. Chem. Phys. 14, 101 (1946).

17. R.H. Dyck and D.S. McClure, J. Chem. Phys. 36, 2326 (1962).

18. C.H. Ting, Ph.D. Thesis, University of Chicago (1965).

19. J.C. Baum, Ph.D. Thesis, Princeton University (1974).

20. C.A. Langhoff and G.W. Robinson, Chem. Phys. 6, 34 (1974).

21. M. Goeppert-Mayer and A.L. Sklar, J. Chem. Phys. 6, 645 (1938).

22. W.F. Radle and C.A. Beck, J. Chem. Phys. 8, 507 (1940).

23. H. Sponer, J. Chem. Phys. 8, 705 (1940).

24. R. Pariser, J. Chem. Phys. 24, 250 (1956).

25. J.R. Platt, K. Ruedenberg, C.W. Scherr, J.S. Ham, H. Labhart and W. Lichten, Free Electron Theory of Conjugated Molecules, A Source Book, Wiley, London and New York (1964).

26. E. Clementi, J. Chem. Phys. 46, 4737 (1967).

27. M. Hackmeyer and J.L. Whitten, J. Chem. Phys. 54, 3739 (1971).

28. W.R. Wadt and W. A. Goddard, III, to be published.

CORE EXCITATION AND ELECTRON CORRELATION IN CRYSTALS

Satoru Sugano

The Institute for Solid State Physics
The University of Tokyo
Roppongi, Minato-ku, Tokyo, Japan

ABSTRACT

The purpose of this article is to emphasize the importance of electron correlation in the excited states of crystals in spectroscopy using high energy photons in the range from 10 eV to 10^4 eV. Electrons in the inner cores of crystals are excited by photons of this energy range. The effects of electron correlation we are discussing are the change of excited-electron orbitals due to the presence of an inner-core hole, the formation of multiplets due to the correlated motion of the hole and the excited electron, and the effects of configuration interactions due to the breakdown of the orbital picture. At the beginning, we point out characteristic properties of core excitons by using a simple model of crystals. Then, citing two examples, vacuum-ultraviolet absorption spectra of alkali halides and X-ray photoelectron and K-emission spectra of transition-metal compounds, we clarify the importance of electron correlation in high energy spectroscopy.

I. FUNDAMENTALS OF CORE-EXCITON THEORY

I.A. Representations by Bloch and Wannier Functions

1. A Ground State of a Model Insulator. Let us denote a Bloch function with wave vector $\vec{k}$ and Wannier function localized around lattice point $\vec{R}$ for an electron in a valence band of a crystal by $u_{\vec{k}}$ and $a_{\vec{R}}$, respectively. They are related to each other as

$$u_{\vec{k}} = N^{-1/2} \sum_{\vec{R}} \exp(i\vec{k}\cdot\vec{R}) a_{\vec{R}}. \quad (1)$$

For simplicity we replace vectors $\vec{k}$ and $\vec{R}$ by scalars k and R assuming a linear lattice in what follows. This does not lose any generality. Then, we use abbreviation u_i for u_k with $k = k_i$ and a_j for a_R with $R = R_j (i,j = 1,2,\cdots,N)$.

If spins are ignored, the ground state of the crystal is given as

$$\Psi_0 = |u_1, u_2, \cdots, u_N| \quad (2)$$

where $|u_1, u_2, \cdots, u_N|$ is a Slater determinant including the normalization factor $(N!)^{-1/2}$. Inserting eq.(1) into eq.(2), one obtains

$$\Psi_0 = |a_1, a_2, \cdots, a_N| \times \det U, \quad (3)$$

where det U is the determinant of matrix U whose elements are given by

$$U_{pq} = N^{-1/2} \exp(ik_p R_q)\ ; \qquad (p,q = 1,2,\cdots,N). \quad (4)$$

It is easy to see that U is a unitary matrix, satisfying the relation $|\det U|^2 = 1$. Then, by choosing a phase so as to give $\det U = 1$, eq.(3) is reduced to

$$\Psi_0 = |a_1, a_2, \cdots, a_N|. \quad (5)$$

Eqs.(3) and (2) tell us that the Heitler-London method is identical to the Molecular-Orbital method as long as the state concerned has a closed shell configuration.

2. <u>States of One-Electron Excitation</u>. We consider the excitation of a valence electron to a non-degenerate conduction band, whose Bloch function is given by u'_k and Wannier function by a'_R. As previously, we ignore spins. We introduce a function corresponding to the excitation of an electron in the Bloch orbital with k_h to the Bloch orbital of the conduction band with k_e as follows;

$$\Phi(k_h, k_e) = |u_1, u_2, \cdots, u_{h-1}, u'_e, u_{h+1}, \cdots, u_N|. \quad (6)$$

Further we introduce another function corresponding to the excitation of an electron in the Wannier orbital localized at R_i to that of the conduction band localized at R_j as follows;

$$\Phi(R_i, R_j) = |a_1, a_2, \cdots, a_{i-1}, a_j', a_{i+1}, \cdots, a_N|. \tag{7}$$

Laplace-expanding $\Phi(k_h,k_e)$ in terms of column u_e' and $\Phi(R_i,R_j)$ in terms of column a_j', and using the relation obtained from eqs.(2) and (5),

$$|u_1, u_2, \cdots, u_{i-1}, u_{i+1}, \cdots, u_N|$$

$$= N^{-1/2} \sum_j (-1)^{i+j} \exp(-ik_iR_j)|a_1,a_2,\cdots,a_{j-1},a_{j+1},\cdots,a_N|, \tag{8}$$

one obtains the relation between $\Phi(k_h,k_e)$ and $\Phi(R_i,R_j)$ as

$$\Phi(k_h, k_e) = N^{-1} \sum_{i,j} \exp(ik_eR_j-ik_hR_i)\Phi(R_i, R_j). \tag{9'}$$

In the following arguments, it is convenient to introduce new indices, $k=k_e$, $K=k_e-k_h$ and $R=R_i$, $\beta=R_j-R_i$, and rewrite functions (6) and (7) as

$$\Phi(k_h, k_e) = \Phi(k-K, k),$$

$$\Phi(R_i, R_j) = \Phi(R, R+\beta).$$

Then, eq.(9') may be reexpressed as

$$\Phi(k-K, k) = N^{-1} \sum_{R,\beta} \exp\{i(k\beta + KR)\}\Phi(R, R+\beta). \tag{9}$$

Evidently function (7) cannot be an eigenfunction of the crystal, as it does not satisfy the symmetry requirement for a periodic lattice. A function satisfying this requirement is given by a linear combination of functions (7),

$$\Phi(K, \beta) = N^{-1/2} \sum_R e^{iKR}\Phi(R, R+\beta). \tag{10}$$

The use of eq.(9) shows that $\Phi(K,\beta)$ is also given by a linear combination of functions (6) as

$$\Phi(K, \beta) = N^{-1/2} \sum_k e^{-ik\beta} \phi(k-K, k). \tag{11}$$

Function $\Phi(K,\beta)$ is called an exciton representation.

I.B. Exciton Wavefunctions

1. Frenkel Excitons. In $\Phi(K,\beta)$ given by eqs.(10) and (11), the index K representing the wavevector associated with the motion of a composite system, an excited electron plus a hole, is a quantum number, but index β is not: β represents the spatial coordinate of the electron relative to the hole. Therefore, exciton wavefunctions with eigenvalues E_{Kp} $(p = 0,1,2,\cdots)$ are given by

$$\Psi_{Kp} = \sum_\beta U_K^p(\beta)\Phi(K\beta). \tag{12}$$

Index p is arranged so that $E_{K0} < E_{K1} < E_{K2} < \cdots$. If the excitation is localized within a lattice point for the lowest exciton state, i.e., if

$$U_K^0(\beta) = \delta_{\beta 0}, \tag{13}$$

the wavefunction of this state is given as

$$\Psi_{K0} = \Phi(K,0) = N^{-1/2} \sum_R e^{iKR} \phi(R, R). \tag{14}$$

The idealized exciton of this limit is called a Frenkel exciton. When $U_k^p(\beta)$ is non-vanishing in a finite range of β, it is easy to extend the picture of a Frenkel exciton.

2. Wannier Excitons. When the interaction between an excited electron and a hole is very weak as compared with the widths of the conduction and valence bands, $U_k^p(\beta)$ is a smooth function extending over a large range of β. In this limit, it has been shown (1) that $U_0^p(\beta)$ is a hydrogen-like wavefunction. A Schrödinger-type differential equation whose eigenfunction is $U_0^p(\beta)$ is called an effective-mass equation. The idealized exciton of this limit is called a Wannier exciton.

I.C. Core-Excitons

1. Energy Matrix in the Wannier-Function Representation. The total Hamiltonian of the system of interest is given as

$$\mathcal{H} = \sum_{i=1}^{N} h_i + \sum_{i>j=1}^{N} g_{ij}, \tag{15}$$

where h_i is a one-electron operator acting on the i-th electron involving operators of the kinetic energy and the periodic potential; g_{ij} is the Coulomb interaction between the i-th and j-th electrons.

The calculation of matrix elements of this Hamiltonian with bases (7) is straightforward. The result will be summarized below. We are using the following simplified notation:

$$\langle R, R+\beta|\mathcal{H}|R', R'+\beta'\rangle \equiv \int d\tau \Phi(R, R+\beta)^* \mathcal{H} \Phi(R', R'+\beta'),$$

$$\langle 0|\mathcal{H}|0\rangle \equiv \int d\tau \Psi_0^* \mathcal{H} \Psi_0,$$

$$\langle a_i a_j \| a_n a_m\rangle \equiv \int d\vec{r}_1 d\vec{r}_2 a_i(\vec{r}_1)^* a_j(\vec{r}_2)^* g_{12} a_n(\vec{r}_1) a_m(\vec{r}_2).$$

The diagonal elements are

(i) $\langle R,R+\beta|\mathcal{H}|R,R+\beta\rangle - \langle 0|\mathcal{H}|0\rangle$

$$= (\varepsilon'-\varepsilon) - [\langle a'_{R+\beta} a_R \| a'_{R+\beta} a_R\rangle - \langle a'_{R+\beta} a_R \| a_R a'_{R+\beta}\rangle]$$

$$\equiv E(\beta), \tag{16}$$

where ε' is the Hartree-Fock energy of the excited electron in the (N+1)-electron system, (-ε) that of the hole in the (N-1)-electron system, and -[···] is the energy of the electron-hole interaction. Elements (16) depend upon β only. The nondiagonal elements are classified into three types. The first one is

(ii) $\langle R, R+\beta|\mathcal{H}|R, R+\beta'\rangle \qquad (\beta\neq\beta')$

$$= t'_{\beta\beta'} - [\langle a'_{R+\beta} a_R \| a'_{R+\beta'} a_R\rangle - \langle a'_{R+\beta} a_R \| a_R a'_{R+\beta'}\rangle]$$

$$\equiv t'(\beta,\beta') \tag{17}$$

where $t'_{\beta\beta'}$ is the matrix of the electron transfer in the excited state in the periodic potentials of cores and N valence electrons, and -[···] is that of the excited-electron transfer due to the hole. Elements (17) depend upon β and β'. The second-type nondiagonal elements are

$$(\text{iii})\quad \langle R, R+\beta|\mathcal{H}|R', R+\beta\rangle \qquad (R\neq R')$$

$$= -t_{RR'} - [\langle a'_{R+\beta}a_{R'}\|a'_{R+\beta}a_R\rangle - \langle a'_{R+\beta}a_{R'}\|a_R a'_{R+\beta}\rangle]$$

$$\equiv t(R-R',\beta), \tag{18}$$

where $-t_{RR'}$ is the matrix of the hole transfer in the valence state in the periodic potentials of cores and N valence electrons, and $-[\cdots]$ is that of the hole transfer due to the excited electron. Elements (18) depend upon R-R' and β. The third-type nondiagonal elements are

$$(\text{iv})\quad \langle R, R+\beta|\mathcal{H}|R', R'+\beta'\rangle \qquad (R\neq R',\ R+\beta\neq R'+\beta')$$

$$= -\langle a_{R'}a'_{R+\beta}\|a_R a'_{R'+\beta'}\rangle + \langle a_{R'}a'_{R+\beta}\|a'_{R'+\beta'}a_R\rangle, \tag{19}$$

where the first term represents the transfer of excitation $\Phi(R',R'+\beta')$ to $\Phi(R,R+\beta)$ by transferring the excited electron at $R'+\beta'$ in $\Phi(R',R'+\beta')$ to R+β and the valence electron at R to R'. This excitation transfer is the exchange transfer which is important in magnetic insulators (2). The second term in eq. (19) represents the excitation transfer by recomining the excited electron at R'+β' and the hole at R', at the same time creating the excited electron at R+β and the hole at R. This transfer is caused by the dipolar interaction and is most popular in molecular crystals. The nondiagonal elements in eq. (19) are not of primary importance in Wannier excitons.

2. Approximations in Core-Excitons. The most characteristic property of core-excitons is that a hole in inner-cores hardly moves from site to site, i.e., the nondiagonal elements in eq. (18) are negligibly small. If the excitation transfer in eq. (19) is neglected for a moment in addition to the hole transfer in eq. (18), the energy matrix of a core-exciton is reduced to a simple form as described on the next page;

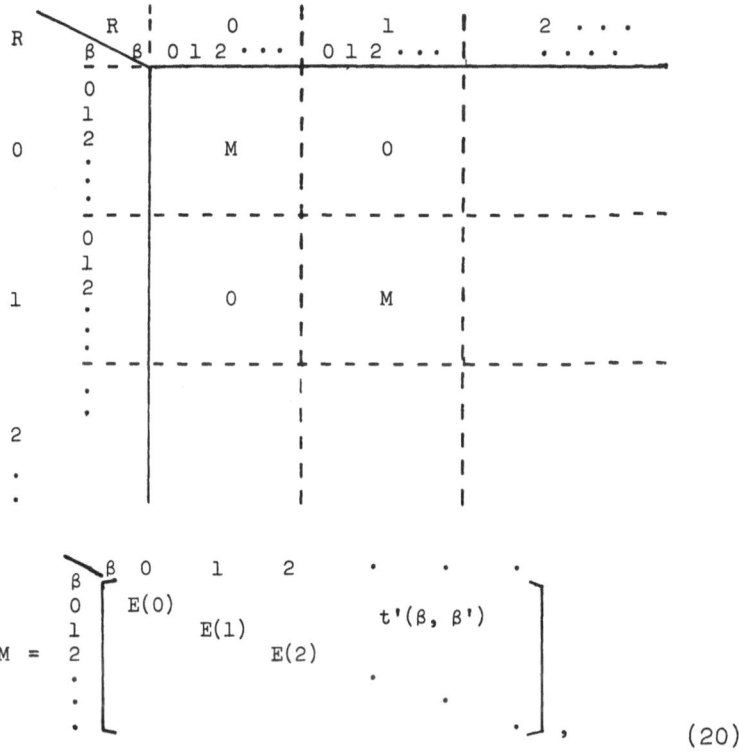

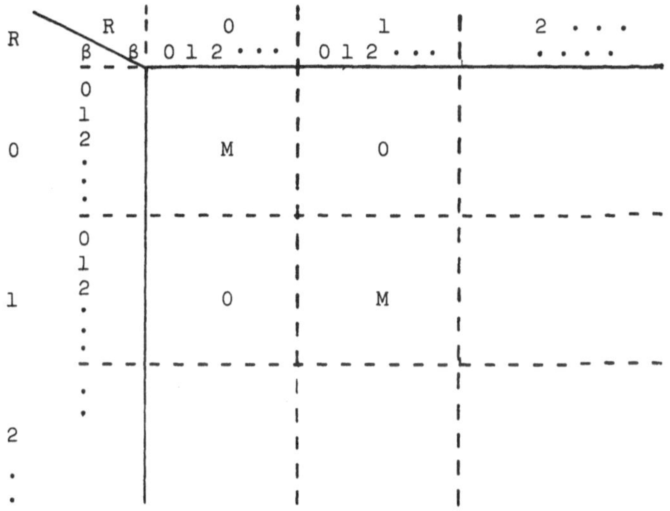

$$M = \begin{array}{c|cccccc} \beta \backslash \beta' & 0 & 1 & 2 & \cdot & \cdot & \cdot \\ \hline 0 & E(0) & & & & t'(\beta,\beta') & \\ 1 & & E(1) & & & & \\ 2 & & & E(2) & & & \\ \cdot & & & & \cdot & & \\ \cdot & & & & & \cdot & \\ \cdot & & & & & & \cdot \end{array}, \qquad (20)$$

where hermitian matrix M is independent of R. The energy matrix in the exciton representation to this approximation is obtained as

$$\langle K, \beta|\mathcal{H}|K', \beta'\rangle = \langle R, R+\beta|\mathcal{H}|R, R+\beta'\rangle\delta_{KK'}. \qquad (21)$$

Therefore, $U_K^p(\beta)$, being independent of K to our approximation, is determined so as to diagonalize matrix M. This procedure is equivalent to that for the localized state of an impurity center in crystals.

So far we have neglected without any argument the excitation transfer given by eq. (19) whose magnitude is at most of the order of 0.1eV as shown in section II.A.: the largest contribution comes from the matrix elements with $\beta = \beta' = 0$. In our problem of core-excitons, where the orbital of the excited electron is also localized due to a strong attractive force of the localized hole as will be discussed later, the largest part of the excitation transfer can be taken into account after solving the localized impurity problems.

I.D. Hole-Electron Interactions

1. The Excited-Electron Orbital. Since a hole in an inner-core is localized at a lattice site and its orbital is almost entirely inside the orbitals of outer electrons, the hole exerts a strong attractive force on the excited electron at the same site. It should be noted that the dielectric screening of the Coulomb force due to polarization of valence electrons does not work for the excited electron at the same site.

This strong force may induce a fairly big change in the atomic part of the excited-electron orbital, consequently a change in the excited-electron distribution as a whole. This change is one of the most important characteristics of core-excitons, and observed most clearly as shake-up satellites in the X-ray photoelectron spectra. This point will be discussed in detail later. The change can also be seen in examining the values of the Coulomb interaction parameters determined so as to fit the calculated multiplet structure of core-excitation with various experimental data. This point will also be discussed in detail in later sections. The change of the excited-electron orbital due to the hole is sometimes called an electronic relaxation or simply relaxation effect.

Since the force exerted by an inner-core hole predominantly affects the excited electron at the same site, the change of the excited-electron orbital may well be represented by mixing Wannier orbitals of higher energies at the same site. On the other hand, if Bloch functions are used to represent the change, one has to mix a larger number of Bloch functions of higher energies: a localized orbital in an energy band is given by a linear combination of Bloch functions with all the possible wavevectors in the Brillouin zone as shown in eq.(1). Simply speaking, if one uses the band theory, a conventional two-band approximation as discussed in subsection I.A. is not valid for core-excitons. One has to take into account many bands, instead. Therefore, the use of the Wannier function representation is suitable for treating core-excitons.

In the force exerted by the hole on the excited electron, a spherically symmetric part is predominant. This part of the force does not change the site-symmetry of the excited-electron orbital. Therefore, one may specify the relaxed orbital of the excited electron by using the same irreducible representation of the site symmetry as that for the unrelaxed one which is known in many cases. Thus, one may proceed a little further without knowing details of the relaxed orbital.

2. Complementary States. In many cases an exciton in an insulator is treated as if it is a system of two particles, an

electron and a hole. This treatment cannot be justified when the hole-electron interaction is taken into account more accurately. The concept of a hole can be introduced as the state complementary to the one where an electron is missing from a closed-shell configuration. It is well known that a simple relation holds between the matrix elements of an operator in the complementary states. For example the operator is a spin-independent, hermitian and real one-electron operator like a crystalline field potential, the matrix elements in the complementary states differ only in sign except a constant term in the diagonal element.

When two shells, γ_1 and γ_2, are dealt with as in the exciton theory, the state complementary to the $\gamma_1^{2[\gamma_1]-1}\gamma_2 S\Gamma$ state is the $\gamma_1\gamma_2^{2[\gamma_2]-1}S\Gamma$ but not the $\gamma_1\gamma_2 S\Gamma$ (3). Here $[\gamma_i]$ $(i = 1,2)$ is the orbital degeneracy of γ_i, S the resultant spin, and Γ specifies the symmetry of the orbital part of a many-electron state. In general there is no simple relation between the matrix elements in the $\gamma_1^{2[\gamma_1]-1}\gamma_2 S\Gamma$ and $\gamma_1\gamma_2 S\Gamma$ states. One may, however, calculate the matrix element of the Coulomb interaction operator G in the $\gamma_1^{2[\gamma_1]-1}\gamma_2 S\Gamma$ state from the $\gamma_1\gamma_2 S\Gamma$ state as follows (4).

First we confine ourselves to the system consisting of two electron-shells, γ_1 and γ_2. For this system the Coulomb interaction operator can be expressed in the form,

$$G = G_1 + G_2 + G_3 + \cdots , \tag{22}$$

where

$$\begin{aligned}
G_1 &= \sum_{S\Gamma M_S M_\Gamma} \langle\gamma_1^2 S\Gamma|g|\gamma_1^2 S\Gamma\rangle A^\dagger(\gamma_1^2 S\Gamma M_S M_\Gamma)A(\gamma_1^2 S\Gamma M_S M_\Gamma),\\
G_2 &= \sum_{S\Gamma M_S M_\Gamma} \langle\gamma_1\gamma_2 S\Gamma|g|\gamma_1\gamma_2 S\Gamma\rangle A^\dagger(\gamma_1\gamma_2 S\Gamma M_S M_\Gamma)A(\gamma_1\gamma_2 S\Gamma M_S M_\Gamma),\\
G_3 &= \sum_{S\Gamma M_S M_\Gamma} \langle\gamma_2^2 S\Gamma|g|\gamma_2^2 S\Gamma\rangle A^\dagger(\gamma_2^2 S\Gamma M_S M_\Gamma)A(\gamma_2^2 S\Gamma M_S M_\Gamma),\\
&\cdots\cdots\cdots
\end{aligned} \tag{23}$$

In eq.(23), $\langle\gamma_1^2 S\Gamma|g|\gamma_1^2 S\Gamma\rangle$ is the diagonal matrix elements of the Coulomb interaction operator in the $\gamma_1^2 S\Gamma$ state, etc., and $A^\dagger(\gamma_1\gamma_2 S\Gamma M_S M_\Gamma)$ the two-electron creation operator defined as

$$A^{\dagger}(\gamma_1\gamma_2 S\Gamma M_S M_\Gamma) = \sum_{\substack{m_1 m_2 \\ \mu_1\mu_2}} \langle sm_1 sm_2|SM\rangle\langle\gamma_1\mu_1\gamma_2\mu_2|\Gamma M_\Gamma\rangle$$

$$\times\, a^{\dagger}_{s\gamma_1 m_1\mu_1} a^{\dagger}_{s\gamma_2 m_2\mu_2}, \quad \text{etc..} \qquad (24)$$

Here, $a^{\dagger}_{s\gamma m\mu}$ is the one-electron creation operator which creates an electron in the $s\gamma m\mu$ spin orbital when operated on a state vector: m and μ are the degenerate components of s and γ, respectively. In eq.(24) $\langle sm_1 sm_2|SM_S\rangle$ are the Wigner coefficients for spins ($s = 1/2$) (5), and $\langle\gamma_1\mu_1\gamma_2\mu_2|\Gamma M_\Gamma\rangle$ the Clebsch-Gordan coefficients for orbitals. It is evident that, when $A^{\dagger}(\gamma_1\gamma_2 S\Gamma M_S M_\Gamma)$ is operated on the vacuum state $|0\rangle$, it gives the two-electron eigenstate $|\gamma_1\gamma_2 S\Gamma M_S M_\Gamma\rangle$, and so on. The two-electron annihilation operator $A(\gamma_1\gamma_2 S\Gamma M_S M_\Gamma)$ is adjoint to $A^{\dagger}(\gamma_1\gamma_2 S\Gamma M_S M_\Gamma)$, etc.. In eq.(22) terms irrelevant to our problem such as those involving $\langle\gamma_1{}^2 S\Gamma|g|\gamma_1\gamma_2 S\Gamma\rangle$, etc., are not explicitly given.

Now we introduce one-hole creation operator $b^{\dagger}_{s\gamma m\mu}$ which creates a hole in the $s\gamma{-m}\mu$ spin orbital for an electron when operated on a state vector. Here all the orbital functions denoted by $\gamma\mu$ are assumed to be real without any loss of generality. Operator $b^{\dagger}_{s\gamma m\mu}$ is related to $a_{s\gamma-m\mu}$ as

$$b^{\dagger}_{s\gamma m\mu} = (-1)^{s+m} a_{s\gamma-m\mu}. \qquad (25a)$$

Similarly,

$$b_{s\gamma m\mu} = (-1)^{s+m} a^{\dagger}_{s\gamma-m\mu}. \qquad (25b)$$

We replace one-electron operators for the γ_1 electron in eq. (22) by hole operators using relations in eq. (25), leaving electron operators for the γ_2 electron unreplaced. By using the anti-commutation relation for the creation and annihilation operators, G_1 and G_2 in eq. (22) are finally reexpressed as follows:

$$G_1 = (1-\tilde{n}_{\gamma_1}/[\gamma_1])(\sum_{S\Gamma}[S][\Gamma]\langle\gamma_1{}^2 S\Gamma|g|\gamma_1{}^2 S\Gamma\rangle) + \tilde{G}_1,$$

$$G_2 = (n_{\gamma_2}/2[\gamma_2])(\sum_{S\Gamma}[S][\Gamma]\langle\gamma_1\gamma_2 S\Gamma|g|\gamma_1\gamma_2 S\Gamma\rangle) + \tilde{G}_2, \qquad (26)$$

where

$$\tilde{G}_1 = \sum_{S\Gamma M_S M_\Gamma} \langle\gamma_1{}^2 S\Gamma|g|\gamma_1{}^2 S\Gamma\rangle A^\dagger(\tilde{\gamma}_1{}^2 S\Gamma M_S M_\Gamma)A(\tilde{\gamma}_1{}^2 S\Gamma M_S M_\Gamma),$$

$$\tilde{G}_2 = -\sum_{\substack{S\Gamma S'\Gamma' \\ M_S' M_\Gamma'}} \langle\gamma_1\gamma_2 S\Gamma|g|\gamma_1\gamma_2 S\Gamma\rangle[S][\Gamma]$$

$$\times \begin{Bmatrix} s & s & S' \\ s & s & S \end{Bmatrix} \begin{Bmatrix} \gamma_1 & \gamma_2 & \Gamma' \\ \gamma_1 & \gamma_2 & \Gamma \end{Bmatrix}$$

$$\times \quad A^\dagger(\tilde{\gamma}_1\gamma_2 A'\Gamma' M_S' M_\Gamma')A(\tilde{\gamma}_1\gamma_2 S'\Gamma' M_S' M_\Gamma'). \qquad (27)$$

In eq. (26) $\tilde{n}_\gamma$ and n_γ are the number operators for the γ holes and the γ electrons, respectively, and $[S] = 2S+1$. In eq. (27) $A^\dagger(\tilde{\gamma}_1{}^2 S\Gamma M_S M_\Gamma)$ and $A^\dagger(\tilde{\gamma}_1\gamma_2 S\Gamma M_S M_\Gamma)$ are the operators which are obtained by replacing $a^\dagger_{s\gamma_1 m_1\mu_1}$ in $A^\dagger(\gamma_1{}^2 S\Gamma M_S M_\Gamma)$ and $A^\dagger(\gamma_1\gamma_2 S\Gamma M_S M_\Gamma)$ by $b^\dagger_{s\gamma_1 m_1\mu_1}$, leaving $a^\dagger_{s\gamma_2 m_2\mu_2}$ as they were. When $A^\dagger(\tilde{\gamma}_1{}^2 S\Gamma M_S M_\Gamma)$ and $A^\dagger(\tilde{\gamma}_1\gamma_2 S\Gamma M_S M_\Gamma)$ are operated on the γ_1-filled but the γ_2-empty state, one obtains states $|\gamma_1{}^{2[\gamma_1]-2}\gamma_2 S\Gamma M_S M_\Gamma\rangle \equiv |\tilde{\gamma}_1{}^2 S\Gamma M_S M_\Gamma\rangle$ and $|\gamma_1{}^{2[\gamma_1]-1}\gamma_2 S\Gamma M_S M_\Gamma\rangle \equiv |\tilde{\gamma}_1\gamma_2 S\Gamma M_S M_\Gamma\rangle$, respectively. The 6-j symbols (6) in eq. (27) are defined as

$$\begin{Bmatrix} s & s & S' \\ s & s & S \end{Bmatrix} = [S]^{-1} \sum_{\substack{m_1 m_2 m_3 \\ m_1' M_S}} (-1)^{2s-(m_1+m_1')}$$

$$\times \langle s m_1 s m_2|S M_S\rangle\langle S M_S|s m_1' s m_3\rangle$$

$$\times \langle s{-m_2}\, s m_3|S' M'\rangle\langle S' M_S'|s{-m_1'}\, s m_2\rangle,$$

$$\left\{\begin{matrix}\gamma_1 & \gamma_3 & \Gamma' \\ \gamma_1 & \gamma_2 & \Gamma\end{matrix}\right\} = [\Gamma]^{-1} \sum_{\substack{\mu_1\mu_2\mu_3 \\ \mu_1'M_\Gamma}} \langle\gamma_1\mu_1\gamma_2\mu_2|\Gamma M_\Gamma\rangle\langle\Gamma M_\Gamma|\gamma_1\mu_1'\gamma_3\mu_3\rangle$$

$$\times \langle\gamma_1\mu_1\gamma_3\mu_3|\Gamma' M_\Gamma'\rangle\langle\Gamma' M_\Gamma'|\gamma_1\mu_1'\gamma_2\mu_2\rangle, \qquad (28)$$

which are independent of M_S' and M_Γ', respectively.

The use of G_1 and G_2 in eq. (26) for calculating the Coulomb matrix elements in the $\tilde{\gamma}_1\gamma_2$ states enables us to express them in terms of those in the $\gamma_1\gamma_2$ states except a common diagonal term which is given in terms of those in the $\gamma_1{}^2$ states: the matrix elements of $\tilde{G}_1$ in eq. (26) is vanishing and the first term in G_1 in eq. (26) gives the common diagonal term. Term G_3 in eq. (23) gives no contribution to the matrix elements in the $\tilde{\gamma}_1\gamma_2$ states.

In the problems of core-excitons, it becomes sometimes necessary to calculate configuration interactions in the excited states, i.e., non-diagonal matrix elements of G between the $\tilde{\gamma}_1\gamma_2 S\Gamma$ state and the $\tilde{\gamma}_1\gamma_3 S\Gamma$ state, where γ_3 is neither γ_1 nor γ_2. A similar procedure gives an additional term, G_4, in G, which is given by

$$G_4 = -\sum_{\substack{S\Gamma \\ S'\Gamma' M_S' M_\Gamma'}} \langle\gamma_1\gamma_2 S\Gamma|g|\gamma_1\gamma_3 S\Gamma\rangle[S][\Gamma]$$

$$\times \left\{\begin{matrix}s & s & S' \\ s & s & S\end{matrix}\right\}\left\{\begin{matrix}\gamma_1 & \gamma_3 & \Gamma' \\ \gamma_1 & \gamma_2 & \Gamma\end{matrix}\right\}$$

$$\times\ A^\dagger(\tilde{\gamma}_1\gamma_2 S'\Gamma' M_S' M_\Gamma')A(\tilde{\gamma}_1\gamma_3 S'\Gamma' M_S' M_\Gamma')$$

$$+\ (\text{hermitian adjoint of the first term}). \qquad (29)$$

3. <u>Multiplets</u>. As already emphasized, the Coulomb interaction between an inner-core hole and an excited electron is quite strong so that this interaction should be treated as accurately as possible. In the conventional Hartree-Fock approximation treating the Coulomb interaction between particles, one replaces the field

acting on a particle due to the other particles by its time-average neglecting the correlation of motion of particles. In our problem this approximation gives a spherically symmetric field of a hole, even if it is a p-hole, acting on an excited electron. In this case, when an orbital of the excited electron is given, the exciton energy is completely degenerate yielding always one degenerate eigenstate of the exciton. This is the case in the effective-mass theory for Wannier excitons: many non-degenerate hydrogenic eigenstates of the Wannier exciton correspond to many excited-electron orbitals but not to one. If one takes into account the correlated motion of the electron and the hole (but neglects the exciton migration), one may generally obtain two-particle eigenstates with different energies even if a specific orbital is assumed for the excited electron. These eigenstates are called multiplet terms or simply multiplets.

The theory of multiplets is well developed for atoms and localized centers in crystals. The latter is called the ligand field theory. The method of calculating multiplets of localized excitons has already been described in subsection I.D.-2. As mentioned in I.D.-1, the orbital of the excited electron is changed or relaxed by introducing an inner-core hole, nevertheless, its symmetry can be specified without knowing its detailed form. Let us specify the symmetry of the excited-electron orbital by γ_2 and that of the hole by γ_1, where γ_1 and γ_2 are irreducible representation of the site-symmetry group. Using the one-electron one-hole creation operators introduced in I.D.-2, one obtains the exciton eigenstates (without migration) as follows:

$$|\tilde{\gamma}_1\gamma_2 S\Gamma M_S M_\Gamma\rangle = A^\dagger(\tilde{\gamma}_1\gamma_2 S\Gamma M_S M_\Gamma)|\gamma_1\text{-filled}, \quad \gamma_2\text{-empty}\rangle. \qquad (30)$$

According to the method of calculating term energies in I.D.-2, the Coulomb matrix elements in the $\tilde{\gamma}_1\gamma_2$ states are given in terms of those in the two-electron states $\gamma_1\gamma_2$, except a common diagonal term giving a shift of multiplets as a whole. In general, matrix elements in many-electron states are expressed in terms of two-electron Coulomb and exchange integrals which are independent of each other. These integrals are often left as adjustable parameters to fit with experiments. Examples of such calculations will be given in later sections.

If energies of different orbitals of the excited electrons are close, mixing of the states with different electron configurations becomes important. The method of calculating the configuration mixing has been given in I.D.-2. The importance of the configuration mixing will be demonstrated in the example treated in section II.

I.E. Spin-Orbit Interactions

In the cases where an inner-core hole has the p-character, the spin-orbit interaction of the hole is quite large. The relation between the spin-orbit coupling constant of the np electron and the effective nuclear charge of Slater in the atoms having electron closed-shells plus an np hole is shown in Fig.1 (4). As seen in the figure, the spin-orbit coupling constant of the inner-core 2p electron is more than 10 eV for the effective nuclear charge larger than 25: it is 11 eV for Ni. Thus, it is important to take the spin-orbit interaction of the inner-core hole into account. An example of the calculation of the spin-orbit interaction will be given in section III.D.-3 for the cases where more than one valence-hole is present besides an inner-core hole.

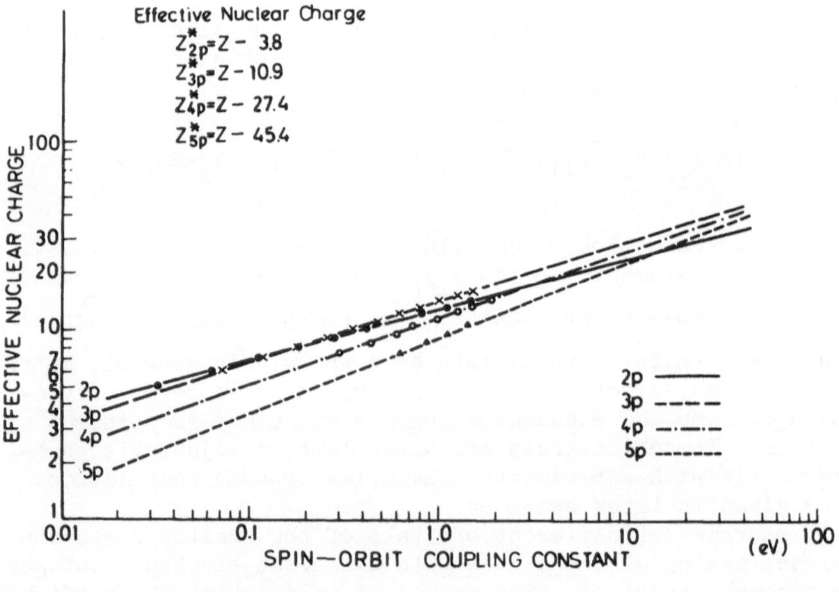

Fig.1. The spin-orbit coupling constant of the np-electron versus Slater's effective nuclear charge in the atoms having electron closed-shells plus an np-hole.

II. VACUUM-ULTRAVIOLET ABSORPTION OF ALKALI HALIDES

II.A. Application of the Core-Exciton Theory

1. The Starting Model. We confine ourselves to the fine structure of absorption spectra around 20 eV in MX alkali halide crystals (M=Cs, Rb, K, Na; X=I, Br, Cl, F). In view of such excitation energies, the fine structure is considered to be related in some way to the excitations $np^6 \rightarrow np^5(n+1)s$ and $np^5n'd$ of alkali ions (n'=n+1 for M=Na; n'=n for M=Cs, Rb, K). It is, however, impossible by assuming purely atomic excitations to explain the observed details such as the chemical shifts, the number, and the relative positions of the fine structure peaks: in RbX five peaks called A through E are observed, and peaks A and C chemically shift towards higher energy upon replacement F → Cl → Br → I while peaks B, D, and E towards lower energy.

As mentioned in section I.C., we start with the assumption that a hole left in the optical excitation is localized in the np shell of an alkali ion. Then, we assume that the relaxed orbitals of the excited electron are those specified by irreducible representations of the cubic site symmetry group, a_{1g}, t_{2g} and e_g which may be obtained by relaxing atomic orbitals (n+1)s and n'd of the alkali ion in a cubic field. As discussed in section I.D., these orbitals are localized, but neither given by linear combinations of Bloch functions associated with a conduction band (see later discussions in II.C.-3) nor by the atomic orbitals of the alkali ion. These orbitals may differ from the corresponding atomic orbitals in both their radial and angular dependence. However, to reduce the number of parameters appearing in the theory, we assume that the angular dependence of the orbitals are the same as those of the corresponding atomic orbitals: for example, the t_{2g} and e_g orbitals are assumed to have the d-character only. We further assume that the radial parts of the t_{2g} and e_g orbitals are the same. These approximations are known to be valid to a good extent in the ligand field theory for transition-metal ions in crystals. To these approximations the one-electron or one-hole orbitals of interest are given as follows.

For the np hole, the orbital function is assumed to be identical to the atomic orbital,

$$|t_{1u}\alpha\rangle = -(1/\sqrt{2})R_p(r)[Y_{11}(\theta\phi) - Y_{1-1}(\theta\phi)],$$

$$|t_{1u}\beta\rangle = (i/\sqrt{2})R_p(r)[Y_{11}(\theta\phi) + Y_{1-1}(\theta\phi)],$$

$$|t_{1u}\gamma\rangle = R_p(r)Y_{10}(\theta\phi), \quad (31)$$

where $Y_{\ell m}(\theta\phi)$'s are the spherical harmonics and $R_p(r)$ the radial function of the np atomic orbital of an alkali ion, M^+. For the relaxed a_{1g}, t_{2g}, and e_g orbitals of the excited electron, the orbital functions are assumed to be

$$|a_{1g}e_1\rangle = R_s(r)Y_{00}(\theta\phi), \quad (32)$$

$$|t_{2g}\xi\rangle = (i/\sqrt{2})R_d(r)[Y_{21}(\theta\phi) + Y_{2-1}(\theta\phi)],$$
$$|t_{2g}\eta\rangle = -(1/\sqrt{2})R_d(r)[Y_{21}(\theta\phi) - Y_{2-1}(\theta\phi)],$$
$$|t_{2g}\zeta\rangle = -(i/\sqrt{2})R_d(r)[Y_{22}(\theta\phi) - Y_{2-2}(\theta\phi)], \quad (33)$$

$$|e_g u\rangle = R_d(r)Y_{20}(\theta\phi),$$
$$|e_g v\rangle = (1/\sqrt{2})R_d(r)[Y_{22}(\theta\phi) + Y_{2-2}(\theta\phi)], \quad (34)$$

where radial functions $R_\mu(r)$ ($\mu = s,d$) generally differ from those of the (n+1)s and n'd atomic orbitals. In eq. (31), α, β, and γ are the components of t_{1u} transforming in the same way as x, y, and z, respectively, ξ, η, and ζ in eq. (33) the components of t_{2g} transforming as yz, zx, and xy, respectively, and u and v in eq. (34) those of e_g transforming as $3z^2-r^2$ and x^2-y^2, respectively.

2. Calculation of Exciton Multiplets. Now, we calculate energies of the multiplet terms arising from configurations $\tilde{t}_{1u}a_{1g}$ ($\equiv t_{1u}^5 a_{1g}$), $\tilde{t}_{1u}t_{2g}$, and $\tilde{t}_{1u}e_g$, to which optical transitions from the $t_{1u}^6\,{}^1A_{1g}$ ground state are allowed. It is evident that these terms including spins should belong to the single-valued irreducible representation $\mathcal{T}_{1u}$ of the cubic double-group. There are four $\mathcal{T}_{1u}$ terms arising from $\tilde{t}_{1u}t_{2g}\,{}^3E_u$, ${}^1T_{1u}$, ${}^3T_{1u}$, and ${}^3T_{2u}$ terms, three $\mathcal{T}_{1u}$ from the $\tilde{t}_{1u}e_g\,{}^1T_{1u}$, ${}^3T_{1u}$, and ${}^3T_{2u}$, and two $\mathcal{T}_{1u}$ from the $\tilde{t}_{1u}a_{1g}\,{}^1T_{1u}$ and ${}^3T_{1u}$. In our problem, the magnitudes of the hole-electron interaction, the spin-orbit interaction and the cubic field splitting are expected to be of the same order, so that we have to diagonalize a nine-dimensional energy matrix of $\mathcal{T}_{1u}$.

Matrix elements of the hole-electron interaction are calculated by using the method given in section I.D. with parent states $\tilde{t}_{1u}\gamma S\Gamma$ as the bases ($\gamma = a_{1g}$, t_{2g}, e_g). We present here an example of calculating a diagonal element of G in the $\tilde{t}_{1u}e_g\,{}^1T_{1u}$ state. Noting that the value of the matrix elements of $\tilde{n}_{t_{1u}}$ in this state is unity, $[t_{1u}] = 3$, and

$$\langle t_{1u}^2\ {}^3T_{1g}|g|t_{1u}^2\ {}^3T_{1g}\rangle = F_0(p,p) - 5F_2(p,p),$$

$$\langle t_{1u}^2\ {}^1T_{2g}|g|t_{1u}^2\ {}^1T_{2g}\rangle = \langle t_{1u}^2\ {}^1E_g|g|t_{1u}^2\ {}^1E_g\rangle$$

$$= F_0(p,p) + F_2(p,p),$$

$$\langle t_{1u}^2\ {}^1A_{1g}|g|t_{1u}^2\ {}^1A_{1g}\rangle = F_0(p,p) + 10F_2(p,p), \quad (35)$$

one obtains from the first term of G_1 in eq.(26)

$$\langle \tilde{t}_{1u}e_g\,{}^1T_{1u}|G_1|\tilde{t}_{1u}e_g\,{}^1T_{1u}\rangle = 10F_0(p,p) - 20F_2(p,p). \quad (36)$$

Similarly, the matrix element of the first term of G_2 in eq.(26) is calculated to be

$$\langle \tilde{t}_{1u}e_g\,{}^1T_{1u}|G_2-\tilde{G}_2|\tilde{t}_{1u}e_g\,{}^1T_{1u}\rangle$$

$$= 6F_0(p,d) - 6G_1(p,d) - 63G_3(p,d). \quad (37)$$

In eqs.(35)-(37), the Slater-Condon parameters (5) $F_k(\mu,\nu)$ and $G_k(\mu,\nu)$ are the Coulomb and exchange integrals between the μ and ν electrons, respectively. The matrix element of $\tilde{G}_2$ is calculated by using eq.(27) as

$$\langle \tilde{t}_{1u}e_g\,{}^1T_{1u}|\tilde{G}_2|\tilde{t}_{1u}e_g\,{}^1T_{1u}\rangle$$

$$= -3\varepsilon({}^1T_{1u})\begin{Bmatrix} 1/2 & 1/2 & 0 \\ 1/2 & 1/2 & 0 \end{Bmatrix}\begin{Bmatrix} t_{1u} & e_g & T_{1u} \\ t_{1u} & e_g & T_{1u} \end{Bmatrix}$$

$$-9\varepsilon({}^3T_{1u})\begin{Bmatrix} 1/2 & 1/2 & 0 \\ 1/2 & 1/2 & 1 \end{Bmatrix}\begin{Bmatrix} t_{1u} & e_g & T_{1u} \\ t_{1u} & e_g & T_{1u} \end{Bmatrix}$$

$$-9\varepsilon({}^3T_{2u})\begin{Bmatrix} 1/2 & 1/2 & 0 \\ 1/2 & 1/2 & 1 \end{Bmatrix}\begin{Bmatrix} t_{1u} & e_g & T_{1u} \\ t_{1u} & e_g & T_{2u} \end{Bmatrix}, \quad (38)$$

where

$$\varepsilon(^1T_{1u}) \equiv \langle t_{1u}e_g\ ^1T_{1u}|g|t_{1u}e_g\,^1T_{1u}\rangle$$

$$= F_0(p,d) + 4F_2(p,d) + 4G_1(p,d) + 27G_3(p,d),$$

$$\varepsilon(^3T_{1u}) \equiv \langle t_{1u}e_g\ ^3T_{1u}|g|t_{1u}e_g\,^3T_{1u}\rangle$$

$$= F_0(p,d) + 4F_2(p,d) - 4G_1(p,d) - 27G_3(p,d),$$

$$\varepsilon(^3T_{2u}) \equiv \langle t_{1u}e_g\ ^3T_{2u}|g|t_{1u}e_g\,^3T_{2u}\rangle$$

$$= F_0(p,d) - 4F_2(p,d) - 15G_3(p,d). \qquad (39)$$

Using the following values of the 6-j symbols,

$$\begin{Bmatrix} 1/2 & 1/2 & 0 \\ 1/2 & 1/2 & 0 \end{Bmatrix} = -1/2, \qquad \begin{Bmatrix} 1/2 & 1/2 & 0 \\ 1/2 & 1/2 & 1 \end{Bmatrix} = 1/2,$$

$$\begin{Bmatrix} t_{1u} & e_g & T_{1u} \\ t_{1u} & e_g & T_{1u} \end{Bmatrix} = 1/3, \qquad \begin{Bmatrix} t_{1u} & e_g & T_{1u} \\ t_{1u} & e_g & T_{2u} \end{Bmatrix} = 0, \qquad (40)$$

one obtains

$$\langle \tilde{t}_{1u}e_g\,^1T_{1u}|\tilde{G}_2|t_{1u}e_g\,^1T_{1u}\rangle$$

$$= -F_0(p,d) - 4F_2(p,d) + 8G_1(p,d) + 54G_3(p,d). \quad (41)$$

Summing up terms in eqs. (36),(37), and (41), one finally obtains

$$\langle \tilde{t}_{1u}e_g\,^1T_{1u}|G|\tilde{t}_{1u}e_g\,^1T_{1u}\rangle$$

$$= 10F_0(p,p) - 20F_2(p,p) + 5F_0(p,d) - 4F_2(p,d)$$

$$+ 2G_1(p,d) - 9G_3(p,d). \quad (42)$$

Matrix elements of the spin-orbit interaction are calculated by using states $|\tilde{t}_{1u}\gamma S\Gamma M_S M_\Gamma\rangle$ as the bases, whose appropriate linear combinations give the $\mathcal{T}_{1u}$ terms. Coefficients of these combinations are the Clebsch-Gordan coefficients for the cubic group, which have already been tabulated (3). The results are given in terms of the spin-orbit interaction constants ζ_p, ζ and ζ' defined as

$$\langle t_{1u}\alpha\ m_s = 1/2|\mathcal{H}_{so}|t_{1u}\beta\ 1/2\rangle = -i\zeta_p/2,$$

$$\langle t_{2g}\xi\ 1/2|\mathcal{H}_{so}|t_{2g}\eta\ 1/2\rangle = i\zeta/2,$$

$$\langle t_{2g}\zeta\ 1/2|\mathcal{H}_{so}|e_g v\ 1/2\rangle = i\zeta'. \tag{43}$$

The difference between the e_g and t_{2g} orbital energies is given by the cubic-field splitting parameter 10 Dq which is positive when the e_g is higher. The crystalline field energies appear only in the diagonal elements for our bases $t_{1u}\gamma S\Gamma$: they are -4 Dq for $\gamma = t_{2g}$, 6 Dq for $\gamma = e_g$, and zero for $\gamma = a_{1g}$.

All the matrix elements thus calculated are given in Tables I and II. Slater-Condon parameters $R^1(pd,sp)$ and $R^2(pd,ps)$ are the nondiagonal exchange and Coulomb integrals, respectively, responsible for the mixing of the terms of the same $S\Gamma$ but of different configurations $\tilde{t}_{1u}\gamma$.

Table I. Diagonal elements of the energy matrix for the $\mathcal{T}_{1u}$ terms

Base number	Parent terms		Matrix elements
1	$\tilde{t}_{1u}t_{2g}$	^3E_u	$E_d{}^{\dagger}-4Dq-F_0(p,d)+7F_2(p,d)$
2		$^1T_{1u}$	$E_d-4Dq-F_0(p,d)-5F_2(p,d)+12G_1(p,d)+36G_3(p,d)$
3		$^3T_{1u}$	$E_d-4Dq-(\zeta_p+\zeta)/4-F_0(p,d)-5F_2(p,d)$
4		$^3T_{2u}$	$E_d-4Dq-(\zeta_p+\zeta)/4-F_0(p,d)+F_2(p,d)$
5	$\tilde{t}_{1u}e_g$	$^1T_{1u}$	$E_d+6Dq-F_0(p,d)-4F_2(p,d)+8G_1(p,d)+54G_3(p,d)$
6		$^3T_{1u}$	$E_d+6Dq-\zeta_p/4-F_0(p,d)-4F_2(p,d)$
7		$^3T_{2u}$	$E_d+6Dq+\zeta_p/4-F_0(p,d)+4F_2(p,d)$
8	$\tilde{t}_{1u}a_{1g}$	$^1T_{1u}$	$E_s{}^{\dagger\dagger}-F_0(p,s)+2G_1(p,s)$
9		$^3T_{1u}$	$E_s+\zeta_p/2-F_0(p,s)$

† $E_d = \varepsilon_d-\varepsilon_p-5F_0(p,p)+10F_2(p,p)+6F_0(p,d)-6G_1(p,d)-63G_3(p,d)$.

†† $E_s = \varepsilon_s-\varepsilon_p-5F_0(p,p)+10F_2(p,p)+6F_0(p,s)-3G_1(p,s)$.

Table II. Off-diagonal elements of the energy matrix for the $\mathcal{T}_{1u}$ term. Only non-vanishing elements $\mathcal{H}_{ij}(j>i)$ are given. Note $\mathcal{H}_{ji} = \mathcal{H}_{ij}{}^*$

$$\mathcal{H}_{12} = (-\zeta_p+\zeta)i/2$$
$$\mathcal{H}_{13} = -(\zeta_p+\zeta)/2\sqrt{2}$$
$$\mathcal{H}_{14} = (-\zeta_p+\zeta)/2\sqrt{2}$$
$$\mathcal{H}_{17} = -\zeta'$$
$$\mathcal{H}_{23} = (-\zeta_p+\zeta)i/2\sqrt{2}$$
$$\mathcal{H}_{24} = -(\zeta_p+\zeta)i/2\sqrt{2}$$
$$\mathcal{H}_{25} = \sqrt{6}[-F_2(p,d)+4G_1(p,d)-18G_3(p,d)]$$
$$\mathcal{H}_{26} = \sqrt{3}\zeta' i/2$$
$$\mathcal{H}_{27} = -\zeta' i/2$$
$$\mathcal{H}_{28} = \sqrt{2}[10R^1(pd,sp)-3R^2(pd,ps)]/5\sqrt{15}$$
$$\mathcal{H}_{34} = (\zeta_p-\zeta)/4$$
$$\mathcal{H}_{35} = -\sqrt{3}\zeta' i/2$$
$$\mathcal{H}_{36} = -\sqrt{6}[F_2(p,d)+\zeta'/4]$$
$$\mathcal{H}_{37} = -\zeta'/2\sqrt{2}$$
$$\mathcal{H}_{39} = -\sqrt{6}R^2(pd,ps)/\sqrt{125}$$
$$\mathcal{H}_{45} = -\sqrt{3}\zeta' i/2$$
$$\mathcal{H}_{46} = \sqrt{3}\zeta'/2\sqrt{2}$$
$$\mathcal{H}_{47} = -3\sqrt{2}F_2(p,d)+\zeta'/2\sqrt{2}$$
$$\mathcal{H}_{56} = -\zeta_p i/2\sqrt{2}$$
$$\mathcal{H}_{57} = \sqrt{3}\zeta_p i/2\sqrt{2}$$
$$\mathcal{H}_{58} = 2[2R^1(pd,sp)/3-R^2(pd,ps)/5]/\sqrt{5}$$
$$\mathcal{H}_{67} = -\sqrt{3}\zeta_p/4$$
$$\mathcal{H}_{69} = -2R^2(pd,ps)/5\sqrt{5}$$
$$\mathcal{H}_{89} = \zeta_p i/\sqrt{2}$$

3. Transition Dipole Moments. Since no polarization is expected in our system of cubic symmetry, transition dipole moments are calculated for the z-component p_z of the electric dipole moment. Non-vanishing transition moments between the initial $t_{1u}^6\ {}^1A_{1g}$ state

and the parent terms $\tilde{t}_{1u}\gamma\,{}^1T_{1u}$ giving rise to $\mathcal{T}_{1u}$ are given as follows:

$$\langle t_{1u}^6\,{}^1A_{1g}|p_z|\tilde{t}_{1u}t_{2g}\,{}^1T_{1u}\gamma\rangle = (4/5)^{1/2}T(p,d),$$

$$\langle t_{1u}^6\,{}^1A_{1g}|p_z|\tilde{t}_{1u}e_g\,{}^1T_{1u}\gamma\rangle = (8/15)^{1/2}T(p,d),$$

$$\langle t_{1u}^6\,{}^1A_{1g}|p_z|\tilde{t}_{1u}a_{1g}\,{}^1T_{1u}\gamma\rangle = (2/3)^{1/2}T(p,s), \qquad (44)$$

where

$$T(p,\mu) = e\int_0^\infty dr\, r^3\, R_p(r)R_\mu(r) \qquad (\mu = s,d).$$

4. Excitation Transfer. So far we have neglected the effect of the migration of excitation corresponding to the terms in eq.(19) in the simplified model treated in section I.C. In the present model, the exciton wavefunctions corresponding to those in eq.(12) are given by

$$\Psi_{\vec{K}}(t_{1u}\gamma S\Gamma M_S M_\Gamma) = N^{-1/2}\sum_i e^{i\vec{K}\cdot\vec{R}_i}\,\Phi_i(\tilde{t}_{1u}\gamma S\Gamma M_S M_\Gamma),$$

$$\Phi_i(\tilde{t}_{1u}\gamma S\Gamma M_S M_\Gamma) = A_i^\dagger(\tilde{t}_{1u}\gamma S\Gamma M_S M_\Gamma)|G\rangle, \qquad (45)$$

where $|G\rangle$ is the ground state of the crystal and operator $A_i^\dagger(\tilde{t}_{1u}\gamma S\Gamma M_S M_\Gamma)$ is the one-hole and one-electron creation operator at site $\vec{R}_i$ introduced in eq.(27). By using eqs.(25), the terms of excitation transfer corresponding to those in eq.(19) are given by

$$\langle G|a_{is\gamma m_2\mu_2}\, b_{ist_{1u}m_1\mu_1}\,\mathcal{H}\, b^\dagger_{jst_{1u}m_1'\mu_1'}\, a^\dagger_{js\gamma' m_2'\mu_2'}|G\rangle$$

$$= (-1)^{2s+m_1+m_1'}[\langle\phi_j(st_{1u}-m_1'\mu_1')\phi_i(s\gamma m_2\mu_2)\|\phi_i(st_{1u}-m_1\mu_1) \times \phi_i(s\gamma' m_2'\mu_2')\rangle$$

$$- \langle\phi_j(st_{1u}-m_1'\mu_1')\phi_i(s\gamma m_2\mu_2)\|\phi_j(s\gamma' m_2'\mu_2')\phi_i(st_{1u}-m_1\mu_1)\rangle]$$

$$=\delta_{m_1'm_1}\delta_{m_2m_2'}\langle\phi_j(t_{1u}\mu_1')\phi_i(\gamma\mu_2)\|\phi_i(t_{1u}\mu_1)\phi_j(\gamma'\mu_2')\rangle$$

$$-\ \delta_{-m_1'm_2'}\ \delta_{-m_1m_2}(-1)^{2s+m_1+m_1'}\langle\phi_j(t_{1u}\mu_1')\phi_i(\gamma\mu_2)\|\phi_j(\gamma'\mu_2')$$

$$\times\ \phi_i(t_{1u}\mu_1)\rangle, \qquad (46)$$

where $a^{\dagger}_{is\gamma m\mu}$ and $b^{\dagger}_{is\gamma m\mu}$ are, respectively, the operators creating one-electron and one-hole in the $s\gamma m\mu$ orbital at site i, which have already been introduced in section I.D.-2, and $\phi_i(s\gamma m\mu)$'s are the electron spin-orbitals $s\gamma m\mu$ at site i whose orbital parts $\phi_i(\gamma\mu)$ are given in eqs.(31) - (34). Here, Hamiltonian $\mathcal{H}$ is assumed to be spin-independent. By using eqs.(24) and (46), the matrix element of the excitation transfer between the $\tilde{t}_{1u}\gamma S\Gamma M_S M_\Gamma$ state at site i and the $\tilde{t}_{1u}\gamma' S'\Gamma' M_S' M_\Gamma'$ state at j ($i\neq j$) is calculated as

$$\langle\Phi_i(\tilde{t}_{1u}\gamma S\Gamma M_S M_\Gamma)|\mathcal{H}|\Phi_j(\tilde{t}_{1u}\gamma' S'\Gamma' M_S' M_\Gamma')\rangle$$

$$= \sum_{\substack{\mu_1\mu_2\\ \mu_1'\mu_2'}} \langle\Gamma M_\Gamma|t_{1u}\mu_1\gamma\mu_2\rangle\langle t_{1u}\mu_1'\gamma'\mu_2'|\Gamma' M_\Gamma'\rangle$$

$$\times\ [\delta_{SS'}\delta_{M_S M_S'}\langle\phi_j(t_{1u}\mu_1')\phi_i(\gamma\mu_2)\|\phi_i(t_{1u}\mu_1)\phi_j(\gamma'\mu_2')\rangle$$

$$-\begin{cases} 2\langle\phi_j(t_{1u}\mu_1')\phi_i(\gamma\mu_2)\|\phi_j(\gamma'\mu_2')\phi_i(t_{1u}\mu_1)\rangle] \\ \qquad\qquad\qquad\qquad \text{for } S = S' = 0, \\ (\text{zero})], \text{ otherwise.} \end{cases} \qquad (47)$$

As mentioned in section I.C., the transfer given by the first term in eq.(47) is caused by the exchange interaction, and that given by the second term, which is effective only for the spin singlet, by the dipolar interaction. Normally the exchange-transfer is of the order of 10^{-2} eV, so that the first term in eq.(47) may be neglected. By using an expansion of the Coulomb interaction operator, $g = e^2/r_{12}$, the non-vanishing second term for $S = S' = 0$ can be reexpressed (1) as

$$\sum_{\substack{\mu_1\mu_2 \\ \mu_1'\mu_2'}} \langle \Gamma M_\Gamma | t_{1u}\mu_1\gamma\mu_2\rangle\langle t_{1u}\mu_1'\gamma'\mu_2' | \Gamma' M_\Gamma'\rangle$$

$$\times \; [(\vec{d}_{\mu_1\mu_2}\cdot\vec{d}_{\mu_1'\mu_2'})R^2 - 3(\vec{d}_{\mu_1\mu_2}\cdot\vec{R})(\vec{d}_{\mu_1'\mu_2'}\cdot\vec{R})]R^{-5},$$

where

$$\vec{d}_{\mu_1\mu_2} = \sqrt{2}e\int\phi_i(\gamma\mu_2)^*\vec{r}\;\phi_i(t_{1u}\mu_1)d\vec{r},$$

$$\vec{d}_{\mu_1'\mu_2'} = \sqrt{2}e\int\phi_j(t_{1u}\mu_1')^*\vec{r}\phi_j(\gamma'\mu_2')d\vec{r}, \qquad (48)$$

and $\vec{R} = \vec{R}_i - \vec{R}_j$. The contribution of this type of excitation transfer to exciton energies has been well studied in simple cases (1). The magnitude of the contibution is estimated to be of the order of 10^{-1}eV.

II.B. Interpretation of the Spectra

1. The Energy Level Diagram. In what follows, we confine ourselves to the absorption spectra of RbX crystals, which have been studied most in detail. In order to obtain the term energies and the eigenvectors of $\mathcal{T}_{1u}$, we have to estimate numerical values of the Slater-Condon parameters, the spin-orbit coupling constant, and the cubic field strength 10 Dq. As the first step, the Slater-Condon parameters for a free Rb^+ ion are determined so as to obtain a reasonable agreement with the observed spectrum of a free Rb^+ ion (8). Then, a set of the parameters for RbX crystals is estimated so as to obtain a reasonable agreement with the observed spectra of all the RbX by reducing the parameters for a free Rb^+ a little bit. The parameters thus determined are listed in Table III. The calculated spectra are quite sensitive to the spin-orbit coupling constant ζ_p in eq.(43). It is determined by Using Fig.1. Since the spectra are not sensitive to ζ and ζ' in eq.(43), they are estimated from the Rb^+ spectrum on a crude basis. These values of the spin-orbit coupling constants are also given in Table III.

By using a set of the parameters for RbX in Table III, energies of the $\mathcal{T}_{1u}$ terms are calculated as a function of the cubic field splitting parameter Dq. The result is shown in Fig.2. At the right side of each curve, parent terms involved in each $\mathcal{T}_{1u}$ state

Table III. Numerical values of the Slater-Condon parameters and the spin-orbit coupling constants.

	Free Rb^+ ion	RbX
$E_d'^{\dagger}$	-17.61 eV	-16.40 eV
E_s'	-20.10	-19.16
$F_0(p,d)$	7.0	6.6
$F_2(p,d)$	0.060	0.049
$G_1(p,d)$	0.070	0.065
$G_3(p,d)$	0.0027	0.0026
$F_0(p,s)$	7.6	7.2
$G_1(p,s)$	0.40	0.15
$R^1(pd,sp)$	0.80	0.50
$R^2(pd,ps)$	1.5	1.0
ζ_p	0.60	0.60
$\zeta = \zeta'$	0.02	0.02

† $E_\nu' = \varepsilon_\nu - \varepsilon_p - 5F_0(p,p) + 10F_2(p,p) \quad (\nu = d,s)$

at the limit of large Dq are indicated in such sequence that the left term is a major component as compared with the right. Three curves going upwards at large Dq correspond to the $\mathcal{T}_{1u}$ terms arising from the $\tilde{t}_{1u}e_g$ configuration, four curves going downwards at large Dq to those from $\tilde{t}_{1u}t_{2g}$, and two curves almost horizontal at large Dq to those from $\tilde{t}_{1u}a_{2g}$. At Dq = 0, transitions to the five terms indicated by arrows are allowed. In the intermediate values of Dq, the configuration mixing is so large that absorption intensities sensitively depend upon the value of Dq as will be seen in what follows.

2. <u>Comparison with the Experiments</u>. By using the eigenvectors numerically obtained at each value of Dq together with numerical values of transition dipole moments in eq.(44), T(p,d)=0.7 and T(p,s) =1.0, the absorption intensities are calculated for the transitions to the nine $\mathcal{T}_{1u}$ terms. Then, the value of Dq for each RbX is determined so as to reproduce the observed spectra (7) as closely as possible. The Dq positions thus determined are indicated in Fig.2. The calculated spectral patterns by using these values of

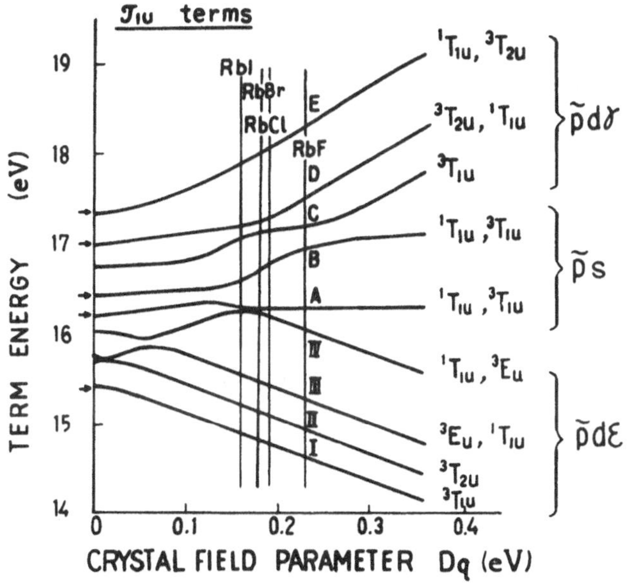

Fig.2. Energies of the $\mathcal{T}_{1u}$ terms as a function of Dq. The energy levels to which optical excitations are allowed at Dq=0 are indicated by arrows.

Dq are shown and compared with the observed ones in Fig.3. In our calculation we have neglected effects of the excitation transfer given in eq. (48), as they merely shift the term energies by a small amount, ∿ 0.1 eV.

As seen in Fig.3, the present calculation reproduces fairly well the observed relative intensities and relative positions of the five absorption peaks called A through E in all the RbX crystals. Peak F seems to arise from a different origin (9). The present theory predicts the presence of additional peaks I through IV, although their intensities are small. The observed chemical shift as mentioned at the beginning of section II.A. can be understood as follows. According to our assignment the main components of the excited states of peaks B, D and E arise from configuration $t_{1u}e_g$, so that they shift towards higher energy when Dq increases. The increase in Dq when one goes from the iodide to the fluoride,

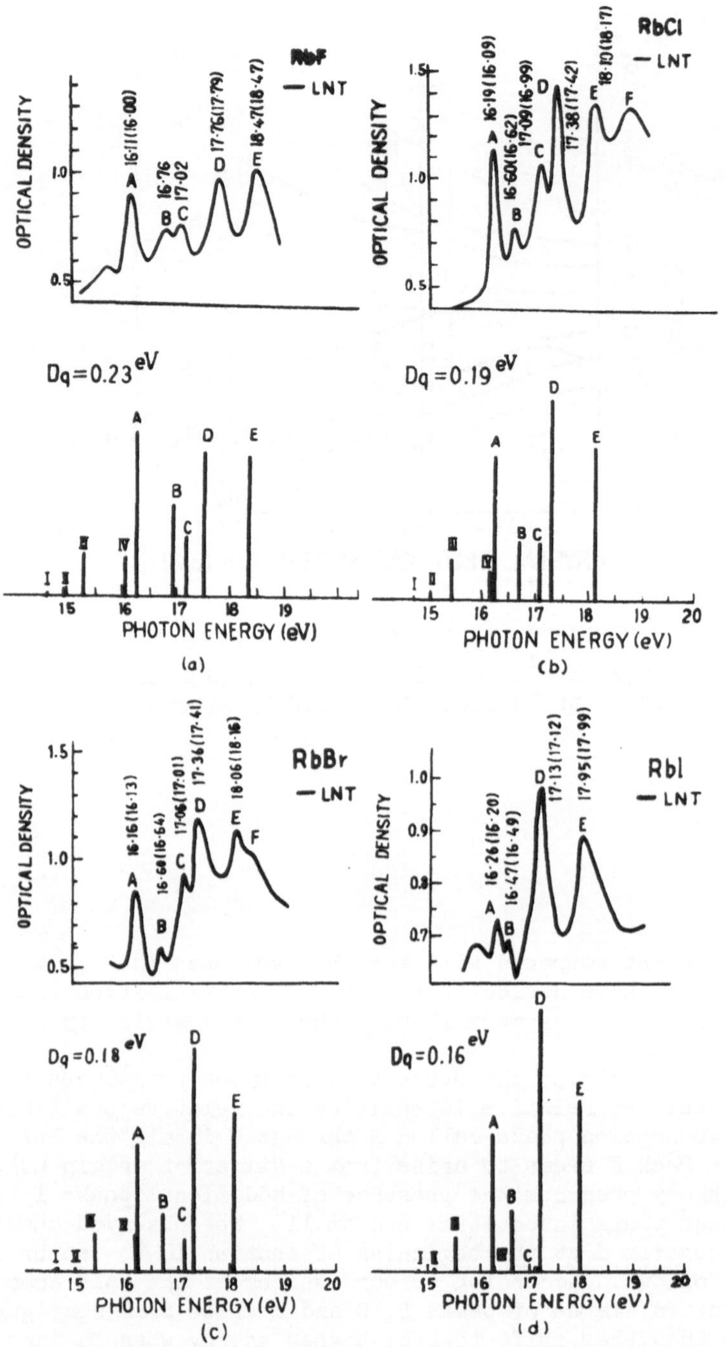

Fig.3. Comparison of the calculated spectral patterns with the observed ones for RbX (X = F, Cl, Br, I) crystals. The observed patterns are given by M. Watanabe et al. (7).

as assumed in the assignment, is in agreement with Tsuchida's spectrochemical series (3) empirically found to be applicable to a wide range of materials. On the other hand, the main components of the excited states of peaks A and C arise from configuration $\tilde{t}_{1u}a_{1g}$, so that their chemical shift may be expected to be much smaller. The same treatment as that mentioned here has been shown to be applicable to interpreting the spectra of KX crystals with success (10).

II.C. Effects of Electron Correlation

1. Comparison with the Wannier Exciton Theory. So far we have shown that a theory of the exciton, in which correlated motion of the localized excited electron and hole is taken into account, is successful in explaining the fine structure of the absorption spectra of alkali halide crystals around 20 eV. On the other hand, the theory of Wannier excitons based on the band theory has been successful in interpreting the ultraviolet absorption spectra exciting a valence electron (11). These two theories are physically quite different when they are applied to the open-shell systems such as excited states of alkali halides. Our theory emphasizes the electron correlation, but the Wannier exciton theory stresses the itinerancy of electrons.

One can, however, easily show that the two theories are formally equivalent, when applied to the excitation to the a_{1g} orbital (the s band at the Γ-point of the Brillouin zone) from the t_{1u} orbital (the p band at Γ), as long as the internal excitation of the Wannier exciton giving rise to the hydrogen series is ignored. Satoko has further discussed in his thesis (11) the interrelationship between our theory and the Wannier exciton theory when applied to the excitation to the t_{2g} orbital (the dε band at the X-point of the Brillouin zone). It is known that the minimum of the conduction band at the X-point is state $X_3(xy)$ given by the dε (t_{2g}) orbital of alkali atoms (12): although state X_3 is non-degenerate, the existence of three equivalent valleys corresponds to the triple-degeneracy of the excited t_{2g} orbital in our theory. The inner-core p band of alkali atoms at X consists of states $X_5'(x,y)$ and $X_4'(z)$, which are degenerate if the band width is ignored. When the spin-orbit interaction is incorporated, X_3 goes to X_7^+, and (X_5', X_4') to (X_7^-, X_{6u}^-, $X_{6\ell}^-$) in which X_7^- and X_{6u}^- are degenerate when the transfer of the p-hole is ignored. Since reduction of direct products $X_7^- \otimes X_7^+$, $X_{6u}^- \otimes X_7^+$, and $X_{6\ell}^- \otimes X_7^+$ give, respectively, two, one, and one $\Gamma_{15}(x, y, z: k=0)$ irreducible representations, one may expect four exciton absorption lines in general. This number of the expected lines is equal to that in our theory for the transition to $\tilde{t}_{1u}t_{2g}$. However, if one takes into account only spherically symmetric parts of the Coulomb and exchange interaction between the hole and the excited electron as Onodera and Toyozawa have done (12), one of the two symmetry-allowed transitions becomes forbidden yielding three lines as a whole. Further, if the hole transfer is ignored, two of the three lines become degenerate yielding two lines.

In our theory we have four parent states, $\Phi_1 \equiv \Phi(\tilde{t}_{1u}t_{2g}\,{}^3E_u)$, $\Phi_2 \equiv \Phi(\tilde{t}_{1u}t_{2g}\,{}^1T_{1u})$, $\Phi_3 \equiv \Phi(\tilde{t}_{1u}t_{2g}\,{}^3T_{1u})$, and $\Phi_4 \equiv \Phi(\tilde{t}_{1u}t_{2g}\,{}^3T_{2u})$ for $\mathcal{T}_{1u}$ as shown in Table I. We have shown that all the transitions to four $\mathcal{T}_{1u}$ terms given by linear combinations of these parent states are allowed. However, if one neglects $F_2(p,d)$, a non-spherically symmetric part of the Coulomb interaction, the energy matrix is given by use of Tables I and II as

$$\begin{bmatrix} -F_0-\zeta_p/2 & 0 & 0 & 0 \\ 0 & -F_0+G & i\zeta_p/\sqrt{2} & 0 \\ 0 & -i\zeta_p/\sqrt{2} & -F_0+\zeta_p/2 & 0 \\ 0 & 0 & 0 & -F_0-\zeta_p/2 \end{bmatrix},$$

$$F_0 \equiv F_0(p,d), \qquad G = 12G_1(p,d) + 36G_3(p,d), \tag{49}$$

for the bases,

$$\Phi_{I} = \frac{1}{2}(\sqrt{2}\Phi_1 + \Phi_3 + \Phi_4),$$

$$\Phi_{II} = \Phi_2,$$

$$\Phi_{III} = \frac{1}{2}(\sqrt{2}\Phi_1 - \Phi_3 - \Phi_4),$$

$$\Phi_{IV} = \frac{1}{\sqrt{2}}(\Phi_3 - \Phi_4).$$

In calculating matrix eq. (43) ζ is assumed to be vanishing. Energy matrix eq. (49) shows that only Φ_{III} is admixed into Φ_{II} to which the optical transition is allowed. Thus, one may expect two absorption lines corresponding to the excitations to two states given by linear combinations of Φ_{II} and Φ_{III}. Since we neglect the hole transfer in our theory, this result is same as that obtained in the Wannier exciton theory.

The example mentioned here emphasizes the importance of the non-spherically symmetric part of the Coulomb interaction for core-excitons arising from the correlated interaction of the electron and the hole as already discussed in section I.D.-3. This kind of electron correlation is not taken into account in the framework of the effective mass theory for the Wannier exciton.

2. Importance of the Configuration Mixing. In Fig.3 we have seen that the intensities of the transitions to the t_{2g} orbital is much reduced compared with those of the a_{1g} and e_g orbitals. Further, the calculated intensity of peak D arising from the transition to e_g relative to that of peak A arising from the transition to a_{1g} increases as Dq decreases in agreement with the experiment. These results are due to the mixing of configurations $\tilde{t}_{1u}t_{2g}$, $\tilde{t}_{1u}a_{1g}$ and $\tilde{t}_{1u}e_g$.

In order to see the effect of configuration mixing on intensities, we consider a simple example (11) in which two states A and B are admixed with each other by interaction I. We assume for simplicity that transition moments μ_A and μ_B to states A and B, respectively, have the same sign. Relative intensity of the transitions to states A and B is plotted against interaction I divided by E in Fig.4. E is the energy of state A minus that of state B.

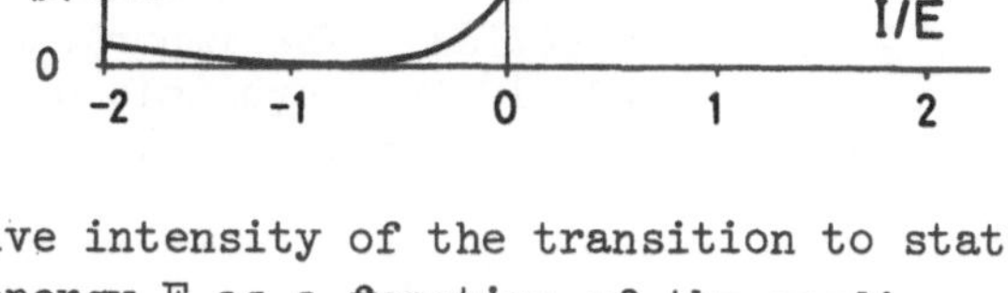

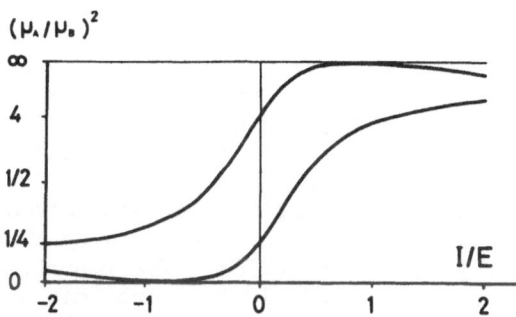

Fig.4. Relative intensity of the transition to states A and B separated by energy E as a function of the nondiagonal interaction I (11).

Except the cases where one of the transitions is forbidden, i.e., $(\mu_A/\mu_B) = 0$ or ∞, the relative intensity (μ_A/μ_B) begins to increase when a positive I is introduced and vice versa. This effect can be interpreted to be caused by the interference of transition moments.

We showed in Table II that the configuration interaction consists of the plus off-diagonal exchange interaction and the minus off-diagonal Coulomb interaction, which are comparable in the present problem. The sum of these interactions turns out to be positive in RbX and KX (10), that makes the intensity of the transition to t_{2g} much weaker than that to e_g. The decrease in Dq means the decrease in E, which increases I/E.

3. Relaxation of the Excited-Electron Orbitals. Very often optical absorption is assigned to a transition between one-electron orbitals which are obtained by solving Hartree-Fock equations. The example is the fundamental absorption of crystals around the visible region which is assigned to the transition from the valence band to the conduction band. The conduction band is normally calculated by assuming the same potential as that used for the calculation of the valence band. This means that, in the calculation of the excited-electron orbital, the system is treated as if it is an (N+1)-electron system where an electron is in the conduction band and N electrons in the closed shells. Roughly speaking, the conduction electron in alkali halides is instantaneously in the (n+1)s or nd orbital of a neutral alkali atom (Cs, Rb, K).

It is, however, more reasonable to consider that, at the presence of an inner-core hole, an excited electron is in the corresponding orbital of a M^+ ion if it is localized. Schematic energy level diagrams illustrating the two cases just mentioned above are given in Fig.5 (11), where the positions of energy levels of free Rb and Rb^+ are determined by using spectroscopic data, those of energy bands by the band calculation, and those of Rb^+ in the crystal by estimating the Madelung energy and the polarization energy. The most remarkable effect due to the presence of the inner-core hole is to lower the 4d relative to the 5s. A similar effect is also expected for CsX and KX. The experimental fact that the transitions to the d and s orbitals have almost the same energy in CsX, RbX and KX suggest the preference of the relaxed excited orbitals: the excited orbitals are not given by the wavefunctions of the conduction band but are given by those of the excited M^+ ion to a good approximation. The importance of using the relaxed orbitals for the excited-electron orbitals has already been pointed out at the beginning of section I.D.

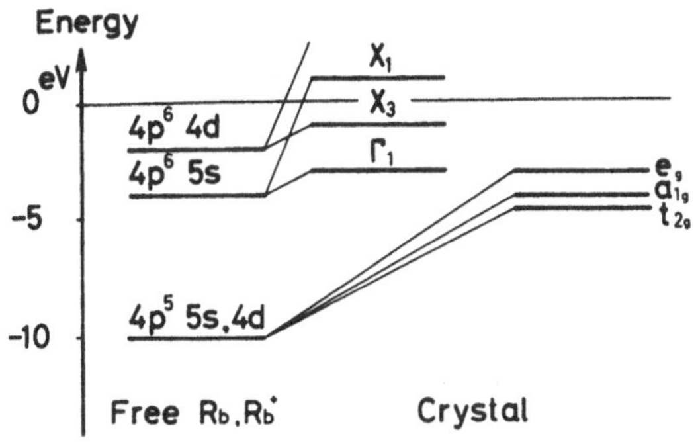

Fig.5. Schematic energy level diagrams of a Rb atom and a Rb^+ ion, and those of a RbCl crystal (11).

III. X-RAY SPECTROSCOPY OF TRANSITION-METAL COMPOUNDS

III.A. Covalency in the Absence of a Hole

1. A Covalency Parameter. It is well known that the optical transitions within an incomplete d shell in transition-metal compounds can successfully be explained by treating a molecular cluster in which a central transition-metal ion is surrounded by ligands, in many cases by six ligands octahedrally. In X-ray spectroscopy where a hole is produced in the inner-core of a metal ion, d-electrons are expected to be more localized, so that the treatment of a cluster for d electrons is easily justified. In discussing the distribution of d-electrons in the cluster the concept of covalency is useful both in the absence and the presence of the core hole. Our final purpose is to discuss in terms of covalency the relaxation of d-orbitals when the hole is produced. Before doing this we first explain in this subsection what covalency means in the absence of the hole.

Let us confine ourselves to the octahedral cluster MX_6 where M is a transition-metal ion and X a ligand. We first assume that the distance between the metal ion and the ligand is large enough to have no overlap of the electron clouds of the metal ion and the ligand. In this case no electron is transferred between the metal ion and the ligand, and the system is purely ionic. Then, the t_{2g} and e_g orbitals of the incomplete shell are given in terms of the d-function of the free metal ion. We next bring the ligands closer to the metal ion keeping the octahedral symmetry of the system. Then, the electron clouds of the metal ion and ligands overlap and some of the electrons are transferred or exchanged between them. Assume that the ligand has a closed-shell configuration like an F^- ion when the metal ion and the ligand are sufficiently far apart. At such a large atomic distance, the energies of the ligand orbitals are lower than those of the d-orbital of the metal ion. When the electron transfer begins to occur at a smaller distance, the d- and ligand orbitals are admixed and the energies of the electron mainly in the ligand orbitals are depressed and those in the d-orbitals are raised as shown in Fig.6, resulting in the decrease of the total energy of the system. The stabilized orbital ψ^b is called a bonding orbital and the destabilized orbital ψ^a an antibonding orbital. The bonding orbital has mainly the ligand character with a small admixture of the d-orbital, and the antibonding orbital mainly the d-orbital character with a small admixture of the ligand orbital. These orbitals are expressed in the following forms;

$$\psi^a = N_a^{-1/2}(\phi - \lambda\chi),$$

$$\psi^b = N_b^{-1/2}(\chi + \gamma\phi), \qquad (50)$$

where ϕ is the atomic d-orbital of the metal ion, χ a suitable linear combination of the atomic orbitals of the ligands, and

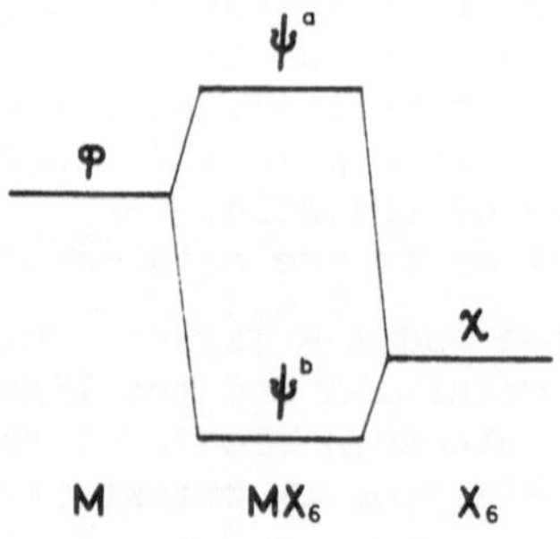

Fig.6. Schematic energy-level diagram showing formation of the bonding and antibonding orbitals.

$N_a^{-1/2}$ and $N_b^{-1/2}$ normalization constants. Orbital functions ψ^a and ψ^b satisfy the symmetry requirement of the system and are labeled, if necessary, with the irreducible representations of the octahedral group like $\psi^a_{t_{2g}}$, $\psi^a_{e_g}$ and so on. Because of the orthogonality between ψ^a and ψ^b, λ and γ are related to each other as

$$\lambda = (\gamma + S)/(1 + \gamma S), \tag{51}$$

where $S = \langle\phi|\chi\rangle$ is the overlap integral between ϕ and χ.

The physical meaning of coefficient γ becomes clear when one constructs the total wavefunction Ψ of the system using one-electron orbitals in eq. 50. For simplicity, we consider a three-electron system in which two electrons are in ψ^b and one in ψ^a. By using eq.(51), the total wavefunction of this system is given by

$$\begin{aligned} \Psi &= |\psi^a\psi^b\bar{\psi}^b| \\ &= [N_b(1 - S^2)]^{-1/2}[\,|\phi\chi\bar{\chi}| + \gamma|\phi\chi\bar{\phi}|\,], \end{aligned} \tag{52}$$

where the orbital without a bar on its top is the up-spin orbital and the one with a bar the down-spin orbital. $|\cdots|$ is the Slater determinant including the normalization factor as used in eq.(2). The first Slater determinant $|\phi\chi\bar{\chi}|$ in eq.(52) represents the state in which two electrons are placed in the ligand orbital χ. This state may be considered to correspond to the purely ionic configuration. The second term $|\phi\chi\bar{\phi}|$ represents the state in which a down-spin electron in the ligand orbital in the ionic configuration is transferred into the metal d-orbital, forming covalent bond with the up-spin electron in the ligand orbital. Parameter γ measuring the amount of the covalent configuration is called a covalency parameter.

2. Hartree-Fock Equation. The covalency parameter γ is determined by the Hartree-Fock equations which are derived from the variation principle as follows. We minimize the total energy by setting

$$\delta\langle\Psi|\mathcal{H}|\Psi\rangle = 0, \tag{53}$$

with subsidiary condition

$$\langle\psi^a|\psi^a\rangle = \langle\psi^b|\psi^b\rangle = 1, \tag{54}$$

$$\langle\psi^a|\psi^b\rangle = 0. \tag{55}$$

It should be noticed that the orthogonality conditions between different orbitals are automatically fulfilled in the Hartree-Fock solutions for a closed-shell system, but not in those for the open-shell system we are considering. The importance of the subsidiary orthogonality condition in eq.(55) will be realized soon. We are still concerned with the three-electron system whose total wave-function is given by eq. (52). The total Hamiltonian is given by

$$\mathcal{H} = \sum_{i=1}^{3} f_i + \sum_{j>i=1}^{3} g_{ij}, \tag{56}$$

where f_i is the kinetic energy plus the potential energy and g_{ij} the Coulomb interaction energy. From eqs. (53),(54) and (55) the following Hartree-Fock equations are derived;

$$\begin{aligned} h_1\psi^a &= \varepsilon_1\psi^a + \lambda_0\psi^b, \\ h_2\psi^b &= \varepsilon_2\psi^b + \frac{1}{2}\lambda_0\psi^a, \end{aligned} \tag{57}$$

where

$$\begin{aligned} h_1 &= f + 2\langle\psi^b\|\psi^b\rangle - \langle\psi^b\|P\psi^b\rangle, \\ h_2 &= f + \langle\psi^a\|\psi^a\rangle + \langle\psi^b\|\psi^b\rangle - \frac{1}{2}\langle\psi^a\|P\psi^a\rangle. \end{aligned} \tag{58}$$

In eq.(57) ε_1 and ε_2 are Lagrange's undetermined multipliers for the conditions expressed in eq.(54) and λ_0 is for the condition expressed in eq. (55). In eq. (58) the following abbreviations are used:

$$\langle\psi\|\psi\rangle = \int d\vec{r}_1\psi^*(1)g_{12}\,\psi(1),$$

$$\langle\psi\|P\psi\rangle = \int d\vec{r}_1\psi^*(1)g_{12}P\psi(1),$$

where P is the permutation operator for electrons 1 and 2. Assuming that ψ^a and ψ^b are real, one can determine λ_0 from eqs.(57) and (58) as

$$\lambda_0 = -\langle\psi^a\psi^a\|\psi^a\psi^b\rangle. \tag{59}$$

The abbreviation in eq.(59) was explained in section I.C.

The Hartree-Fock equations in eq.(57) are not Schrödinger-type equations for one-electron orbitals because of the presence of the terms, $\lambda_0\psi^a/2$ and $\lambda_0\psi^b$. It can, however, be shown by using Koopmans' theorem (3) that ε_1 is the orbital energy of ψ^a and ε_2 is the appropriate average of the orbital energies of the up-spin and down-spin electrons in ψ^b.

3. Calculation of the Covalency Parameter. Inserting the explicit forms of ψ^a and ψ^b given in eq.(50) into the Hartree-Fock equations in eq.(57), one obtains the equation to determine λ and γ as follows:

$$(N_aN_b)^{-1/2}[(B_1 - SA_1) + \lambda(A_1 - C_1) - \lambda^2(B_1 - SC_1)$$

$$= \lambda_0(1 - \lambda S), \qquad (60)$$

$$2(N_aN_b)^{-1/2}[(B_2 - SC_2) + \gamma(A_2 - C_2) - \gamma^2(B_2 - SA_2)]$$

$$= \lambda_0(1 + \gamma S), \qquad (61)$$

where

$$A_\nu = \langle\phi|h_\nu|\phi\rangle, \qquad B_\nu = \langle\phi|h_\nu|\chi\rangle, \qquad C_\nu = \langle\chi|h_\nu|\chi\rangle,$$

$$(\nu = 1, 2).$$

Replacing λ_0 in eqs. (60) and (61) by eq.(59) and using the explicit expressions for N_a and N_b and the relation in eq.(51), one finally obtains the equations to determine γ as follows:

$$(B - SA) + \lambda(A - C) - \lambda^2(B - SC) = 0, \qquad (62)$$

$$(B - SC) + \gamma(A - C) - \gamma^2(B - SA) = 0, \qquad (63)$$

where

$$A = \langle\phi|h|\phi\rangle, \qquad B = \langle\phi|h|\chi\rangle, \qquad C = \langle\chi|h|\chi\rangle,$$

and

$$h = f + \langle\psi^a\|\psi^a\rangle + \langle\psi^b\|\psi^b\rangle. \qquad (64)$$

It should be noted that eqs.(62) and (63) are not independent of each other: one is obtained from another by using the relation in eq. (51).

When γ and S are small, neglecting the higher order terms proportional to γ^2 and S^2, one obtains

$$\gamma \approx \frac{-B + SC}{A - C}, \tag{65}$$

and

$$\lambda \approx \gamma + S. \tag{66}$$

It is very important to examine the physical meaning of eq.(65). To simplify the following argument, we set γ and S equal to zero in the one-electron Hamiltonian in eq. (64) and denote it as h_0;

$$h_0 = f + \langle\phi\|\phi\rangle + \langle\chi\|\chi\rangle. \tag{67}$$

Then, $\langle\phi|h_0|\phi\rangle$ in the denominator may be interpreted as the orbital energy of the down-spin electron on the metal in the configuration where one electron is transferred from the ligands to the metal ion: in the example of a $Ni^{2+}(F^-)_6$ cluster, this configuration is $Ni^+(F_6)^{5-}$. On the other hand, $\langle\chi|h_0|\chi\rangle$ may be interpreted as the orbital energy of the down-spin electron on the ligands in the ionic configuration, which is $Ni^{2+}(F^-)_6$ in the example just mentioned above.

In the molecular orbital theory we assume fixed forms of ϕ and χ in eq. (50) and use them in the calculation of both A and C in eq.(65). According to the non-empirical calculation (13) of γ for $KNiF_3$ where the Hartree-Fock atomic wavefunctions of Ni^{2+} and F^- are assumed for ϕ and a component of χ, respectively, the denominator of eq.(65) turns out to be 20 eV. This calculation yields a value of 0.074 for γ for the e_g orbital (14); this is much smaller than the experimental value of 0.23 (15).

As shown in eq.(52), our molecular orbital theory is formally equivalent to the Heitler-London theory in which mixing of the excited covalent configuration is taken into account. According to the Heitler-London theory the denominator of the expression determining γ similar to eq.(65) is given by the energy difference of the excited covalent configuration and the ionic configuration, in harmony with the statement given just below eq.(67). Hubbard et al. (16) estimated the denominator in the Heitler-London theory by using the observed ionization potential of a Ni^+ ion. They found that the denominator was reduced to $\sim$12 eV. Furthermore, they took into account the effect of polarization induced by the transferred electron in the covalent configuration, and estimated that this effect further reduces the denominator to ~8 eV. Once the reduced value of the denominator is used, γ for e_g is calculated to be 0.21 in agreement with the experimental value.

III.B. Hole-Induced Covalency

1. Mixing of Two Configurations. In this section we follow the argument given in our recent work (17). To discuss the effects of an inner-core hole on the bonding, we consider two configurations of a four-electron system as shown in Fig.7. The total wavefunctions of these configuration with $S=1$, $M_S=1$ are given as

$$\Psi_1 = |\psi^a\psi^b\bar{\psi}^b\phi^c|,$$

$$\Psi_2 = |\psi^a\psi^b\bar{\psi}^a\phi^c|, \qquad (68)$$

where ϕ^c is the atomic orbital of the inner-core of the transition-metal ion. Note that Ψ_1 and Ψ_2 are orthogonal to each other because of eq.(55). There exists, however, a non-diagonal matrix element of the total Hamiltonian between Ψ_1 and Ψ_2 in the presence of the hole. By using abbreviations, $\langle\psi^a\phi^c||\psi^b\phi^c\rangle \equiv \langle ac||bc\rangle$, etc., it is given as

$$\langle\Psi_2|\mathcal{H}|\Psi_1\rangle \qquad (\equiv E')$$

$$= \langle ac||bc\rangle - \langle ac||cb\rangle, \qquad (69)$$

in which use has been made of Hartree-Fock equations

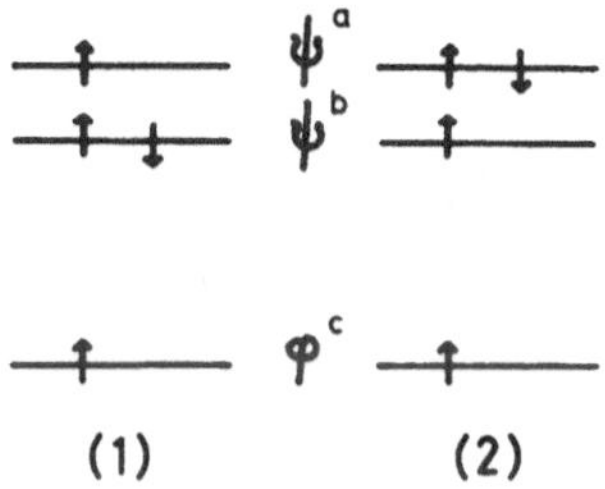

Fig.7. Two configurations of a four-electron system at the presence of an inner-core hole.

similar to those given in eq.(57) by assuming that ψ^a and ψ^b are the molecular orbitals in the absence of the hole. The matrix element in eq.(69) is vanishing at the absence of the hole.

The energy difference of the states Ψ_1 and Ψ_2 is calculated in a similar way as follows:

$$\langle\Psi_2|\mathcal{H}|\Psi_2\rangle - \langle\Psi_1|\mathcal{H}|\Psi_1\rangle \quad (\equiv 2\Delta E)$$

$$= (\varepsilon_1 - \varepsilon_2) - (\langle ac\|ac\rangle - \langle ac\|ca\rangle) + (\langle bc\|bc\rangle - \langle bc\|cb\rangle) + (\langle aa\|aa\rangle - \langle ab\|ab\rangle + \frac{1}{2}\langle ab\|ba\rangle), \tag{70}$$

where ε_1 and ε_2 are identical to those given in eq.(57). The terms in the second bracket represent the lowering in energy of the antibonding orbital by the attractive force of the hole. Those in the third bracket represent the amount of lowering in the bonding orbital due to the hole. The terms in the final bracket represent the energy change due to the electron transferred from the ligand to the metal ion.

In what follows the linear combinations of Ψ_1 and Ψ_2 as given by

$$\Psi_m = (1 + \gamma'^2)^{-1/2}(\Psi_1 + \gamma'\Psi_2),$$

$$\Psi_s = (1 + \gamma'^2)^{-1/2}(\Psi_1 - \gamma'\Psi_2), \tag{71}$$

are assumed to diagonalize the total Hamiltonian. If $E' \ll 2\Delta E$, γ' is approximately given by $E'/2\Delta E$.

2. <u>Increase of the Covalency Parameter</u>. Inspection of eq.(52) shows that Ψ_m and Ψ_s in (71) can be reexpressed in the form,

$$\Psi_m = |\psi^{a'}\ \psi^{b'}\ \bar{\psi}^{b'}\ \phi^c|,$$

$$\Psi_s = |\psi^{a'}\ \psi^{b'}\ \bar{\psi}^{a'}\ \phi^c|, \tag{72}$$

where

$$\psi^{a'} = (1 + \gamma'^2)^{-1/2}(\psi^a - \gamma'\psi^b),$$

$$\psi^{b'} = (1 + \gamma'^2)^{-1/2}(\psi^a + \gamma'\psi^b). \tag{73}$$

By using eq.(50) these new molecular orbitals are expressed in the terms of ϕ and χ as follows;

$$\psi^{a'} = [(1 - S^2)(1 + \gamma'^2)N_b]^{-1/2}$$
$$\times \{(1+\gamma S-\gamma'\gamma\sqrt{1-S^2})\phi - (\gamma+S+\gamma'\sqrt{1-S^2})\chi\},$$
$$\psi^{b'} = [(1 - S^2)(1 + \gamma'^2)N_b]^{-1/2} \tag{74}$$
$$\times \{(\sqrt{1-S^2} - (\gamma+S)\gamma')\chi + (\gamma\sqrt{1-S^2} + (1+\gamma S)\gamma')\phi\}.$$

If γ, S and γ' are first-order small quantities, neglecting higher-order small quantities one may simplify the expressions in eq.(74) as

$$\psi^{a'} \approx \{\phi - (\lambda + \gamma')\chi\},$$
$$\psi^{b'} \approx \{\chi + (\gamma + \gamma')\phi\}. \tag{75}$$

Eq.(75) shows that γ' is the increment of the covalency parameter due to the presence of an inner-core hole.

III.C. Shake-Up Satellites

1. Sudden Approximation. The creation of a core hole by X-ray excitation produces a sudden change of the Hamiltonian which may induce non-adiabatic electron jumps. These jumps are observed as satellites in photoelectron emission spectra. We call the satellites due to the sudden change of the Hamiltonian "shake-up satellites".

The approximation method dealing with a sudden change of the Hamiltonian is well known as the sudden approximation. To explain this approximation in our problem, we first consider a discontinuous change of the electron Hamiltonian: $\mathcal{H} = \mathcal{H}_0$ for $t < 0$ and $\mathcal{H} = \mathcal{H}_1$ for $t > 0$. Further, the equations,

$$\mathcal{H}_0\Psi_n = E_n\Psi_n, \qquad \mathcal{H}_1\Phi_m = E_m\Phi_m, \tag{76}$$

are assumed to be valid, where Ψ_n and Φ_m are complete orthonormal sets of many-electron functions. For simplicity, we assume that their eigenvalues are discrete. The general wavefunctions of our electron system are given by

$$\Psi = \sum_n a_n \Psi_n e^{-iE_n t/\hbar}, \qquad t < 0$$

$$\Psi = \sum_m b_m \Phi_m e^{-iE_m t/\hbar}, \qquad t > 0 \qquad (77)$$

where a_n and b_m are numerical coefficients independent of the time. Since the Schrödinger equation is of first order in the time, the wavefunctions at all spatial points must be a continuous function of the time at $t = 0$, although its time derivative is not. Then, equating the two wavefunctions in eq. (77) at $t = 0$, one obtains

$$b_m = \sum_n a_n \int \Phi_m^* \Psi_n \, d\tau. \qquad (78)$$

When the change in the Hamiltonian is not discontinuous but takes place in a finite interval of time t_0, the criterion for the validity of eq. (78) is that t_0 be small in comparison with the periods associated with the initial motion of the electrons. If this criterion is satisfied, the probability that the electron jumps to the state given by Φ_m for $t > 0$ is given by the absolute square of the overlap integral of Φ_m and the initial-state function Ψ_0; here $a_n = 0$ $(n \neq 0)$ and $a_0 = 1$ is assumed for the initial state. Throughout the following arguments we always assume the validity of the criterion without giving any explanation. The sudden approximation seems to apply well to the satellites in the X-ray photoelectron spectra (18).

Now let us assume that the initial-state function Ψ_0 is given by a single Slater determinant,

$$\Psi_0 = |u_1 u_2 \cdots u_N|, \qquad (79)$$

where u's are orthonormal spin orbitals. In the problem of photoelectron emission, Φ's in eq.(76) are functions of N-1 electrons involving electron coordinates $\vec{x}_2, \vec{x}_3, \cdots, \vec{x}_N$, which may be expanded in terms of orthonormal Slater determinants as follows;

$$\begin{aligned}\Phi_m &= c_{m0}|u_2 u_3 \cdots u_N| \\ &+ c_{m1}|u_2 u_3 \cdots u_N'| \\ &+ c_{m2}|u_2 u_3 \cdots u_N''| + \cdots, \end{aligned} \qquad (80)$$

where u'_N is the excited one-electron spin orbital orthogonal to u_2, u_3, ⋯, u_N, and so on. In writing eq.(80) u_1 is assumed to be the core spin orbital from which an electron is emitted. It is also assumed that, if the orbital relaxation in the presence of the u_1 hole is ignored, $c_{00} = 1$, $c_{0n} = 0$ $(n = 1,2,\cdots)$ and $c_{11} = 1$, $c_{1n} = 0$ $(n = 0,2,3,\cdots)$: if the orbital relaxation is taken into account, c_{0n} $(n = 1,2,\cdots)$ and c_{1n} $(n = 0,2,3,\cdots)$ are not necessarily vanishing.

Laplace-expanding Ψ_0 with respect to the first column, one has

$$\Psi_0 = \sum_{i=1}^{N} (-1)^{i-1} u_1(\vec{x}_i) |u_2\, u_3 \cdots u_N|, \tag{81}$$

where the Slater determinant is a function of electron coordinates $\vec{x}_1$, $\vec{x}_2$, ⋯ $\vec{x}_{i-1}$, $\vec{x}_{i+1}$, ⋯ $\vec{x}_N$. Inserting Ψ_0 and Φ_m into eq.(78) with $a_n = \delta_{n0}$, one obtains

$$\begin{aligned} b_0 &= c_{00}\, u_1(\vec{x}_1), \\ b_1 &= c_{10}\, u_1(\vec{x}_1), \\ &\cdots \end{aligned} \tag{82}$$

Then, the probabilities for the electron to jump to Φ_0 and Φ_1 are, respectively, given by

$$\int |b_0|^2 d\vec{x}_1 = |c_{00}|^2,$$

and

$$\int |b_1|^2 d\vec{x}_1 = |c_{10}|^2. \tag{83}$$

Since Φ_1 is the state in which both the emission of the u_1 electron and the excitation of the u_N electron may be considered to occur simultaneously, to the approximation of neglecting the relaxation of orbitals, the transition to this state is called the shake-up of the u_N electron in the u_1 photoelectron emission. It is important to note that the non-vanishing intensity of this transition is entirely due to the relaxation of orbitals as shown in eq.(83), as $c_{10} = 0$ in the absence of the relaxation.

2. Configuration Interaction Satellites. In the expansion in eq.(80), an inner-core hole was assumed to be in a fixed orbital, u_1. There are, however, cases in which the orbital picture of the inner-core hole breaks down and the hole state is given by a linear combination of some hole configurations. This kind of configuration interaction, which is a characteristic property of the core hole, is due to the ability of the hole to jump into outer orbitals without loss of energy. In these cases, the expansion corresponding to eq.(80) may be given by

$$\Phi_m = c'_{m0}|u_2 u_3 \cdots u_N| + c'_{m1}|u_1 u'_3 \cdots u_N| + \cdots , \tag{84}$$

where u_1 is again the core orbital from which an electron is emitted and u_2 and u_3' are outer core orbitals. Then one sees that

$$\int |b_1|^2 d\vec{r}_1 = |c'_{10}|^2 \tag{85}$$

is non-vanishing in the presence of the configuration interaction, resulting in the appearance of a satellite in photoelectron emission spectra called a configuration interaction satellite. The origin of this satellite is quite different from that due to the orbital relaxation which one can think of in the framework of the orbital picture.

A clear-cut example of the configuration interaction satellite due to a core-hole has been found in the X-ray photoelectron spectroscopy (XPS) of a MnF_2 crystal where a 3s-electron of Mn^{2+} is emitted (19). Since the ground state of MnF_2 is the orbital singlet with $S = 5/2$, the XPS spectrum consists of two lines corresponding to the transitions to two orbital singlets with $S = 3$ and 2 when the 3s-hole is created. Theoretically, the energy separation ΔE of these lines is given by the exchange interaction between the 3s-hole and the 3d-electron, and the intensity ratio R of the $S = 3$ line to the $S=2$ line by the ratio of their spin multiplicities. Such a theory is compared with the experiment as follows:

	Theory	Exp.
ΔE	14.2	6.3 eV
R	1.4	2.0 .

(86)

It has been found that the large discrepancy between the theory and the experiment may be removed by taking into account the mixing of hole configurations within the M shell. This finding also predicts the appearance of the configuration interaction satellites as confirmed by experiments.

A large mixing of hole configurations in this problem owes to the situation that the energy separation, ~35 eV, between the 3p orbital and the lower 3s is relatively close to that, ~50 eV, between the 3p-orbital and the higher 3d. Owing to this situation the configuration mixing of $3s^2 3p^4 3d^6$ into $3s3p^6 3d^5$ does not need much energy (~15 eV), that makes the mixing relatively large. This mixing is the jump of two electrons from the 3p-orbital, one to the 3s and another to the 3d. Since the jumping probability, which is determined by the non-diagonal matrix element of the Coulomb interaction operator, is large when the two 3p-electrons jumping are in the same orbital, (and consequently have antiparallel spins), the configuration mixing is expected to be ineffective for the $S = 3$ final state where no space to accommodate an up-spin is available in the 3s and 3d orbitals. The detailed calculation of the mixing for a free Mn^{2+} ion (19) shows that the mixing of $3s^2 3p^4 3d^6$ actually does not exist for the $S = 3$ state while the mixing coefficients for $3s3p^6 3d^5 (^6S)^5S$ and $3s^2 3p^4 (^1D) 3d^6 (^5D)^5S$ are as large as 0.794 and -0.550, respectively. The large mixing for ^5S reduces ΔE to ~5 eV and enhances R to $1.4 \times (1/0.794)^2 = 2.2$, in agreement with the observation. The mixing also gives rise to a satellite, as seen from the argument leading to eq.(83), which has already been observed. The satellite may be considered to be caused by a combination of the shake-up and the shake-down of the 3p-electrons.

This sort of configuration mixing is small when a 2s-electron is emitted, as no 2d-orbital is available. When the orbital of a different principal quantum number is involved, the mixing requires much energy; this makes the matrix coefficients negligibly small. It has been pointed out that a similar configuration mixing is important in the XPS of rare earth compounds when a 4s-electron of a rare earth ion is emitted, but not when a 5s-electron is emitted. The configuration interaction discussed in this subsection is operative within a principal shell, the M shell in the XPS emitting a 3s-electron. This indicates the validity in specifying a core-hole by the shell even when its specification by orbitals becomes inadequate.

3. Multiplet Satellites. When a core hole is in a degenerate orbital and valence electrons in an open shell, Φ_m in eq.(76) generally cannot be given by a single Slater determinant even in the absence of the orbital relaxation. Then, following the arguments given for the satellites arising from the orbital relaxation

and the configuration interaction, one may predict appearance of satellite lines in the XPS. The energy positions and the intensities of these satellites are determined by the hole-electron interaction and the interaction between the valence electrons. Since the final states of these satellites are the multiplet terms in the system consisting of a hole and valence electrons, the satellites are called multiplet satellites. In the next section we are going to give a detailed example of the multiplet satellites where the orbital relaxation is incorporated at the same time.

III.D. Interpretation of Photoelectron and K-Emission Spectra

1. The Theoretical Scheme. We perform in this section a detailed analysis of the X-ray p-photoelectron and the K-emission spectra of nickel compounds (20). The final states of the transitions in both the cases are the same, $np^5 3d^8$, so that a unified treatment may be done. Our primary interest lies in the multiplet satellites and the shake-up satellites due to the orbital relaxation which is connected with the hole-induced covalency discussed in III.B. It will be shown that some quantitative interrelationship may be found between these two kinds of satellites.

We start with the assumption that d-electrons are localized within a molecular cluster in which a central transition-metal ion is surrounded by six ligands octahedrally. Then, the orbitals for d-electrons are the antibonding orbitals of symmetries t_{2g} and e_g. Since a core-hole is present on the metal ion, these antibonding orbitals are those given in eq.(74) where the covalency parameter is, roughly speaking, increased by γ'.

In what follows we use the complementary states, which are expressed as $|t_{1u}, e_g^{2-n} t_{2g}^{n}(\bar{S}\bar{\Gamma})S\Gamma>$ $(n = 0,1,2)$ for the Ni^{2+} compounds having a $p(t_{1u})$-hole on the metal ion: The ground state of the Ni^{2+} ion is the $e_g^2\,{}^3A_{2g}$ state, but there are many excited multiplet terms within the d-shell approximately given by $e_g^{2-n} t_{2g}^{n}\,\bar{S}\bar{\Gamma}$ $(n = 0, 1,2)$. These excited terms are mixed into the ground term by the hole-electron interaction (p-d interaction) when a p-hole is created in an inner-core. The final terms including the hole are specified by symmetry $S\Gamma$. Note that this kind of mixing is absent when the core hole is s: the final terms to which the transitions occur are restricted to $|a_{1g}, e_g^2({}^3A_{2g})^4A_{2g}$ and ${}^2A_{2g}>$.

Now our problem is reduced to the calculation of the multiplet structure arising from configuration pd^2 in a cubic field. In the calculation we take into account a cubic ligand field acting on d-electrons, the spin-orbit interaction of p-electrons, and the Coulomb and exchange interactions between the p- and the d-electron as well as between the d-electrons. Among these various interactions, the Coulomb and exchange interactions between the p- and the d-electron seem to be much larger than the cubic field strength 10 Dq. This fact may lead us to the thought that the natural scheme of the calculation is to start from the free-ion terms of configuration pd^2 taking into account the cubic field interaction afterwards. We, however, use the strong-field scheme, as just used for describing the complementary states, in which the cubic field interaction is diagonalized. The reason is two-fold: (1) This scheme is convenient to study the interrelationship between the states in the presence of a p-hole and those in the absence of a p-hole, the latters of which are well described in the strong-field scheme. One of such merits is found in the calculation of transition intensities; (2) This is convenient to study the effect of chemical bond between the metal ion and the ligands in the presence of a core-hole in terms of molecular orbitals as discussed in section III.B.

2. Calculation of Term Energies. In our simple system of the pd^2 configuration, the matrices of the Coulomb and exchange interactions can be directly calculated for each set of SΓ by using the wavefunctions $|t_{1u}, e_g^{2-n} t_{2g}^{n} (\bar{S}\bar{\Gamma})\, S\Gamma\rangle$ which are constructed by adding a p-electron (t_{1u}-electron) to the already known wavefunctions of the $e_g^{2-n} t_{2g}^{n} \bar{S}\bar{\Gamma}$ states (3) and antisymmetrizing the total wavefunctions with respect to any permutations of electrons. The matrix elements are given in terms of those for the two-electron system of configuration pd, which are given in Table IV. In the systems of pd^n ($n > 2$) it is more convenient to express the matrix elements in terms of those for the two-electron system by using the coefficients of fractional parentage (21).

3. Calculation of the Spin-Orbit Interaction Matrices. The matrices of the spin-orbit interaction $\mathcal{H}_{so}$ are conveniently calculated by using $S\Gamma\Gamma_J$ scheme where Γ_J is the irreducible representation of the cubic double group arising from the SΓ representation. The matrix elements in the $S\Gamma\Gamma_J$ scheme are given in terms of those in the $S\Gamma M_S M_\Gamma$ scheme as

Table IV. The matrix elements of the p-d interaction in the two-electron system, which are independent of each other.

$t_{1u}t_{2g}-t_{1u}t_{2g}$

$J(\gamma\eta) = F_0+2F_2$ $K(\gamma\eta) = 3G_1+8G_3'$

$J(\gamma\zeta) = F_0-4F_2$ $K(\gamma\zeta) = 5G_3'$

$\langle\alpha\xi\|\beta\eta\rangle = 3F_2$ $\langle\alpha\xi\|\eta\beta\rangle = 3G_1-2G_3'$

$\langle\alpha\eta\|\xi\beta\rangle = 5G_3'$

$t_{1u}e_g-t_{1u}e_g$

$J(\gamma u) = F_0+4F_2$ $K(\gamma u) = 4G_1+9G_3'$

$J(\gamma v) = F_0-4F_2$ $K(\gamma v) = 5G_3'$

$t_{1u}t_{2g}-t_{1u}e_g$

$\langle\alpha\eta\|\gamma u\rangle = \sqrt{3}F_2$ $\langle\alpha\eta\|u\gamma\rangle = -\sqrt{3}(G_1-4G_3')$

$\langle\alpha\eta\|\gamma v\rangle = 3F_2$ $\langle\alpha u\|\eta\gamma\rangle = \sqrt{3}(2G_1-3G_3')$

$\langle\alpha\eta\|v\gamma\rangle = 3G_1-2G_3'$

$\langle\alpha v\|\eta\gamma\rangle = 5G_3'$

$$\langle ab\|cd\rangle = \int d\vec{r}_1 d\vec{r}_2 \psi_a^*(\vec{r}_1)\psi_b^*(\vec{r}_2)\frac{e^2}{r_{12}}\psi_c(\vec{r}_1)\psi_d(\vec{r}_2)$$

$F_0 = F^0(n'p, nd)$

$F_2 = (1/35)F^2(n'p, nd)$

$G_1 = (1/15)G^1(n'p, nd)$

$G_3' = (3/245)G^3(n'p, nd)$

$$\langle S\Gamma\Gamma_J|\mathcal{H}_{so}|S'\Gamma'\Gamma_J\rangle = \sum_{\substack{M_S M_\Gamma \\ M_S' M_\Gamma'}} \langle\Gamma_J M_J|SM_S\Gamma M_\Gamma\rangle$$

$$\times \langle SM_S\Gamma M_\Gamma|\mathcal{H}_{so}|S'M_S'\Gamma'M_\Gamma'\rangle$$

$$\times \langle S'M_S'\Gamma'M_\Gamma'|\Gamma_J M_J\rangle, \tag{87}$$

which is independent of M_J. Clebsch-Gordan coefficients $\langle SM_S\Gamma M_\Gamma|\Gamma_J M_J\rangle$ for the trigonal bases are tabulated in reference (20). The matrix elements of $\mathcal{H}_{so}$ with the $SM_S\Gamma M_\Gamma$ trigonal bases are calculated by using the Wigner-Eckart theorem as

$$\langle SM_S\Gamma M_\Gamma|\mathcal{H}_{so}|S'M_S'\Gamma'M_\Gamma'\rangle = \langle S\Gamma\|V(1T_1)\|S'\Gamma'\rangle$$

$$\times \{[S][\Gamma]\}^{-1/2}\sum(-1)^q\langle SM_S|S'M_S'1q\rangle$$

$$\times \langle\Gamma M_\Gamma|\Gamma'M'T_1-q\rangle, \tag{88}$$

where $V(1T_1)$ is the tensor operator for the spin-orbit interaction. Tables of Wigner coefficients $\langle SM_S|S'M_S'1q\rangle$ and Clebsch-Gordan coefficients $\langle\Gamma M_\Gamma|\Gamma'M'1-q\rangle$ for the trigonal bases are found in reference (3). To calculate reduced matrix elements of the spin-orbit interaction we take into account only the interaction of p-electrons neglecting that of 3d-electrons which is much smaller than any interaction we are considering. Writing $V(1T_1)$ as $V_p(1T_1)$ to this approximation and indicating the electron configurations associated with the $S\Gamma$ terms explicity, one obtains

$$\langle t_{1u}, e_g^{2-n}\, t_{2g}(\bar{S}\bar{\Gamma})S\Gamma\|V_p(1T_1)\|t_{1u}, e_g^{2-n'}\, t_{2g}\,(\bar{S}'\bar{\Gamma}')S'\Gamma'\rangle$$

$$= \delta_{nn'}\,\delta_{\bar{S}\bar{S}'}\,\delta_{\bar{\Gamma}\bar{\Gamma}'}\,\langle t_{1u}\|V_p(1T_1)\|t_{1u}\rangle$$

$$\times \left(\tfrac{1}{2}\bar{S}S[1]\,\tfrac{1}{2}\bar{S}S'\right)(T_1\bar{\Gamma}\Gamma[T_1]T_1\bar{\Gamma}\Gamma'), \tag{89}$$

where $\langle t_{1u}\|V_p(1T_1)\|t_{1u}\rangle = 3i\zeta_p$ and ζ_p is the coupling constant of the spin-orbit interaction operator for a p-electron. The definition of coefficients $(\tfrac{1}{2}\bar{S}S[1]\,\tfrac{1}{2}\bar{S}S')$ and $(T_1\bar{\Gamma}\Gamma[T_1]T_1\bar{\Gamma}\Gamma')$ are given by Tanabe and Kamimura (22) and the numerical values are tabulated in references (22) and (20).

Now, adding the cubic field energies to the diagonal elements, we have the energy matrices characterized by $\Gamma_J = E_1, E_2$, and G, whose dimensions are 23, 22, and 45, respectively.

4. Calculation of Transition Intensities. We first consider the case of the X-ray K-emission. In this case the initial states are $|a_{1g}, e_g^2(^3A_{2g})^2A_{2g}$ and $^4A_{2g}>$ whose energy separation is given by $(3/5)G^2(1s, 3d) \sim 0.04$ eV. Since the process of emitting the K-shell electron is considered to be much faster than the thermalization between these initial states of different spins, we assume that the population of the initial states is proportional to their multiplicities. The final states specified by $k\Gamma_J$, where k distinguishes the eigenstates with the same Γ_J, are given by linear combinations of states $|t_{1u}, e_g^{2-n}(S_1\Gamma_1)t_{2g}^n(S_2\Gamma_2)\overline{S\Gamma}; S\Gamma\Gamma_J>$ with various sets of n, $S_1\Gamma_1$, $S_2\Gamma_2$, $\overline{S\Gamma}$ and $S\Gamma$. Simply denoting these sets as α, one can express the final states as

$$|k\Gamma_J> = \sum_\alpha a(\alpha, k\Gamma_J)\ |\alpha\Gamma_J>. \qquad (90)$$

Since the transitions connect the initial $a_{1g}, e_g^2(^3A_{2g})S_oA_{2g}$ states ($S_o = 1/2, 3/2$) with only the components $|\alpha_o(S_o)\Gamma_J>$ of the final states, where $\alpha_o(S_o) = t_{1u}, e_g^2(^3A_{2g})\ S_oT_{2u}$, the intensities of the transitions to the $k\Gamma_J$ states are given by

$$I(k\Gamma_J) = A\cdot[\Gamma_J] \sum_{S_o=1/2,3/2} |a(\alpha_o(S_o), k\Gamma_J)|^2, \qquad (91)$$

where A is a factor independent of $k\Gamma_J$ and $[\Gamma_J]$ the multiplicity of Γ_J.

In the XPS case the initial state is $|e_g^2\,{}^3A_{2g}>$. When the X-ray is absorbed, a core p-electron has to be excited to a higher bound or unbound orbital of the a_{1g} symmetry. If an electron in such excited a_{1g} orbital is assumed to be non-interacting with the rest electrons, the multiplet structure of this excited state is given by that of the rest electrons which is equivalent to the multiplet structure of the K-emission given by $k\Gamma_J$. In terms of complementary states including the excited a_{1g} orbital, the transition to the excited state is described as

$$\begin{array}{ccc} \text{(I)} & & (\text{F: } S_o) \\ e_g^{2}\,(^3A_{2g})a_{1g}^{2}\,{}^3A_{2g} & \rightarrow & t_{1u},\ e_g^{2}\,(^3A_{2g})S_o T_{2u};\ a_{1g}\,{}^3T_{2u}. \end{array} \tag{92}$$

Reexpressing the initial state as

$$(\text{I}) = \sum_{S_o} |e_g^{2}\,(^3A_{2g})a_{1g}\ S_o A_{2g};\ a_{1g}\,{}^3A_{2g}> \times\ (-1)(2S_o+1)^{1/2}\begin{Bmatrix} 1 & 1/2 & S_o \\ 1/2 & 1 & 0 \end{Bmatrix} \tag{93}$$

where

$$\begin{Bmatrix} 1 & 1/2 & 3/2 \\ 1/2 & 1 & 0 \end{Bmatrix} = -1/\sqrt{2\cdot 3},$$

$$\begin{Bmatrix} 1 & 1/2 & 1/2 \\ 1/2 & 1 & 0 \end{Bmatrix} = 1/\sqrt{2\cdot 3},$$

one can show that the intensities of the transitions to the states corresponding to $k\Gamma_J$ are also given by eq.(90). Finally the excited a_{1g} electron is assumed to escape into vacuum with the probability in dependent of $k\Gamma_J$.

5. Comparison with Experiments. Now, by using the energy matrices and the intensity formula already obtained, the 2p-,3p-photoelectron emission and Kα-,Kβ-emission spectra are calculated. In the calculation we use the values of the spin-orbit coupling constants, ζ_{np}, of a free Ni^{2+} ion determined from Fig.1. The values of Racah parameters B and C (3) for the d-d interaction are assumed to be those for a free Ni^{2+} ion; $B = 0.127$ eV and $C = 4.7$ B. For the p-d interaction, we use several sets of values of the Slater-Condon parameters, which are obtained by multiplying reduction factors to those of the free ion. The Slater-Condon parameters of the free ion given by Watson (26) are shown in Table V. For the cubic field strength, we use the values ranging from 0 up to 4 eV.

The calculated spectra for the 3p-XPS and the Kβ-emission are shown in Figs.8 and 9, and those for the 2p-XPS and the Kα-emission in Figs.10 and 11 in the following subsection. In the figures, vertical lines show the calculated relative intensities of the transitions. The total intensity is normalized to unity. The origin of the energy scale is arbitrarily chosen. In order to facilitate

Table V. Numerical values of the spin-orbit coupling constants and the Slater-Condon parameters for a free Ni^{2+} ion (in eV). For comparison, those for a free Cu^{2+} ion are given in parentheses.

np	2p	3p
ζ_{np}	11.0	1.81
$F_f^2(np,3d)$	7.065 (8.174)	13.596 (14.978)
$G_f^1(np,3d)$	5.188 (6.127)	16.850 (18.474)
$G_f^3(np,3d)$	2.948 (3.484)	10.202 (11.276)

the comparison with experimental spectra, computer plots of the spectra assuming a Gaussian line shape with a 1 eV width for each line are included. The plotted spectra are shown by solid curves in the figures.

Fig.8 shows the change of the calculated spectra when the cubic field strength 10 Dq is changed. In the figure the reduction factors κ_F and κ_G of the Slater-Condon parameters for the p-d interaction are fixed at 0.7 and 0.6, respectively; $\kappa_F = F^2(p,d)/F_f^2(p,d)$ and $\kappa_G = G^{1,3}(p,d)/G_f^{1,3}(p,d)$ where F_f^2 and $G_f^{1,3}$ are those of a free Ni^{2+} ion. The main spectral peaks of Fig.8(a) with 10 Dq = 0 may be labeled with the SL terms of the free ion having the pd^2 electron configuration, as the spin-orbit interaction of a 3p-electron is relatively small. The spin doublet and quartets of D, F and G indicated in Fig.8(a) are the terms obtained by adding a p-electron to the ground state term ^3F of d^2.

As seen in Fig.8, the increase in 10 Dq causes a significant relative increase in intensities on the high energy side. The reason is as follows: when 10 Dq increases, the lowest terms of the pd^2 configuration become almost the $t_{1g}, e_g^2(^3A_{2g})^{2,4}T_{2u}$ terms to which the transitions are allowed. Transitions to these $^{2,4}T_{2u}$ terms are indicated in Fig.8(c) for 10 Dq = 4 eV.

Fig.9 shows the change of the calculated spectra when reduction factors κ_F and κ_G are varied but 10 Dq is fixed to be 2 eV. It is seen that the energy spread of the multiplet structure gets

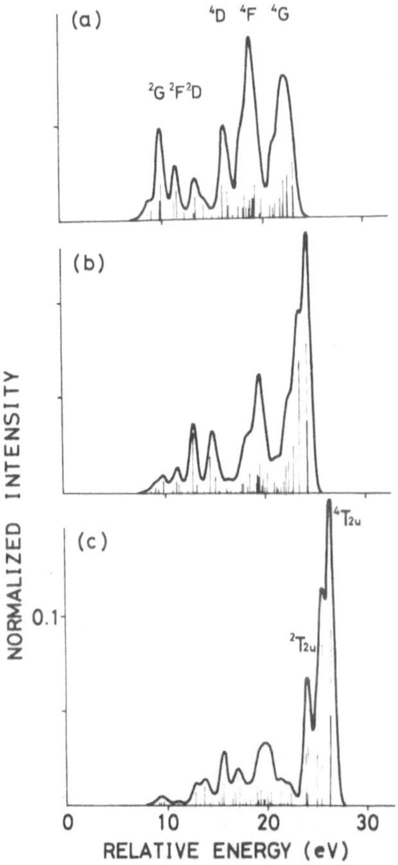

Fig.8. Calculated spectra of the 3p-XPS and the Kβ-emission of Ni^{2+} compounds: (a) 10 Dq = 0 eV, (b) 10 Dq = 2 eV, (c) 10 Dq = 4 eV. The magnitudes of the p-d interaction are fixed at $\kappa_F = 0.7$ and $\kappa_G = 0.6$. For the definition of κ_F and κ_G, see the text. Solid curves are obtained by assuming a Gaussian line shape with 1 eV width for each line.

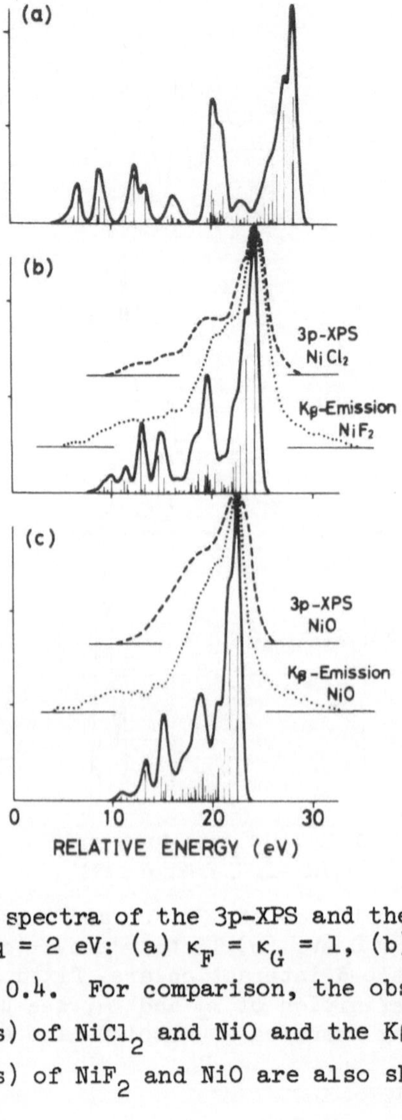

Fig.9. Calculated spectra of the 3p-XPS and the Kβ-emission of Ni^{2+} compounds for 10 Dq = 2 eV: (a) $\kappa_F = \kappa_G = 1$, (b) $\kappa_F = 0.7$, $\kappa_G = 0.6$, (c) $\kappa_F = 0.5$, $\kappa_G = 0.4$. For comparison, the observed 3p-XPS spectra (23) (broken curves) of $NiCl_2$ and NiO and the Kβ-emission spectra (24) (dotted curves) of NiF_2 and NiO are also shown.

smaller when the p-d interaction is reduced. Another tendency, that intensities of the transitions on the high energy side increase as the reduction factors decrease, may be understood as the result of a relative increase of 10 Dq to the p-d interaction.

In Fig.9 the observed spectra of the 3p-XPS and the Kβ-emission are shown by broken and dotted curves, respectively. The gross feature of the computed spectrum is in good agreement with the experimental one, if 10 Dq = 2 eV, $\kappa_F = 0.7$, and $\kappa_G = 0.6$ are assumed. This value of 10 Dq is twice as large as that of divalent nickel compounds, and approximately equal to that of trivalent transition-metal compounds in the absence of a core hole. The large reduction of the Slater-Condon parameters together with the increase of 10 Dq indicates a large relaxation of the d-orbitals in the presence of a core hole. We may imagine two kinds of orbital relaxation. One is the localization of d-electrons by the attractive force of the hole, which increases the Slater-Condon parameters. This effect is seen by comparing the parameters of a free Ni^{2+} ion with those of a free Cu^{3+} ion which also are shown in Table V. The comparison shows that the increase of the parameter values is relatively small.

Another kind of relaxation, which seems most important, is due to the hole-induced covalency already discussed in section III.B. In this case the relaxed d-orbitals are given by the antibonding molecular orbital in eq.(74), or in eq.(75). Detailed explanations of the effects of this relaxation will be given in the next subsection which discusses the 2p-XPS and Kα-emission spectra.

6. <u>Satellites due to the Hole-Induced Covalency</u>. The computed spectra of the 2p-XPS and the Kα-emission are given in Figs.10 and 11. Fig.10 shows the change in the spectra when 10 Dq is changed. In the strong field limit one expects only two peaks, $P_{3/2}$ and $P_{1/2}$; peak $P_{3/2}$ on the high energy side is a group of the transitions to four Γ_J states, $t_{1u}(G)e_g{}^2({}^3A_{2g}T_2)\ \Gamma_J = E_1$, E_2 and 2G, and peak $P_{1/2}$ a group of those two Γ_J, $t_{1u}(E_1)e_g{}^2({}^3A_{2g}T_2)\ \Gamma_J = E_2$ and G. This situation is already seen in Fig.10(c) for 10 Dq = 4 eV, although the spectrum still has a small peak at the low energy side of each main peak. The intensities of these small peaks further decrease with an increase in 10 Dq. The ratio of the intensity of $P_{3/2}$ to that of $P_{1/2}$ is approximately 2.

Fig.11 shows the change of the computed spectra when the reduction factors are changed but 10 Dq is fixed at 2 eV. When the reduction factors are reduced, the energy spread of each main peak, including small structure on its low energy side, becomes smaller. Further, a decrease of the Slater-Condon parameters produces the same effect as that which an increase of 10 Dq induces.

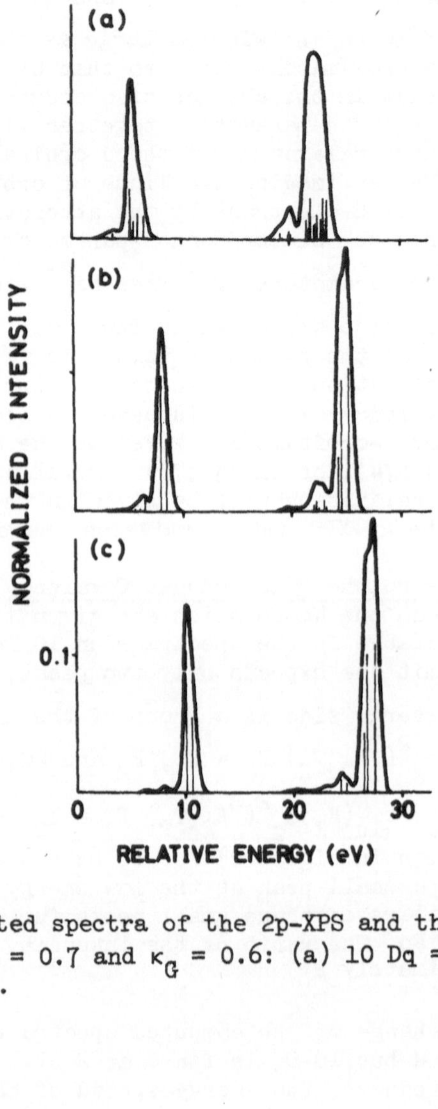

Fig.10. Calculated spectra of the 2p-XPS and the Kα-emission of Ni^{2+} compounds for $\kappa_F = 0.7$ and $\kappa_G = 0.6$: (a) 10 Dq = 0 eV, (b) 10 Dq = 2eV, (c) 10 Dq = 4 eV.

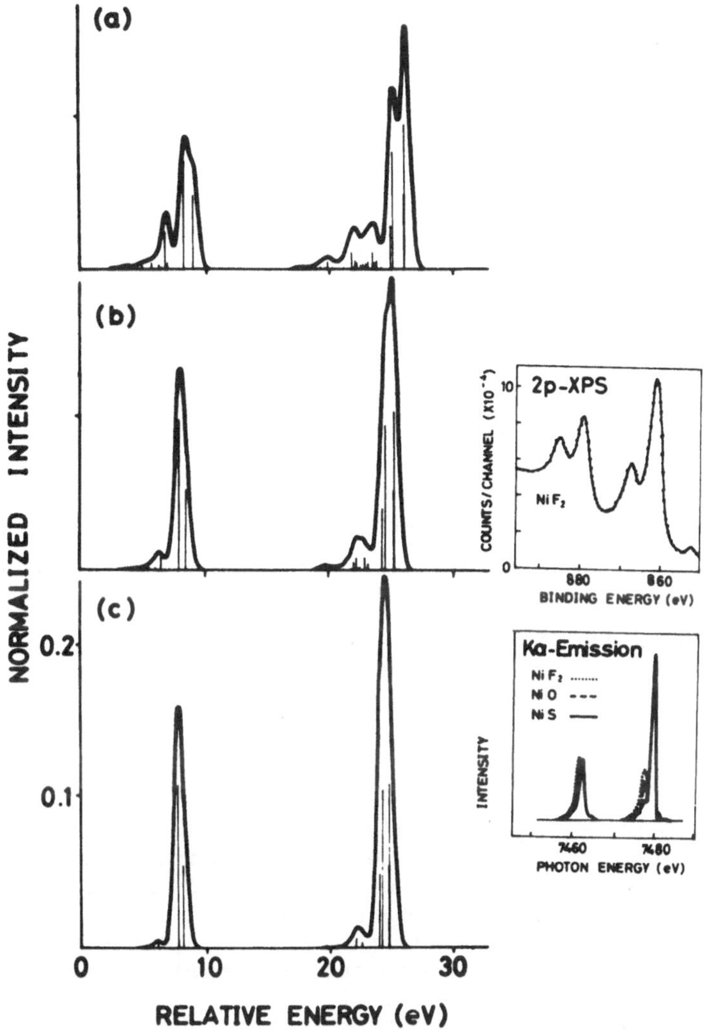

Fig.11. Calculated spectra of the 2p-XPS and the Kα-emission of Ni^{2+} compounds for 10 Dq = 2 eV: (a) $\kappa_F = \kappa_G = 1$, (b) $\kappa_F = 0.7$, $\kappa_G = 0.6$, (c) $\kappa_F = 0.5$, $\kappa_G = 0.4$. For comparison, the observed 2p-XPS spectrum (25) of NiF_2 and the Kα-emission spectra (24) of NiF_2, NiO and NiS are also inserted.

The experimental spectra of the 2p-XPS (25) and the Kα-emission (24) are also given in Fig.11 for comparison. There is a qualitative difference between these two kinds of experimental spectra: a satellite line with remarkable intensity is observed in the 2p-XPS on the ∿6 eV lower energy side of each main line, while it is absent in the Kα-emission. A similar situation has also been found in other transition-metal compounds. The origin of this difference may be ascribed to the difference in the initial states. In the Kα-emission a 1s-hole is initially present which induces the orbital relaxation of 3d-electrons. Although the hole is transferred to the 2p-orbital in the final state, the additional change in the already relaxed 3d-orbital induced by this hole transfer seems to be small. On the other hand, no hole is present in the initial state of the 2p-XPS and a 2p-hole is suddenly created in the final state inducing the orbital relaxation of 3d-electrons. In the latter case the shake-up satellite due to the orbital relaxation may be expected to appear.

Once the satellite in the 2p-XPS is ignored, the calculated spectrum with 10 Dq = 2 eV, $\kappa_F = 0.5$ and $\kappa_G = 0.4$ seems to be in fair agreement with the observed spectra of both the 2p-XPS and the Kα-emission. The observed intensity decrease of a small peak associated with each main peak of the 2p-XPS when one goes from NiF_2 to NiS seems to be reasonable, as a large covalency is expected in NiS (compared to NiF_2).

The large reduction of the Slater-Condon parameters and the remarkable increase of 10 Dq found in the comparison with experiments are explained by taking into account the hole-induced covalency which was discussed in section III.B. As already mentioned in III.D.-1, the t_{2g} and e_g orbitals, with which we started, are the antibonding orbitals given in eq.(74) or eq.(73). The covalency parameter in these orbitals is roughly $(\gamma+\gamma')$ where γ is the covalency parameter in the absence of a core hole and γ' its increment in the presence of the hole. The increase of the covalency may decrease the Slater-Condon parameter and increase 10 Dq.

It is more important to note that the multiplet structure we have calculated is associated with the configuration given by Ψ_m in eq.(72). The excited states associated with that given by Ψ_s in eq.(72) can be observed as satellites when the hole is suddenly created. The multiplet structure of these excited states has been omitted from consideration. However, the integrated intensity of the satellites relative to that of the main lines is easily seen to be γ'^2, if one looks at the form of Ψ in eq.(71), and recalls that the initial state in the absence of the hole is Ψ_1 in eq.(68). Then, we notice that a quantitative relation exists between the satellite intensity and the reduction factor for the Slater-Condon parameter. Actually we have confirmed the existence of such a relation in our recent work (17).

The present analysis of the p-XPS and the K-emission spectra has pointed out a possibility of changing the strength of the chemical bond in transition-metal compounds by creating core holes by X-ray excitation. We hope that further studies of this possibility may lead us to a new field of photochemistry.

In conclusion the author thanks Dr. T. Yamaguchi for reading the manuscript, Mr. S. Asada for checking mathematical expressions in this article and Miss T. Oto for careful typewriting.

REFERENCES

1. R.S.Knox, Theory of Excitons, Solid State Physics suppl.5 (F. Seitz and D. Turnbull,eds.), Academic Press (1963).
2. R.Loudon, Advances in Physics 17, 243 (1968).
3. S.Sugano, Y.Tanabe and H.Kamimura, Multiplets of Transition-Metal Ions in Crystals, Academic Press (1970).
4. C.Satoko and S.Sugano, J.Phys.Soc.Japan 34, 701 (1973).
5. E.U.Condon and G.H.Shortley, The Theory of Atomic Spectra, Cambridge Univ. Press (1935).
6. M.Rotenberg, R.Bivins, N.Metropolis and J.K.Wooten,Jr., The 3-j and 6-j Symbols, The Technology Press, MIT (1959).
7. M.Watanabe, A.Ejiri, H.Yamashita, H.Saito, S.Sato, T.Shibaguchi and H.Nishida, J.Phys.Soc.Japan 31, 1085 (1971).
8. C.E.Moore, Atomic Energy Levels, Vol.II, National Bureau of Standards, pp.184, 169 (1952).
9. Y.Iguchi, T.Sasaki, H.Sugawara, S.Sato, T.Nasu, A.Ejiri, S.Onari, K.Kojima and T.Oya, Phys.Rev.Letters 26, 82 (1971).
10. C.Satoko, Sol.State Comn. 13, 1851 (1973).
11. C.Satoko, Doctoral Thesis submitted to The Univ. of Tokyo (1974).
12. Y.Onodera and Y.Toyozawa, J.Phys.Soc.Japan 22, 833 (1967).
13. S.Sugano and R.G.Shulman, Phys.Rev. 130, 517 (1963).
14. S.Sugano and Y.Tanabe, J.Phys.Soc.Japan 20, 1155 (1965).
15. R.G.Shulman and S.Sugano, Phys.Rev. 130, 506 (1963).
16. J.Hubbard, D.E.Rimmer and F.R.A.Hopgood, Proc.Phys.Soc.(London) 88, 13 (1966).
17. S.Asada and S.Sugano, to be published.
18. T.Åberg, Phys.Rev. 156, 35 (1967).
19. P.S.Bagus, A.J.Freeman and F.Sasaki, Phys.Rev.Letters 30, 850 (1973);
S.P.Kowalczyk, L.Ley, R.A.Pollak, F.R.McFeely and D.A.Shirley, Phys.Rev. B7, 4009 (1973).
20. S.Asada, C.Satoko and S.Sugano, J.Phys.Soc.Japan 38, 855 (1975).
21. Y.Tanabe and S.Sugano, J.Phys.Soc.Japan 9, 753 (1954).
22. Y.Tanabe and H.Kamimura, J.Phys.Soc.Japan 13, 394 (1958).
23. M.Okusawa, T.Ishii and T.Sagawa, Proc.Int.Conf. on X-ray Processes in Matter, Helsinki, p.298 (1974).

24. Y.Gohshi and J.Kashiwakura, Proc.Int.Conf. on X-ray Processes in Matter, Helsinki, p.330 (1974).
25. A.Rosencwaig, G.K.Wertheim and H.J.Guggenheim, Phys.Rev.Letters 27, 479 (1971);
S.Hüfner and G.K.Wertheim, Phys. Rev. B8, 4857 (1973).
26. R.E.Watson, MIT Technical Report No.12 (1959).

MOLECULAR EXCITONS IN SMALL AGGREGATES*

M. Kasha

Department of Chemistry and Institute of Molecular Biophysics, Florida State University

Tallahassee, Florida

ABSTRACT

The molecular exciton model, which deals with the excited state resonance interaction in weakly coupled electronic systems, is described as an interpretative tool for the study of the spectra and photochemistry of composite molecules. Under composite molecules are grouped loosely bound groups of light-absorbing units, held together by hydrogen bonds or by van der Waals forces. Another group of composite molecules included in the study consists of covalently bound light-absorbing units.

A skeletal outline of the simplest quantum mechanical framework for the description of the model is presented. Dimers of various geometries, cyclical higher aggregates, linear chain polymers, helical polymers, and molecular lamellar arrays are reviewed. The exciton splitting diagrams and electric dipole selection rules are discussed quantum mechanically and by means of a transition dipole vector model.

Applications to absorption and luminescence spectroscopy of molecular aggregates are cited. Photochemical sensitization and photobiological applications are suggested, and areas of new research are enumerated.

I. ATOMIC AND MOLECULAR EXCITONS

The exciton concept was introduced by J. Frenkel (1) in 1931 as a general excitation delocalization mechanism to account for the capacity of solids to absorb high energy quanta without local melting. Since that time the concept has been applied to diverse physical systems, so that it is necessary to classify (2,3) the exciton model used according to the physical situation. A diagrammatic classification of excitons (3) is offered in Fig. 1 to indicate the general features of the various exciton cases. This diagram has been revised in accordance with the elegant mathematical comparison given by Förster (4) of the various molecular exciton coupling cases.

Exciton phenomena are generally considered to apply to crystalline solid states. However, we must distinguish sharply between ionic crystals and molecular crystals in discussing exciton phenomena (5). For ionic solids, with strong electronic interaction between components of the lattice, what we call the atomic exciton model (or Wannier exciton) is applicable. This is called the weakly-bound exciton case (2) because the exciton can be treated as a positive-center-electron pair travelling through the crystal, with a significant charge separation (beyond nearest neighbors) between the electron and its positive hole. Rydberg-like exciton cases are observed in the spectrum of the solid, converging to a conduction band continuum. This model has been mistakenly applied to molecular solids (as chlorophyll aggregates) by some researchers.

The molecular exciton model (or Davydov exciton) is applicable to molecular crystals in which the intermolecular electronic interaction is weak (van der Waals crystals), so that the component molecules of the crystal 'preserve their individuality' (6). This is called the tightly-bound case (2) because the electrons are associated with specific molecular centers and do not migrate; the molecular exciton is a migrating excitation center whose characteristic migration mechanism depends on the specific strength of excited intermolecular coupling.

The molecular exciton model was first developed by A.S. Davydov (7) to explain why the spectrum of a molecular crystal (naphthalene) could consist of very sharp line-like spectra at low temperatures in spite of intermolecular interaction. He described the resonance splitting in the crystal of excited states, non-degenerate in the component molecules, and showed that although a quasi-continuous exciton band was generated, the possible optical transitions by electric dipole radiation were severely limited by symmetry, with sharp narrow band absorption spectra still observable for the solid state. In his monograph (6) Davydov showed the generality of the approach by applying the molecular exciton model to electronic and vibrational transitions for molecules in crystals. He also applied the model to intramolecular interaction between molecular units whose π-electron systems could be considered to be only weakly coupled by valence forces (e.g. excited states of para-

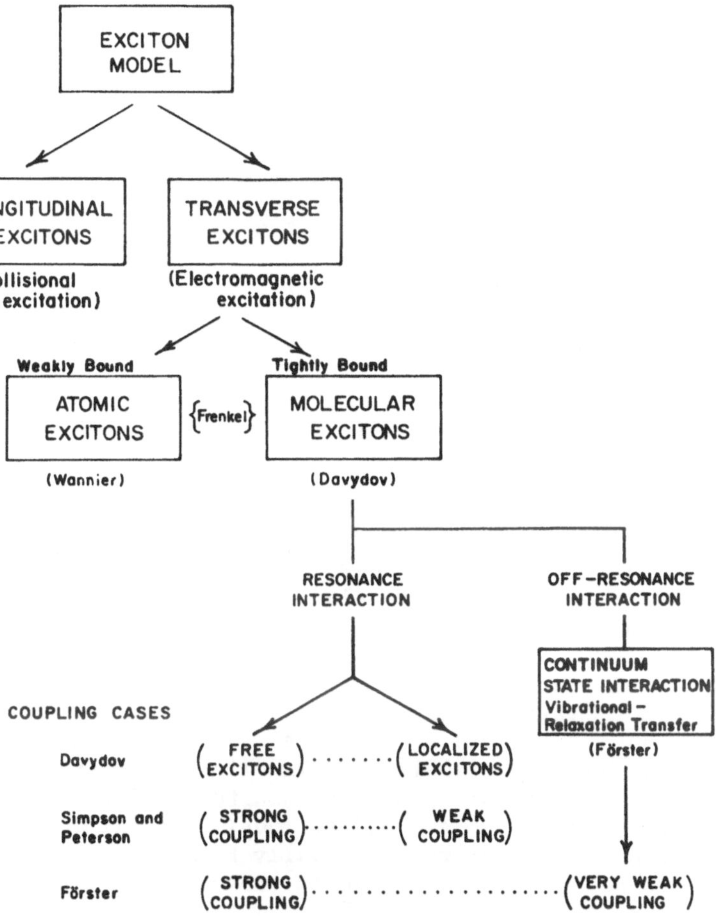

Fig. 1 Classification of excitons.

polyphenyls), in spite of the covalent binding between molecular units.

The molecular exciton model may be defined as describing resonance interaction between excited states of composite molecule systems, in which non-degenerate states of the individual molecular components become split in the composite molecule, usually considered to be caused by dipole-dipole interaction. We may formulate a Davydov rule for electric dipole transitions allowed to the components of the exciton band of states: the number of allowed electric dipole transitions from the ground state to the states of the exciton band will be equal to the number of molecules per unit cell in the molecular aggregate. Thus, if there is one molecule per unit cell in an ordered molecular aggregate, then for an N-component exciton band for N molecules in the aggregate, only a single exciton state can be observed by radiative electric dipole transitions; two molecules per unit cell result in the observation of two observed electronic transitions for the exciton band, etc. The precise quantitative characteristics of the exciton band structure will depend on the intermolecular orientation, center-to-center distance, and intensity of the component molecule optical transition involved in the state resonance interaction.

There is now a highly evolved literature on molecular exciton theory and applications (8). Most of the research has been applied specifically to molecular crystals. However, the molecular exciton model serves very powerfully to elucidate the spectral properties of smaller molecular aggregates. The molecular exciton formalism has been applied to the study of the absorption and emission spectra of van der Waals dimers of organic dye molecules, of hydrogen-bonded dimers in various molecular geometries, of long-chain molecular polymers, of helical molecular arrays including polypeptides and nucleic acids, and of molecular lamellar aggregates. Here will be given a brief survey of the results obtained from the application of the molecular exciton model to small molecular aggregates of defined geometry, especially because many of the publications appear in sources exotic to the molecular spectroscopist's world.

II. THE STRONG-COUPLING MOLECULAR EXCITON: DIMERS

Davydov (6) considered the free exciton case as one involving collective excitation of an aggregate of molecular units, since the rate of transfer of excitation energy between molecules could in the extreme case far exceed the relaxation time for the excited molecules to a new equilibrium orientation and position in a molecular lattice. Simpson and Peterson (9) considered the extreme strong coupling case as being associated with Born-

Oppenheimer separability of electronic and vibrational wave functions. Their criterion for the strong coupling case required that the exciton band displacement exceed the vibronic bandwidth of the molecular electronic transition involved in the coupling. In other words, the effect of the strong coupling exciton interaction would be, e.g. to shift electronic-vibrational bands with the Franck-Condon envelope following the electronic displacement (10). McRae and Siebrand (11) have suggested a modification of the Simpson and Peterson criteria.

We shall examine the strong coupling case in a skeletal outline as applied to molecular dimers. This case will be especially useful to consider because it gives a pellucid example for the effect of exciton state generation from molecular state components. Secondly, the case applies literally to a physically visualizable and realizable example, the van der Waals organic-dye parallel (card-pack) dimer. In such dimers, the monomer $S_1 \leftarrow S_0$ absorption oscillator strength is very high (frequently of the order of unity), and the physical observations generally universal: (a) the lowest electronic absorption band strongly blue-shifts in the dimer, (b) there is a slight red-shifted absorption band shoulder in the dimer, (c) the fluorescence of the molecule is quenched in the dimer, and (d) phosphorescence ($T \rightarrow S$ emission) is greatly enhanced in the dimer (12,13).

Since the strong-coupling exciton case represents that in which the Born-Oppenheimer approximation is assumed to be valid, and the electronic and vibrational wave functions can be factored, we shall examine the formalism for the electronic wave functions only (14). Since we are interested in spectroscopic consequences of exciton state formation, we shall be interested primarily in <u>stationary state exciton theory</u>; the <u>non-stationary state</u> theory for <u>energy transfer dynamics</u> follows naturally (1,4,5). We use the <u>molecular exciton</u> terminology even for the dimer case because of the parallels in quantum mechanical formalism to molecular exciton theory of crystals.

For a molecular dimer the ground-state pure electronic wave function may be written uniquely for a van der Waals dimer (electron localization)

$$\Psi_G = \psi_1 \psi_2 \quad , \tag{1}$$

where ψ_1 represents the <u>ground state of molecule 1</u>, and ψ_2 the <u>ground state of molecule 2</u>; these symbolisms are thus great abstractions (molecules 1 and 2 identical).

The Hamiltonian operator for the dimer is taken as

$$H = H_1 + H_2 + V_{12} \quad . \tag{2}$$

with H_1 and H_2 the energy operators for isolated molecules 1 and 2 separately, and V_{12} represents the intermolecular perturbation. Since we are considering a weak interaction across a van der Waals dimer, we are justified in treating the intermolecular coulombic term V_{12} as a small perturbation term. The energy of the ground state is obtained from the Schrodinger equation

$$E_G = \iint \psi_1\psi_2 H\psi_1\psi_2 dt_1 dt_2 \quad , \tag{3}$$

which reduces to

$$E_G = E_1 + E_2 + \iint \psi_1\psi_2 V_{12}\psi_1\psi_2 dt_1 dt_2 \ . \tag{4}$$

The last term represents the van der Waals energy (symbol D) lowering in the ground state of the dimer, and E_1 and E_2 are the ground state energies of the isolated molecules.

In the non-stationary excited states of the dimer a degeneracy exists, with either molecule 1 or molecule 2 excited (indicated by dagger)

$$\psi_1^\dagger\psi_2 \ , \quad \psi_1\psi_2^\dagger \quad . \tag{5}$$

The excited state dimer pure electronic wave functions (exciton state wave functions) may be written

$$\Psi_E = r\psi_1^\dagger\psi_2 + s\psi_1\psi_2^\dagger \ , \tag{6}$$

where r and s are written as arbitrary coefficients to be determined. The Schrodinger equation for the excited state is

$$H(r\psi_1^\dagger\psi_2 + s\psi_1\psi_2^\dagger) = E(r\psi_1^\dagger\psi_2 + s\psi_1\psi_2^\dagger). \tag{7}$$

Multiplying both sides of eq. (7) by $\psi_1^\dagger\psi_2$, and integrating over coordinates for molecules 1 and 2, and repeating this process for $\psi_1\psi_2^\dagger$ leads to two simultaneous equations containing the terms symmetrical in 1 and 2 (identical molecules). Setting the determinant of the coefficients r and s in the simultaneous equations equal to zero for non-trivial solutions, leads (15) to the roots

$$\begin{aligned} E' = E_1^\dagger + E_2 &+ \iint\psi_1^\dagger\psi_2 V_{12}\psi_1^\dagger\psi_2 dt_1 dt_2 \\ &+ \iint\psi_1^\dagger\psi_2 V_{12}\psi_1\psi_2^\dagger dt_1 dt_2 \\ E'' = E_1^\dagger + E_2 &+ \iint\psi_1^\dagger\psi_2 V_{12}\psi_1^\dagger\psi_2 dt_1 dt_2 \\ &- \iint\psi_1^\dagger\psi_2 V_{12}\psi_1\psi_2^\dagger dt_1 dt_2 \ . \end{aligned} \tag{8}$$

These energy expressions represent first-order perturbation energies, with the interaction term averaged over the zeroth-order wave functions (eqs. 5,6; and 12 with coefficients evaluated). The van der Waals energy-lowering term thus also appears in the exciton state expressions E' and E" as

$$D' = \iint \psi_1^{\dagger}\psi_2 V_{12}\psi_1^{\dagger}\psi_2 dt_1 dt_2 \quad , \tag{9}$$

representing the interaction of the charge distribution of the excited state of molecule 1 with the ground state of molecule 2 (or vice versa).

In addition, the exciton displacement term appears:

$$\varepsilon = \iint \psi_1^{\dagger}\psi_2 V_{12}\psi_1\psi_2^{\dagger} dt_k dt_2 \ , \tag{10}$$

representing energy of interaction arising from the delocalization of excitation between molecules 1 and 2.

The electronic transition energies for the dimer exciton states can be written symbolically:

$$\Delta E'_{dimer} = \Delta E_{mono.} + \Delta D + \varepsilon$$

$$\Delta E''_{dimer} = \Delta E_{mono.} + \Delta D - \varepsilon \quad , \tag{11}$$

and the stationary state exciton wave functions for the dimer are

$$\Psi_{E'} = \frac{1}{\sqrt{2}} (\psi_1^{\dagger}\psi_2 + \psi_1\psi_2^{\dagger})$$

$$\Psi_{E''} = \frac{1}{\sqrt{2}} (\psi_1^{\dagger}\psi_2 - \psi_1\psi_2^{\dagger}) \quad . \tag{12}$$

These are the symmetry-adapted zeroth-order pure electronic wave functions for a van der Waals dimer. The node in the exciton wave function is an excitation node (not an electron node as in orbital wave functions); the physical meaning of the node depends on the arbitrary assignment of phase relations of transition dipoles as will be seen in the following.

The exciton displacement term cannot be easily evaluated if the full intermolecular coulombic potential for all particle interactions V_{12} is used. However, if the point-dipole approximation is made (6,12,16) (from the multipole expansion of the coulombic potential),

$$V_{12(coulombic)} \approx V_{12(dipole\text{-}dipole)} \ ,$$

with

$$V_{12(di-di)} = -\frac{e^2}{r_{12}^3}\sum_{i,j}(2z_1^i z_2^j - x_1^i x_2^j - y_1^i y_2^j), \tag{13}$$

for Cartesian coordinates in a dimer (with z taken as the inter-dimer axis), the ε term immediately can be factored into single molecule coordinates (and for transition dipoles directed uniquely along the coordinate x)

$$\varepsilon = \frac{1}{r_{12}^3}(\int\psi_1 \sum_i ex_1^i \psi_1^{\dagger} dt_1)(\int\psi_2^{\dagger} \sum_j ex_2^j\psi_2 dt_2) . \tag{14}$$

This result for the exciton displacement energy seems designed to delight the spectroscopist, since the energy of exciton displacement can be determined by transition moment integrals (intensity) for the individual molecule transition! The phase relation is perfectly arbitrary, so to make the energy for $\Psi_{E'}$ (eq. 12) correspond to lowering of energy, we choose the transition dipole relation for molecules 1 and 2 with $M_1 = -M_2$, so that eq. (14) becomes

$$\varepsilon = + \frac{|M_1 \cdot M_2|}{r_{12}^3} = - \frac{|M_1|^2}{r_{12}^3} . \tag{15}$$

A schematic energy level diagram for this result is shown in Fig. 2. For a dimer with arbitrary orientation of molecular axes with respect to an x,y,z coordinate frame,

$$\varepsilon = - \frac{|M_1|^2}{r_{12}^3} (2\cos\Theta_1^z\cos\Theta_2^z - \cos\Theta_1^x\cos\Theta_2^x - \cos\Theta_1^y\cos\Theta_2^y), \tag{16}$$

with M_1 representing the transition moment in an isolated molecule and e.g., $\cos\Theta_1^x$, representing the cosines of the angle which M_1 makes with the x-axis.

Of equal interest to the exciton displacement energetics is the question of exciton band transition selection rules. For the dimer these are obtained from, e.g.,

$$M' = \frac{1}{\sqrt{2}} \iint\Psi_G(\sum_i ex_1^i + \sum_j ex_2^j)\Psi_{E'} dt_1 dt_2, \tag{17}$$

which for the two exciton states of the parallel (card-pack) dimer become (16)

$$M' = \frac{1}{\sqrt{2}} (M_1 + M_2) = \frac{1}{\sqrt{2}} (M_1 - M_1) = 0 \tag{18a}$$

$$M'' = \frac{1}{\sqrt{2}} (M_1 - M_2) = \frac{1}{\sqrt{2}} (M_1 + M_1) = \frac{2M_1}{\sqrt{2}} , \tag{18b}$$

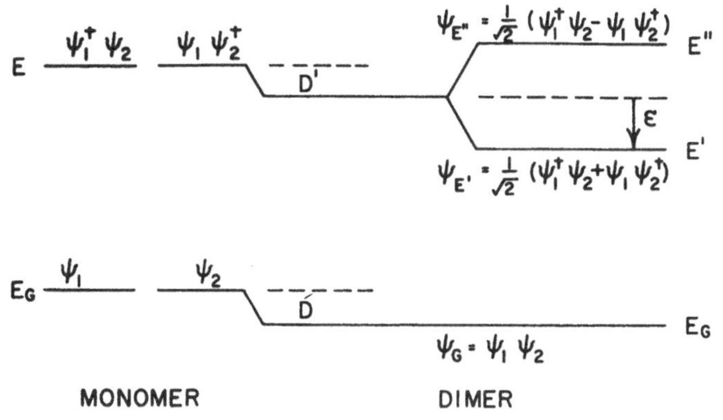

Fig. 2 Exciton resonance splitting in molecular dimers.

since for the parallel or card-pack dimer we had defined $M_1 = -M_2$ as the phase convention. Thus, if we square the transition moment M" we obtain $|M''|^2 = 2|M_1|^2$ or f" = 2f, i.e., the oscillator strength of the dimer transition is twice (per two molecules) that of the corresponding transitions in the monomer. Thus, in this approximation, no change in intensity is predicted. We shall observe later that a second-order diminution in f" would be expected for this case.

III. PHENOMENOLOGY OF EXCITON STATES IN DIMERS

The molecular exciton model permits a simple vector model analysis of the exciton band structure, since the exciton state (energy) displacement term is directly related to the transition moment (a vector integral defining intensity) of the individual molecule transitions. In addition to the simple parallel or card-pack dimer for van der Waals dye molecule dimers (12,13,16), a variety of dimer geometries has been considered for covalently bonded and hydrogen-bonded composite molecules or dimers (14,17,15).

The exciton vector model diagrams (17) of Fig. 3 (van der Waals terms omitted for simplification) will serve to indicate how the exciton displacement energy term and the electric dipole transition moment, for exciton states of the dimer, may be deduced from vector diagrams representing interaction between component molecule transition moments for the electronic state in question.

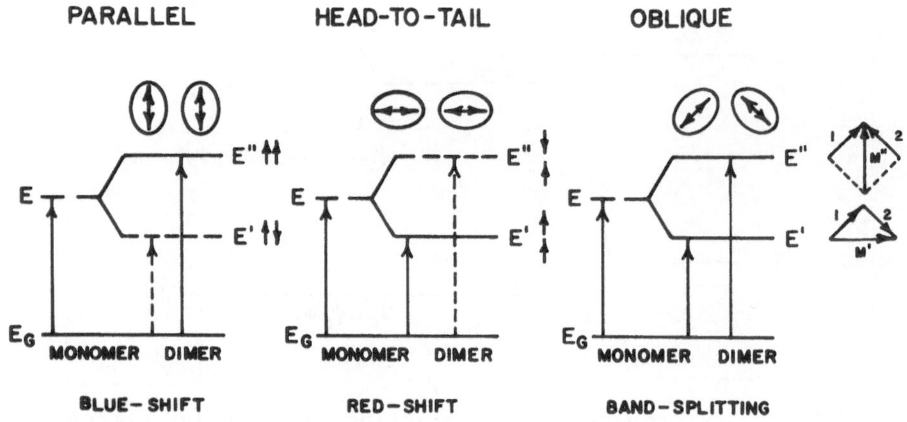

Fig. 3. Exciton splitting in dimers of various geometries.

Consider the parallel dimer case of Fig. 3. The exciton energy displacement is electrostatically repulsive for two transition moment vectors in phase, so E" is raised in energy in the dimer; the out-of-phase transition moment array is attractive, so E' represents a lowering of energy in the dimer.

For the parallel dimer case, the electric dipole selection rules indicate a vector sum for E" which is finite and effectively equal to the sum of the two transition moments (eq. 18b) of the individual molecules; whereas for E' the vector sum of the two transition moments (eq. 18a) is zero. Thus, the monomer $S_1 \leftarrow S_0$ absorption spectrum should appear <u>blue-shifted</u> in the dimer, which is a common observation for van der Waals dye dimers (12,13,16,18). An interesting observation concerning these selection rules can be made by comparison with deductions from the classical theory of light interaction. If we imagine a pair of molecules in van der Waals proximity, both molecules will be simultaneously under the influence of virtually an identical electric field perturbation $\vec{E}$. Therefore, the E" state would be predicted to be excitable in the parallel dimer, whereas the E' state could not be excited by the electric dipolar field of the light wave (opposed induced transition moments being inconceivable). This seems to correspond to the quantum mechanical deduction. However, both the E" and E' states are <u>real</u> energy states of the dimer (not virtual states) and both are <u>required</u> by quantum mechanical theory, and not classical theory. The E' state is of especial interest to spectroscopists: it is a symmetry-forbidden singlet state. The importance of such a state lies in its role in radiationless transitions, to be discussed. If the geometry of the dimer is slightly skew, instead of parallel, the E' state selection rule is relaxed, and a small intensity may be observed at long wavelengths (compared with the monomer absorption) (12,18).

For the head-to-tail dimer (Fig. 3) we can deduce easily that the lower exciton state of the dimer is allowed, and the upper state forbidden. In order to observe such a case a dimer geometry must be fixed by covalent bonds or hydrogen bonding (17,15).

The oblique dimer goemetry indicates that both exciton states now will become allowed in the dimer, with intensities depending on the angular relation between transition moments (13,16). A band doubling is expected in such cases relative to the single molecule electronic transition. This corresponds to the Davydov "Two molecules per unit cell" case in the crystal: as we shall see, even with N-molecule interaction in large aggregates, only two exciton components can be observed by electric dipole radiation if the oblique pattern is repeated N/2 times.

Formulas for the exciton splitting and transition moments of various dimer geometries have been published (15). As in all cases of dipole-dipole interaction, at the angle of inclination of $\alpha = \text{arc cos } 1/\sqrt{3} = 54.7^{\circ}$ of transition moments to a common axis the exciton splitting becomes equal to 0 in spite of close proximity or magnitude of transition moments (strength of coupling). This zero interaction effect (cf. Fig. 4) persists for polymeric and lamellar aggregates, and is an effect not expected intuitively by many experimentalists working with spectral properties of molecular aggregates.

Thus we see that the molecular exciton model offers a state interaction perturbation theory which starts with the frequency and intensity of a single molecule transition, and permits a semi-quantitiative and qualitatively accurate deduction of the absorption spectrum of the molecular aggregate, using the dimer geometry, intermolecular center-to-center distance, and oscillator strength for single molecule transitions. The point-dipole point-dipole approximation (which can be modified by extended dipole calculations) constitutes a limitation on quantitative accuracy. For the strong-coupling model to apply, the oscillator strengths must be high and the intermolecular separation small, so that the displacement vs. bandwidth criteria are applicable (9,11).

The weak coupling case with vibronic state interaction has been treated with great attention in the literature, cf. McClure (14), and especially Förster's review (4). This case is theoretically

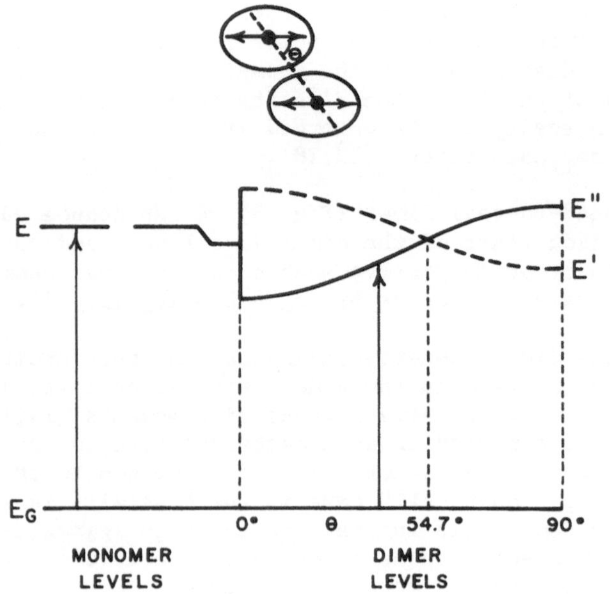

Fig. 4. Exciton splitting for dimers with co-planar inclined transition dipoles.

interesting because it is one of the (somewhat rare) cases where the Born-Oppenheimer separability of electronic and vibrational motion cannot be assumed. McClure (14) has shown the effect of this lack of separability on the vibronic states of a composite double molecule. Nevertheless, the presence of a forbidden lowest exciton vibronic state can still occur (19), affecting the excitation behavior of the molecule.

The luminescence behavior of molecules, especially dye molecules with their strong absorption bands, can be profoundly affected by the generation of exciton states in molecular aggregates (12,13,16). Fig. 5 indicates that exciton band formation

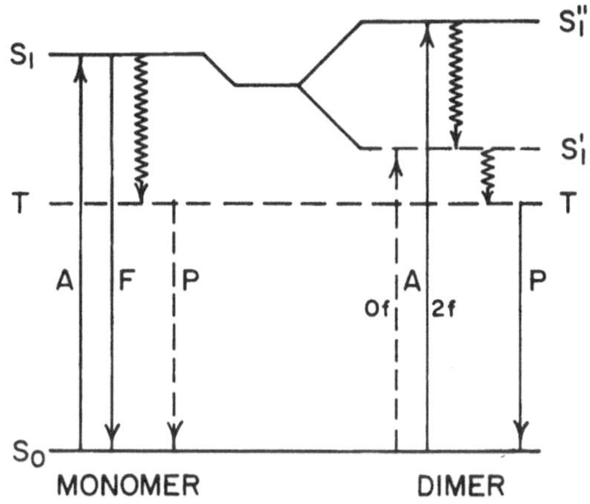

Fig. 5. Absorption and luminescence transitions for parallel dye dimer and dye monomer.

in a dimer (or higher aggregate)can materially affect the S-T split for a monomer; this would directly affect the spin-orbital mixing of states of the molecule. A greater effect is expected, however, if the lowest state of an exciton manifold is forbidden. Then the radiationless intersystem crossing from the forbidden singlet exciton state to the lower-lying triplet is expected. It is interesting that this is a kinetic competition effect which is always present in the intersystem crossing in polyatomic molecules (20), and in this case constitutes a non-spin-orbital enhancement of the intersystem crossing for the monomer when present in the dimer (or aggregate). In other words, the singlet exciton state in the dimer becomes so metastable that the rate of fluorescence may change by six orders of magnitude (even in the distorted or skew dimer), so that the rate of intersystem crossing (fixed by spin-orbital parameters) becomes dominant, leading to enhanced triplet state population.

We have not considered in this treatment (cf. Fig. 5) the splitting of the triplet state dimer degeneracy by the molecular exciton model. If we limited the model to the dipole-dipole coupling represented by eq. (13), the oscillator strength criterion for exciton displacement (eq. 14) would predict zero shift, since the oscillator strength is effectively 0 for singlet-triplet transitions. However, it is found in crystals that finite triplet state exciton splittings can be observed and can be accounted for through an electron exchange mechanism. We have neglected such consideration for triplet exciton effects for small aggregates, and regard this as a worthy field of research in these cases.

Biological applications of dye dimer spectroscopic properties are very extensive. Many of the observed phenomena are used empirically but can be interpreted and extended by the molecular exciton model. Metachromatia, the change in color of dyes upon staining in biological tissue, has been classified by Michaelis (21) as arising from absorption as monomers (α band), dimers (β band) and higher aggregates (γ band). Bradley and Wolf have investigated the interaction of macromolecules with dye molecules quantitatively (22). The change in luminescence properties of acridine orange with dimerization (18), green to red, has been used to distinguish between RNA-and DNA-rich cells in neoplasmic growth vs. normal cell tissue. There is continuing discussion (23) of chlorophyll aggregation and associated absorption and luminescence behavior in relation to the two chlorophyll systems of the photosynthetic apparatus.

IV. LINEAR CHAIN POLYMERS

The quantum theoretical treatment of the linear chain model of molecular aggregates in two geometries has been given by McRae and Kasha (13,16). Although this model is not simply realizable in practice, the exciton brand structure deduced for the model is indicative of the nature of the molecular exciton model as one goes to higher molecular aggregates. There are some, though limited, examples of the application of the model to real molecular systems.

Assuming N identical molecules (N very large), the ground state wave function is the unique product of independent component molecule functions,

$$\Psi_G = \psi_1\psi_2\psi_3 \cdots\cdots \psi_N = \prod_{n=1}^{N} \psi_n \, . \tag{19}$$

The non-stationary (non-symmetry adapted) excited state wave function for an excited state of the aggregate is

$$\Phi_a = \psi_1\psi_2\psi_3 \cdots\cdots \psi_a^{\dagger} \cdots\cdots \psi_N = \psi_a^{\dagger} \prod_{\substack{n=1 \\ n\neq a}}^{N} \psi_n . \tag{20}$$

The k-th stationary state exciton wave functions can be described as symmetry-adapted linear combinations

$$\Psi_k = \frac{1}{\sqrt{N}} \sum_{a=1}^{N} c_{ak} \Phi_a , \tag{21}$$

where $|c_{ak}|^2$ determines the probability that the a-th molecule is excited. If the molecules in the chain are distributed with simple periodicity and N is very large so that each molecule will have equal probability for excitation (absence of end effects), the various coefficients will differ only in their phase factors, so that the wave functions may be written

$$\Psi_k = \frac{1}{\sqrt{N}} \sum_{a=1}^{N} e^{2\pi i k a/N} \Phi_a$$

$$(k = 0, \pm 1, \pm 2, \ldots . N/2). \tag{22}$$

The stationary state wave functions described by eq. 21 and eq. 22 show (a) collective excitation of molecular units in the aggregate; (b) excitation nodes, from a nodeless function, to an N-noded wave function; and (c) orthogonality of the stationary states.

Using the zeroth-order wave functions the first-order interaction energies can be calculated, using the point-dipole point-dipole potential term (eq. 13) as a perturbation, and summing over molecular pair interactions. As an initial simplification, nearest neighbor interactions may be considered; the exciton displacement term per pair is calculated by

$$\varepsilon_{a,a+1} = \int \Phi_a V_{di-di} \Phi_{a+1} dt. \tag{23}$$

The first-order exciton state energies for the linear chain polymer are given by

$$E_k = E_a^o + 2\left(\frac{N-1}{N}\right)\cos\left(\frac{2\pi k}{N}\right) \varepsilon_{a,a+1}$$

$$(k = 0, \pm 1, \pm 2, \ldots N/2). \tag{24}$$

We may consider two geometries for the linear chain polymer, shown in Fig. 6.

The results obtained for the nearest neighbor approximation exciton model applied to the linear chain model (13,16) are given in Table I.

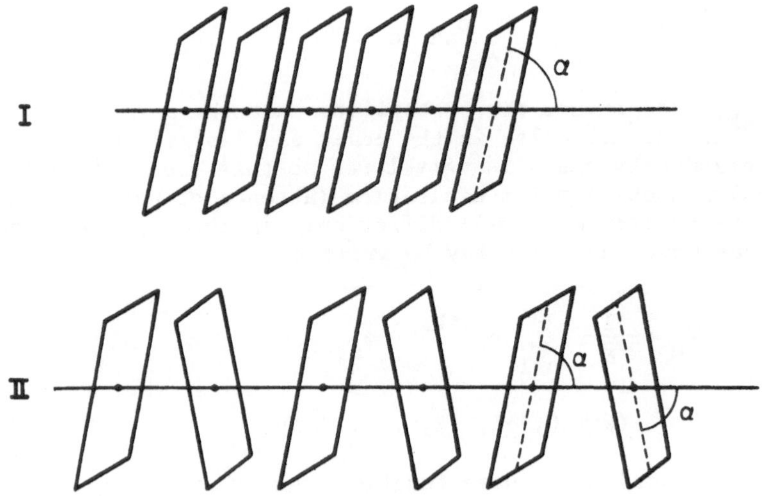

Fig. 6. Geometries for linear chain polymers.

It appears from Table I that the exciton band width (twice the exciton displacement term) for the linear chain model (N >> 2) is exactly twice the exciton band width for the dimer of corresponding structure (N = 2). This result is a fiction of the nearest neighbor approximation: obviously each molecule has two neighbors, so 2 x the dipole-dipole interaction relative to the dimer occurs in the nearest neighbor approximation. The dipole-dipole interaction is quite a long-range interaction, so summation beyond the nearest neighbor interaction must be considered for a more realistic numerical result. Thus, if the pair interactions are summed over the nearest eight neighbors on each side, the exciton band displacement increases to 2.39 times the single pair (dimer) interaction. This is a typical exciton band result: limited band widths are always predicted in the molecular exciton model because of the inverse cube dependence of the dipole-dipole first-order interaction. The formulas of Table I are valid only for N very large, since end effects are neglected ('cyclic boundary conditions').

The linear chain model even in the nearest neighbor approximation does, however, allow a simple physical picture to be obtained for the severe restriction by symmetry of the electric dipole radiative transition selection rules. Fig. 7 shows the exciton band structure (16) for the two linear chain model structures of Fig. 6

TABLE I

EXCITON SELECTION RULES AND BAND WIDTHS FOR LINEAR CHAIN POLYMERS

Geometry	Transition Moment to Exciton State: Lowest	Transition Moment to Exciton State: Highest	Exciton Band Width
Case 1. Translational Chain (one molecule/unit cell)			
$\pi/2 \geq \alpha > \text{arc cos}(1/\sqrt{3})$	0	$(N^{1/2})M$	$4\left\lvert\left(\frac{N-1}{N}\right)\frac{M^2}{r^3}(1-3\cos^2\alpha)\right\rvert$
$0 \leq \alpha < \text{arc cos}(1/\sqrt{3})$	$(N^{1/2})M$	0	
Case II. Alternate Translational (two molecules/unit cell)			
	$(N^{1/2})M\cos\alpha$	$(N^{1/2})M\sin\alpha$	$4\left(\frac{N-1}{N}\right)\frac{M^2}{r^3}(1+\cos^2\alpha)$

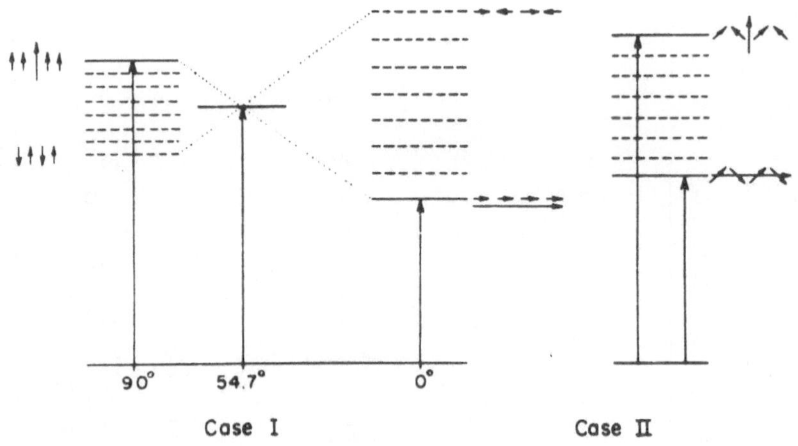

Fig. 7. Exciton band structures for linear chain polymers.

and Table I. The transition moment vector diagrams for the exciton states at the band extremities are shown. The analogy of these exciton band cases to those for the analogous dimer geometries of Figs. 3 and 4 will be apparent. All intermediate states are forbidden, as can be easily verified by study of such vector diagrams for one excitation node, two nodes, etc. (the orthogonality conditions for the stationary state exciton wave functions determine nodal positions (16)). Thus, these cases again illustrate the Davydov molecular exciton selection rule.

Exciton state splittings exceeding singlet-triplet splits of dye molecules would be expected for certain cases (high oscillator strength and intrinsically small S-T splits, < 1500 cm^{-1}) and would result in an anomalous excited state situation with a forbidden lowest singlet!

Quite often dye molecules at high concentration will aggregate to very high molecular weight. Concentrated solutions of some dyes in water, e.g. many symmetrical cyanines, will upon cooling a warm solution to which sodium chloride has been added, form exceedingly high molecular weight polymers (~1,000,000 AWU, estimated by rotational diffusion coefficient). Such polymers develop (24,25) an exceedingly narrow long wavelength absorption band ('J-band') which undoubtedly represents the Davydov free exciton pure electronic transition (not an electron delocalization). The coupling strength is so high, and the excitation delocalization so complete, that the excited molecules exhibit negligible vibrational relaxation: these are excited states of the molecule with ground state nuclear configuration, so the free exciton band is observed as a shifted pure electronic 0,0 band.

Probably such high molecular weight aggregates are helical in configuration as required for hydrodynamic stability. We shall now turn our attention to helical exciton arrays.

V. EXCITON STATES OF HELICAL POLYMERS

Moffitt (26) applied the molecular exciton model to the <u>polypeptides</u> in his study of the optical rotatory dispersion theory of these aggregates. As an example of the physical result expected for the exciton bands of helical aggregates, we present in Fig. 8 a diagrammatic representation of the exciton states by the vector model. The helical array may be studied for dichroism parallel or perpendicular to the polymer (fiber) axis. Parallel to the axis the transition moment vector components, for a regular helical array, constitute a simple linear chain polymer. As discussed in the previous section, the selection rule permits only one exciton state of the N-fold linear exciton band components to absorb electric dipole radiation. Since the allowed (in-phase) transition moment array is the most highly electrostatically attractive dipole array,

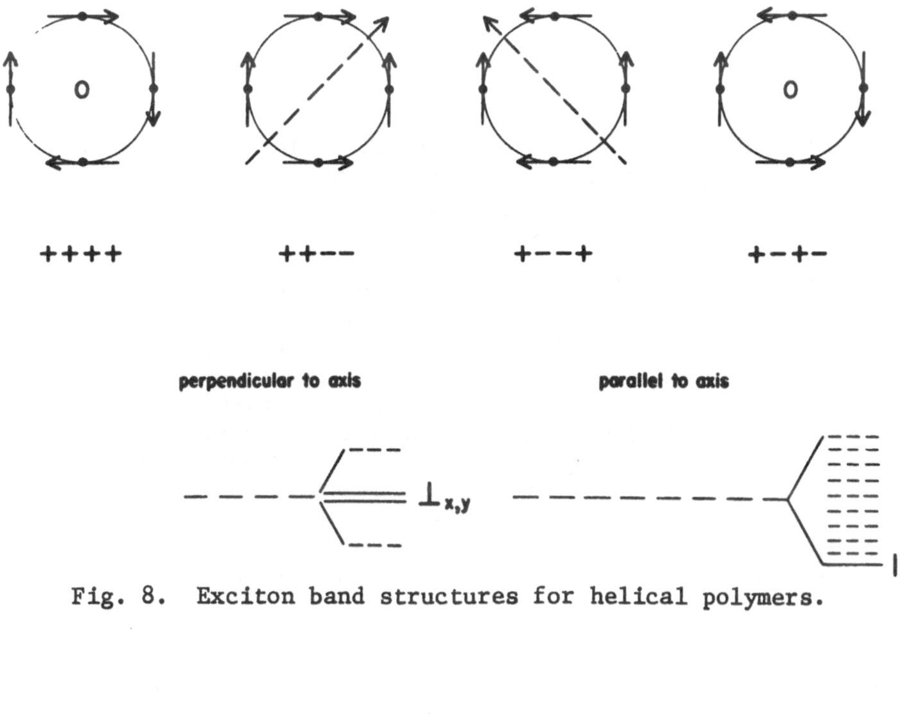

Fig. 8. Exciton band structures for helical polymers.

the longitudinally polarized exciton band has only one allowed component, the bottom or deepest exciton state of the helical exciton band.

The projection of transition moment vectors (per turn of the helix) in a plane perpendicular to the helical axis necessarily leads to degeneracies of 1,2,2,2,2,2 ..... for the exciton states because of orthogonality requirements of wave function components about a cylindrical axis. This degeneracy is observed (10,15) for small cyclical exciton arrays (trimers, tetramers), and is familiar to spectroscopists as λ-type doubling of electronic states in diatomic molecules. Thus, if a helical polymer has four molecule (chromophore) units per turn (electronically isolated by saturated covalent bonds as in the peptide chain), we may consider the exciton band structure of a cyclical tetramer (Fig. 8); summation over the helix will expand the exciton band width in the usual manner but will preserve the (symmetry) selection rule. As we can see from the projections in Fig. 8, the lowest exciton state is forbidden in perpendicular projection. The middle state, with a single excitation node is doubly degenerate and is allowed for dipole radiation. The highest state is forbidden (it is not intrinsically doubly degenerate for 4 chromophores because the rotation of nodes for orthogonality would place nodes on each chromophore, negating excitation).

Optical spectroscopic and rotatory dispersion studies of the helical polypeptides have confirmed the predictions of the molecular exciton model (27,28). Rhodes and Barnes (29) have extended the theoretical treatment of helical polymer optical properties.

A spectroscopically interesting result could be expected for dyes adsorbed onto extruded polymer fibers of helical macromolecules. The possibility of observing dye molecule helical exciton effects would be anticipated. Induced dye molecule Cotton effects for dyes adsorbed to helical polypeptide macromolecules in solution have been reported (30).

Nucleic acid and polynucleotide macromolecules present in the famous Watson-Crick double helix another helical geometry for molecular exciton effects. The hydrogen-bonded base pairs are considered to be arrayed with molecular-pair planes perpendicular to a helical axis, analogous to the treads of a spiral staircase. The lowest singlet-singlet absorption band in the purines and pyrimidines is not sufficiently intense to correspond to the strong coupling molecular exciton case. The random-coil to ordered-helix ('native') transition is however accompanied by a large hypochromism (30-40% diminution in absorption intensity),even though the spectral shifts in the lowest absorption band are small or negligible. Tinoco (31) attempted to use the molecular exciton first-order transition moment integral of Moffitt et al. (26) to account for the hypochromism of polynucleotides and DNA. Rhodes (32) discovered that the hypochromism can be attributed to a dispersion force interaction rather than to an exciton interaction (cf. vector diagrams of ref. 17).

A spectroscopically unique result for helical polymers has been obtained theoretically by Rhodes (33) in his demonstration of an expected $n, \pi^* - \pi,\pi^*$ state mixing for helical Space Groups. Normally electronic states arising from such orbitally contrasting configurations are mixed through out-of-plane vibrational coupling, resulting in a small interaction. This pure electronic mixing in helical arrays could result in unusual enhancement effects, e.g., in spin-orbital interaction.

VI. EXCITON STATES IN MOLECULAR LAMELLAR ARRAYS

Biological systems exhibit molecular lamellar structure in a wide variety of circumstances. Not only do membranes of varied biological function, but also many subcellular particles and other organelles, exhibit a high degree of molecular organization. How do molecules behave in lamellae as compared with their isolated molecule behavior? As far as the spectroscopic behavior and photochemical sensitization properties are concerned, the molecular exciton model offers an ideal and relatively tractable methodology for relating the observed photophysical and photochemical properties of molecular lamellae to their structure (geometry and intermolecular spacing) and the optical properties of the component molecules.

Hochstrasser and Kasha (34) carried out a comprehensive analysis of the theoretical molecular exciton band structure for a series of (monomolecular) lamellar arrays. The studies included (a) the simple square (rectangular) planar lattice, (b) in-plane inclined dipole planar lattices, (c) the normal (pin-cushion) lattice, (d) arbitrary out-of-plane inclined dipole planar lattices, (e) the zero-interaction out-of-plane lattice, and (f) in-plane and out-of-plane lattices with two molecules per unit cell.

For the various monomolecular lamellar arrays, explicit exciton band splitting formulas and electric dipole selection rules were derived. The molecular lamellar array is interesting because, unlike the case of molecular crystals, the molecular-pair interactions can be expressed in closed analytical form for summation over the entire array.

The molecular lamellar array exciton formulas show strikingly the effect of molecular organization. For example, the pin-cushion or normal array of regularly spaced transition dipoles perpendicular to the lamellar plane (Fig. 9) gives rise to a single allowed transition

$$\Delta E_{lam} = \Delta E_{mol} + \Delta D + 8.4 \frac{|M|^2}{r^3}, \qquad (25)$$

or over eight times the dimer interaction. As expected, this allowed exciton state comes at the top of the exciton band, Fig. 10, since the in-plane-dipole-array is most electrostatically repulsive. Obviously a very large blue shift would be observed for such a stacking, and the forbidden level at the bottom of the band would act as an efficient sensitizer in photochemical processes.

Another interesting aspect of the molecular lamellar exciton theory is that with an alternate translational array (two molecules/unit cell), two allowed exciton states (Davydov rule) are predicted, but for the lamellar structure the allowed states do not bracket the exciton band limits but occur close together with one component at the top or bottom of the exciton band.

Experimental spectroscopic studies of molecular lamellar arrays in relation to the exciton model would be very worthwhile. The extension of the model to bilayered systems is suggested by biological fine-structure research.

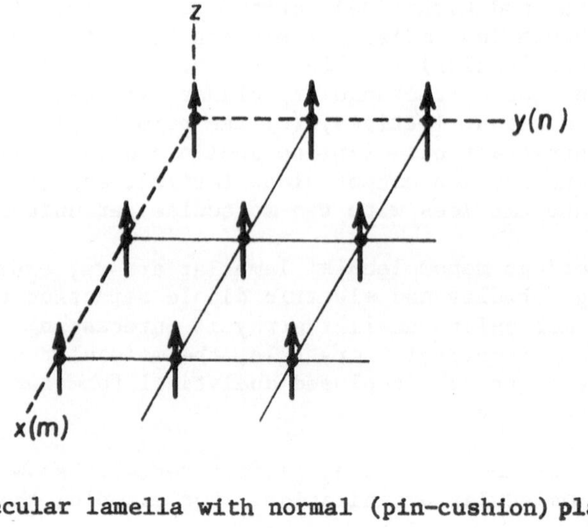

Fig. 9. Molecular lamella with normal (pin-cushion) planar lattice.

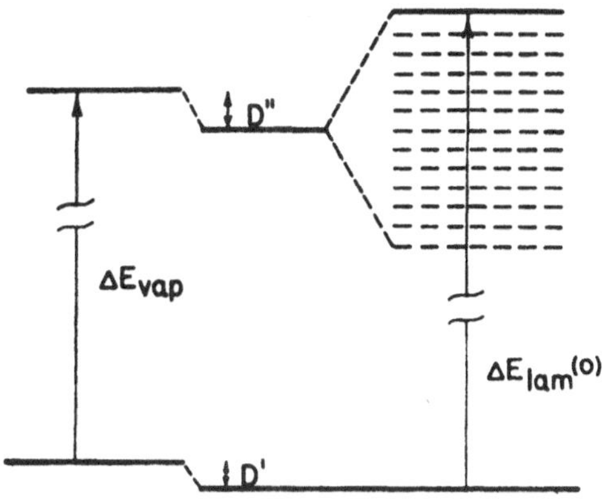

Fig. 10. Exciton band structures for normal (pin-cushion) molecular lamella.

VII. MOLECULAR EXCITONS IN SPHERICAL ARRAYS

Monomolecular dye molecule layers on spherical particles of small size occur in a very interesting context in photochemical and photobiological systems.

The chromatophore, the spherical particle isolated from purple photosynthesizing bacteria, has been described as representing a spherical monolayer of bacteriochlorophyll. This suggests the study of the effect of spherical aggregation of dye molecules on the spectroscopic and photosensitizing properties of the dye.

Two other chemical systems offer accessible experimental tools for possible tests of molecular exciton strong-coupling theory for spherical arrays. Dye molecules may adsorb onto the surface of micelles in solution, with changes in photosensitization behavior of the dye expected to parallel changes in the absorption and emission spectra. Secondly, the adsorption of dye molecules by colloidal particles, as studied by Fajans et al. (35), exhibits striking changes in fluorescence behavior and spectral absorption upon colloidal adsorption. In the well-known titration method, e.g. for Ag ion by Cl ion, fluorescein is adsorbed at the endpoint onto the AgCl colloidal particles, with a dramatic quenching of fluorescence and a change in color of the dye from yellow to pink.

It is possible that a molecular exciton effect can exist for the adsorbed dye and that both the absorption change and luminescence behavior could be explained thereby. However, an external spin-orbital effect on the dye by adsorption to the silver halide surface could also be present (36).

In the context of the discussion just presented, the subject of organic dye lake or mordant effects should be worth a reexamination from the point of view of the molecular exciton model for aggregated dyes, and the external spin-orbital perturbation effect for adsorption of dyes on high-Z atom surfaces (36). Both of these influences can alter the absorption and luminescence spectroscopy in a fundamentally discernible fashion. The plant flower anthocyanidin pigments are drastically altered in spectral properties in the plant cell compared with the dye molecule in dilute solution. The nature of the spectral changes as associated with adsorption on macromolecules (22), mordanting on inorganic colloidal particles (35), or molecular aggregation at high concentrations (25) has not been deciphered.

The molecular exciton state wave functions for spherical monomolecular arrays would be expected to have the spherical harmonic degeneracies 1,3,5,7,9,.... . A simple meridional stacking of transition moments would appear to give an allowed lowest exciton state, with forbidden exciton states for all degenerate wave functions. It is more likely that the Kitaigorodskii requirement for close-packing of ellipsoids would prevail, with a staggered array (2 molecules, or more, per unit cell) of the molecular units. The molecular exciton band pattern and consequences on excited state kinetics and spin-orbital mixing deserve some study because of the experimental correlations which could be expected for small particle surface adsorbed dye layers.

VIII. DYNAMICAL ASPECTS AND PHOTOSENSITIZATION

In this review we have been preoccupied with the stationary state strong-coupling molecular exciton model because of its direct spectroscopic consequences. On the other hand, an equally useful and important aspect of the theory is the non-stationary state exciton model and its deductions for energy transfer phenomena. These phenomena have occupied the attention of many researchers in experimental aspects of photobiology. Th. Förster had developed a phenomenological theory for dipole-dipole energy transfer, and later was able to demonstrate the quantum theoretical basis of his theory as a very weak coupling case molecular exciton model (4) (cf. Fig. 1). Förster's comparative study presents the dynamical aspects of the various exciton cases and offers a rational quantitative starting point for most of the useful energy transfer mechanisms.

Photosensitization by organic dyes as modulated by molecular aggregation was first described by Kautsky et al. (37). These authors recognized the formation of metastable excited states of dyes conditioned by aggregation, but failed to distinguish between intrinsic intersystem-crossing in the dye and intersystem-crossing enhanced (15) by molecular aggregation. The more recent studies of the molecular exciton model in relation to excitation phenomena (12,13,16) have now clarified the complex situation. Undoubtedly, molecular exciton effects contribute distinctive features to photosensitization phenomena and cannot be ignored. Instead, the full consideration of the nuances of photosensitization by molecular exciton effects, with or without intersystem-crossing consequences, offers the experimentalist a greater richness of excitation mechanisms. Such deduction of photosensitization mechanisms not only are of great interest to photochemists, but also can be invaluable to photobiologists. The molecular exciton effects may play a role in intrinsically photobiological phenomena, and also in applications of photosensitization as a research tool, as in the dye-sensitized stimulation of nerve axon studied by Chalazonitis (38) and Arvanitaki (39).

* Work done under a contract between the Division of Biomedical and Environmental Research, Energy Research and Development Administration, and the Florida State University.

REFERENCES

1. J. Frenkel, Phys. Rev. 37, 7, 1276 (1931); Physik. Z. Sowjetunion 9, 158 (1936) (in English).

2. R. S. Knox, Theory of Excitons (Solid State Physics Series), Academic Press, New York (1963).

3. M. Kasha, in Physical Processes in Radiation Biology (Augenstein, Rosenberg, and Mason, eds.), Academic Press, New York, pp. 17-22 (1964). We consider the Frenkel treatment as offering the general excitation delocalization model, including the Davydov treatment as the tightly-bound case, and the Wannier treatment as the loosely-bound case (cf. Knox).

4. Th. Förster, in Modern Quantum Chemistry, Vol. III (O. Sinanoğlu, ed.), Academic Press, New York, pp. 93-137 (1965).

5. M. Kasha, Rev. Modern Phys. 31, 162 (1959).

6. A. S. Davydov, Theory of Molecular Excitons (trans. from first Russian (1951) ed. by M. Kasha and M. Oppenheimer, Jr.), McGraw-Hill Book Co., New York (1962).

7. A. S. Davydov, Zhur. Eksptl. Teoret. Fiz. 18, 210 (1948).

8. Cf. first 12 chapters in Section III, Modern Quantum Chemistry, Vol. III (Sinanoğlu, ed.), Academic Press, New York (1965).

9. W. T. Simpson and D. L. Peterson, J. Chem. Phys. 26, 588 (1957).

10. Cf. M. Kasha, Radiation Research 20, 55 (1963).

11. E. G. McRae and W. Siebrand, J. Chem. Phys. 41, 905 (1964).

12. G. L. Levinson, W. T. Simpson and W. Curtis, J. Am. Chem. Soc. 79, 4314 (1957).

13. E. G. McRae and M. Kasha, J. Chem. Phys. 28, 721 (1958).

14. Cf. D. S. McClure, Canadian J. Chem. 36, 59 (1958).

15. M. Kasha, H.R. Rawls, and M. Ashraf El-Bayoumi, Pure and Applied Chemistry 11, 371 (1965).

16. E. G. McRae and M. Kasha, in Physical Processes in Radiation Biology (Augenstein, Rosenberg and Mason, eds.), Academic Press, New York, pp.23-42 (1964).

17. M. Kasha, M. Ashraf El-Bayoumi and W. Rhodes, J. chim. Phys. 58, 916 (1961).

18. V. Zanker, Z. physik Chem. 199, 225 (1952); ibid. 200, 250 (1952).

19. M. Kasha, in International Conference on Luminescence, Hungarian Acad. Sci., Akadémiai Kiadó, Budapest, pp. 166-182 (1968).

20. M. Kasha, Faraday Soc. Discussion No. 9, 14 (1950).

21. L. Michaelis, Cold Spring Harbor Symposium on Quantitative Biology 12, 131 (1947); J. Phys. Colloid Chem. 54 1 (1950).

22. D. F. Bradley and M. K. Wolf, Proc. Nat. Acad. Sci. U.S. 45, 944 (1959).

23. S. S. Brody and M. Brody, Nature 189, 547 (1961); Trans. Faraday Soc. 58, 416 (1962).

24. E. E. Jelley, Nature 138, 1009 (1936); ibid. 139, 631 (1937).

25. Cf. S.E. Sheppard, Rev. Modern Phys. 14, 303 (1942).

26. W. Moffitt, Proc. Nat. Acad. Sci. U.S. 42, 736 (1956); W. Moffitt, D.D. Fitts and J.G. Kirkwood, Proc. Nat. Acad. Sci. U.S. 43, 723 (1957).

27. K. Rosenheck and P. Doty, Proc. Nat. Acad. Sci. U.S. 47, 1775 (1961).

28. W. B. Gratzer, G. M. Holzwarth and P. Doty, Proc. Nat. Acad. Sci. U.S. 47, 1785 (1961).

29. W. Rhodes and D. G. Barnes, J. chim. Phys. 65, 78 (1968).

30. L. Stryer and E. R. Blout, J. Am. Chem. Soc. 83, 1411 (1961).

31. I. Tinoco, J. Am. Chem. Soc. 82, 4785 (1960).

32. W. Rhodes, J. Am. Chem. Soc. 83, 3609 (1961).

33. W. Rhodes, J. Chem. Phys. 37, 2433 (1962).

34. R. M. Hochstrasser and M. Kasha, Photochem. Photobiol. 3, 317 (1964).

35. K. Fajans and O. Hassel, Z. Elektrochem. 29, 495 (1923); Fajans and Wolff, Z. anorg. Chem. 137, 221 (1924).

36. E. Clementi and M. Kasha, J. Chem. Phys. 26, 956 (1957).

37. Kautsky, Hirsch, and Baumeister, Ber. deutsch. chem. Ges. 64, 2053 (1931); H. Kautsky and H. Merkel, Naturwissenschaften 27, 195 (1939).

38. N. Chalazonitis, Photochem. Photobiol. 3, 539 (1964).

39. A. Arvanitaki and N. Chalazonitis, in Nervous Inhibition, Pergamon Press, Oxford, p. 194 (1961).

SINGLET MOLECULAR OXYGEN: FROM A SCIENTIFIC CURIOSITY TO A UBIQUITOUS CHEMICAL SPECIES* (Abstract only)

M. Kasha

Institute of Molecular Biophysics, Florida State University
Tallahassee, Florida

Spectroscopists have recognized the existence of infrared electronic states of molecular oxygen through observations of an astrophysical nature. In the 40 years since careful spectroscopic observations on singlet molecular oxygen were made, no research on its terrestrial chemical and biological importance was carried out. Suddenly it was discovered about ten years ago that singlet molecular oxygen could be produced quantitatively in simple chemical reactions. The last ten years have seen a renascence of research in singlet molecular oxygen indicating its general importance as an organic reactant, as a chemical species occurring under widely different chemical reaction conditions, and, more recently, in extensive applications to biochemical systems, some of which are directly involved in human physiology. This lecture reviewed the study of singlet molecular oxygen and some interesting recent developments.

---

*Work done under a contract between the Division of Biomedical and Environmental Research, Energy Research and Development Administration, and the Florida State University.

TRIPLET STATE EXCITATION PHENOMENA* (Abstract only)

M. Kasha

Institute of Molecular Biophysics, Florida State University
Tallahassee, Florida

The nature of spin-orbital interaction in molecules permits the study of triplet-state properties as a means of orbital characterization of the excited state. Electronic transitions in aza-N-heteroaromatics, in pyrrolo-heteroaromatics, and in hydrocarbon aromatics, e.g., can thus be classified into three separate groups according to the nature of spin-orbital interaction. The application of these results to diverse excitation phenomena of photochemical and photobiological interest was outlined in this lecture.

*Work done under a contract between the Division of Biomedical and Environmental Research, Energy Research and Development Administration, and the Florida State University.

SPECTROSCOPIC THEORY OF THE SOLVENT CAGE* (Abstract only)

M. Kasha and B. Dellinger

Florida State University

Tallahassee, Florida

The solvent cage concept of photochemistry is translated into spectroscopic language by an extension of the Born-Oppenheimer approximation. The intermolecular perturbation of the solute modes is taken as the barrier to viscous flow in the solvent cage. It was shown that novel phenomenological consequences in the spectroscopy of polyatomic molecules in rigid solvent matrices can be predicted. Explicit consideration was given to molecular torsional potentials, the Dushinsky potential, the hydrogen-bonding potential, and a variety of molecular potentials in the region of large intermolecular distortions.

*Work done under a contract with the Division of Biomedical and Environmental Research, and the U.S. Energy Research Development Administration.

MULTIPLE EXCITATION IN COMPOSITE MOLECULES*† (Abstract only)

M. Kasha

Institute of Molecular Biophysics, Florida State University
Tallahassee, Florida

The study of single-photon, single-molecule interaction reveals basic information about electronically excited states of isolated molecules, but may reveal little about the excitation processes of molecules in the aggregate such as usually occur in nature. The present study explored the range of multi-excitation phenomena in multi-molecule systems. The interaction of one photon with a cluster of two or more molecules was considered. The list of phenomena includes Charge Transfer, Contact CT, Excimer, Exciton, Simultaneous Transition, Successive Excitation, Triplet-Triplet Annihilation, Exciton Fission, Non-Linear (Biphotonic) Absorption, and Biprotonic Phototautomerism. The purpose of the study was to examine the phenomenology, potential diagrams, and quantum theoretical formalism for the phenomena with the purpose of correlating what have been usually treated as separate phenomena, as accidentally discovered. The search for unifying elements of theory and the search for intermediate phenomena were both suggested. Numerous applications to photochemistry and photobiology were pointed out in this lecture.

*Work done under a contract between the Division of Biomedical and Environmental Research, Energy Research and Development Administration, and the Florida State University.

†Study done in collaboration with M. Ashraf El-Bayoumi, Department of Biophysics, Michigan State University, East Lansing, Michigan.

LONG SEMINARS (ABSTRACTS)

OPTICAL SPECTROSCOPY OF MOLECULAR IONS

S. Leach

Laboratoire de Photophysique Moléculaire du C.N.R.S.

Université Paris-Sud, Orsay, France

Experimental techniques for obtaining optical spectra of molecular ions are summarized, and a brief review of the spectra of polyatomic ions is given. Renner-Teller interactions and spectral perturbations provide interesting problems concerning the structure and spectra of the 15 valence electron ions N_2O^+, CO_2^+, COS^+ and CS_2^+. Examples are given where optical high-resolution spectra of molecular ions supply useful information for interpreting photoelectron spectra and also observations on planetary atmospheres and comets.

Optical spectroscopy has always been a fundamental source of information on the geometrical structure, internal dynamics and electron configurations of molecular species. Such information for ions is often required for interpreting experimental results in collision physics and chemistry, including flames and shock tubes, plasma physics, radiation chemistry as well as structural and fragmentation aspects of mass spectrometry. In recent years, optical spectroscopy of molecular ions has been seen to be important in two other fields: 1) as an aid in the identification and interpretation of photoelectron spectra of molecules; 2) in the identification of molecular ions and determination of their properties in order to assess the role of these species in planetary atmospheres, comets, stellar and interstellar media.

Laboratory studies of the optical spectra of molecular ions are not easy to perform by classical emission and absorption techniques. This is due essentially to the difficulty of creating molecular ion densities that are sufficiently high for times long enough for spectroscopic measurements to be carried out. Molecular ions are easily destroyed by ion-electron recombination processes, ion-molecule reactions and, for complex molecular ions, spontaneous dissociation. Under suitable experimental conditions one can avoid or reduce recombination and ion-molecule reactions. The spectra and structure of molecular ions have been reviewed previously (1).

The emission spectra of about 50 diatomic ions have been obtained using, for the most part, various electron impact or discharge sources (2). The spectra of 8 polyatomic ions are known; their generation requires special experimental techniques. A number of special sources have been developed in our laboratory at Orsay over the past years. These are of the crossed-molecule and electron-beam type (3,4), with extension to multiple beams (5), as well as various forms of a.c. and d.c. electrical discharges, including hollow cathode discharges (6). A low voltage discharge source has recently been developed at Ottawa (7). These sources are sufficiently intense for studying high-resolution spectra of the molecular ions observed.

Low-resolution spectra of molecular ions have been obtained with other, less intense, spectral sources. The fluorescence of molecular ions has been excited by far ultraviolet photons absorbed by neutral molecules (8). This technique has recently been extended by the use of synchrotron radiation as a far u.v. source (9). Collisions between neutral molecules and ions or fast neutrals can give rise to molecular ion emission (10). At suitable projectile velocities, the newly formed ions have abnormal populations of their vibrational states (11). The possibility of achieving non-Franck-Condon vibrational distributions is potentially of help in analyzing complex vibronic spectra. Penning ionization is another promising source of ion fluorescence generation (12).

The photoexcitation of molecular ions trapped in potential wells has been achieved and used to study their photodissociation spectra and cross-sections (13); the maximum ion densities at present obtainable by this trapping technique is about 10^7 ions/cm^3 (14) and are probably too low for high-resolution optical spectroscopy experiments.

Spectroscopic information on ions has also been obtained through the study of the photodissociation spectra of beams of positive ions (15) and photodetachment spectra of negative ions (16), monitoring fragment ions or photoelectrons, respectively. In these experiments one can study threshold phenomena as a function of excitation wavelength or, at a fixed wavelength, measure the kinetic energy spectrum of the photoejected species. In similar fashion, ion (17) and electron (18) spectroscopy can be used to give information about the internal energy states of molecular ions formed by inelastic collisions of neutral molecules with photons or electrons. Photoelectron spectroscopy is indeed a powerful and rapidly expanding source of (low-resolution) information on molecular ion states.

Absorption spectra of a few molecular ions have been obtained using the flash discharge (19) and pulsed electron (Febetron) (20) techniques. Classically, absorption spectra of neutral molecules to Rydberg states, especially superexcited electronic states, was one of the earliest sources of information on ionization limits and on limited aspects of the structure and electronic states of molecular ions (21). It should also be mentioned that absorption spectra, mainly infrared, have been measured of ions trapped in rigid matrices (22).

High-resolution optical spectra are known of the following polyatomic ions:

- N_2O^+ (23); CO_2^+ (24,25); COS^+ (26); CS_2^+ (27);
- H_2O^+ (7,28); H_2S^+ (29,30); D_2S^+ (29,30);
- $C_4H_2^+$ (31).

It is interesting to note that the existence of the ions HCO^+ (32) and N_2H^+ (33) in the interstellar medium has been inferred from their microwave spectra; their electronic spectra have not been observed as yet.

The ions N_2O^+, CO_2^+, COS^+ and CS_2^+ are 15 valence electron species and have linear ground and first few excited doublet states in accordance with the Walsh rules (21). These states have the following configurations:

$$\ldots\ \sigma_{(g)}^2\ \sigma_{(u)}^2\ \pi_{(u)}^4\ \pi_{(g)}^3 \qquad \tilde{X}\ ^2\Pi_{(g)}$$

$$\ldots\ \sigma_{(g)}^2\ \sigma_{(u)}^2\ \pi_{(u)}^3\ \pi_{(g)}^4 \qquad (\tilde{A})^2\Pi_{(u)}$$

$$\ldots\ \sigma_{(g)}^2\ \sigma_{(u)}\ \pi_{(u)}^4\ \pi_{(g)}^4 \qquad (\tilde{B})^2\Sigma_{(u)}^+$$

$$\ldots\ \sigma_{(g)}\ \sigma_{(u)}^2\ \pi_{(u)}^4\ \pi_{(g)}^4 \qquad C\ ^2\Sigma_{(g)}^+ .$$

The inversion symmetry characters g and u are only defined for the centrosymmetrical CO_2^+ and CS_2^+ species. The adjacent orbitals σ_u and π_u have very similar energies; their order and consequently that of the $^2\Pi_{(u)}$ and $^2\Sigma_{(u)}^+$ states are reversed in N_2O^+. The structure as well as the radiative and nonradiative transitions of the 15 valence electron species have been discussed in detail (34,35). It should be noted that some of the vibronic transitions in these ions provide interesting examples of Renner-Teller (23,25) and Fermi resonance (36) interactions. Spectroscopic data on CO_2^+, in particular the spectral perturbations indicating a coupling between the $\tilde{B}\ ^2\Sigma_u^+$ and $\tilde{A}\ ^2\Pi_u$ states of this ion (25,37) may be important in the interpretation of the Mariner 6, 7 and 9 observations of the $\tilde{B}\ ^2\Sigma_u^+ \rightarrow \tilde{X}\ ^2\Pi_g$ and $\tilde{A}\ ^2\Pi_u \rightarrow \tilde{X}\ ^2\Pi_g$ emission bands in the upper atmosphere of the planet Mars (38).

The $C_4H_2^+$ ion is also centrosymmetric and linear in its ground $\tilde{X}\ ^2\Pi_g$ and first excited $\tilde{A}\ ^2\Pi_u$ states. The $\tilde{A}\ ^2\Pi_u \rightarrow \tilde{X}\ ^2\Pi_g$ transition observed in emission in the diacetylene ion appears to be analogous to that of CO_2^+ (31).

The electronic spectrum of H_2S^+ (39) was the first to be investigated of a bent triatomic ion (29,30). Another example is the H_2O^+ ion which was recently observed both in emission in the laboratory (7) and in the tail of the comet Kohoutek (28). The emission transition observed in both H_2S^+ and H_2O^+ is $\tilde{A}\ ^2A_1 \rightarrow \tilde{X}\ ^2B_1$, the two states being derived from an electronic Π state split by static Renner-Teller interaction. Vibronic transitions involving levels both below and above the upper state barrier to linearity were observed in $H_2S^+(D_2S^+)$ (29,30). A comparison between the photoelectron spectra of $H_2S(D_2S)$ and the much higher resolved optical emission spectrum of $H_2S^+(D_2S^+)$ made it possible to unravel the complex vibronic structure involving the $\tilde{A}\ ^2A_1$ state of $H_2S^+(D_2S^+)$ in both types of spectra. Other examples where optical spectroscopy of molecular ions has provided useful information to obtain or test assignments of photoelectron spectra concern the ions CS^+ (40), COS^+ (41) and CS_2^+ (41).

REFERENCES

1. G. Herzberg, Quart. Rev. Chem. Soc. 25, 201 (1971).

2. B. Rosen, editor, Spectroscopic Data Relative to Diatomic Molecules, Tables Internationales de Constantes, Vol. 17, Pergamon, Oxford (1970); R. F. Barrow, editor, Diatomic Molecules: A Critical Bibliography of Spectroscopic Data, Tables Internationales de Constantes, C.N.R.S., Paris (1973).

3. M. Horani and S. Leach, Comptes-rendus Ac. Sci. (Paris) 248, 2196 (1959).

4. M. Horani and S. Leach, J. Chim. Phys. 58, 825 (1961).

5. M. Horani, J. Chim. Phys. 64, 331 (1967).

6. D. Cossart, Thèse d'Etat, Orsay (1974); D. Gauyacq-Abad, Thèse 3ème Cycle, Orsay (1975).

7. H. Lew and I. Heiber, J. Chem. Phys. 58, 1246 (1973).

8. D. L. Judge, G. S. Bloom and A. L. Morse, Can. J. Phys. 47, 489 (1969).

9. W. Sroka and R. Zietz, Phys. Letters 43A, 493 (1973).

10. M. C. Poulizac and M. Dufay, Astrophys. Letters 1, 17 (1967); M. J. Haugh and J. H. Birely, J. Chem. Phys. 60, 264 (1974).

11. J. P. Doering, The Physics of Electronic and Atomic Collisions, VII ICPEAC (1971), (T. R. Govers and F. J. de Heer, eds.), North-Holland, pg. 341 (1972); H. Bregmann-Reisler and J. P. Doering, Chem. Phys. Letters 27, 199 (1974).

12. J. A. Coxon, M. A. A. Clyne and D. W. Setser, Chem. Phys. 7, 255 (1975).

13. R. C. Dunbar, J. Am. Chem. Soc. 95, 6191 (1973); E. E. Ensberg and K. B. Jefferts, Astrophys. J. 195, L89 (1975).

14. F. L. Walls and G. H. Dunn, Physics Today 27, 30 (1974).

15. J. B. Ozenne, D. Pham and J. Durup, Chem. Phys. Letters 17, 422 (1972); N.P.F.B. Van Asselt, J. G. Maas and J. Los, Chem. Phys. 5, 429 (1974); J. Durup, Proceedings 21st Ann. Conference Mass Spectrometry, San Francisco (1973), ppg. 109-116.

16. W. C. Lineburger, in Chemical and Biochemical Applications of Lasers (C. Bradley Moore, ed.), Academic Press, N.Y., pg. 71 (1974).

17. e.g., K. E. Mc Culloh, J. Chem. Phys. 59, 4250 (1973).

18. J. H. Eland, Photoelectron Spectroscopy, Butterworths, London (1974); K. D. Sevier, Low Energy Electron Spectrometry, Wiley-Interscience, N.Y. (1972).

19. G. Herzberg and A. Lagerqvist, Can. J. Phys. 46, 2363 (1968).

20. M. Clerc and B. Lesigne, J. Chim. Phys. 67, 701 (1970); M. Clerc and B. Lesigne, Canad. Spectroscopy 16, 17 (1971).

21. G. Herzberg, Electronic Spectra of Polyatomic Molecules, Van Nostrand, Reinhold, N.Y. (1966).

22. D. E. Milligan and M. E. Jacox, J. Chem. Phys. 51, 1952 (1969); M. E. Jacox and D. E. Milligan, J. Mol. Spectroscopy 52, 363 (1974); D. E. Milligan and M. E. Jacox, in Spectroscopy, MTP Review of Physical Chemistry, Series 1, Vol. 3 (D. A. Ramsay, ed.), Butterworths, London, Chapter 1 (1972).

23. J. H. Callomon and F. Creutzberg, Phil. Trans. Roy. Soc. (London) 277, 157 (1974).

24. F. Bueso-Sanllehi, Phys. Rev. 60, 556 (1941); S. Mrozowski, Phys. Rev. 60, 730 (1941); ibid. 62, 270 (1942); ibid. 72, 682 (1947).

25. D. Gauyacq, M. Horani, S. Leach and J. Rostas, Canad. J. Phys. (1975), in press.

26. M. Horani, S. Leach, J. Rostas and G. Berthier, J. Chim. Phys. 63, 1015 (1966).

27. J. H. Callomon, Proc. Roy. Soc. (London) A244, 220 (1958).

28. P. A. Wehinger, S. Wyckoff, G. H. Herbig, G. Herzberg and H. Lew, Astrophys. J. 190, L43 (1974).

29. R. N. Dixon, G. Duxbury, M. Horani and J. Rostas, Mol. Phys. 22, 977 (1971).

30. G. Duxbury, M. Horani and J. Rostas, Proc. Roy. Soc. (London), A331, 109 (1972).

31. J. H. Callomon, Canad. J. Phys. 34, 1046 (1956).

32. D. Buhl and L. E. Snyder, Astrophys. J. 180, 791 (1973); U. Wahlgren, B. Liu, P. K. Pearson and H. F. Schaefer III, Nature 246, 4 (1973).

33. B. E. Turner, Astrophys. J. 193, L83 (1974); S. Green, J. A. Montgomery, Jr. and P. Thaddeus, ibid. 193, L89 (1974).

34. S. Leach, J. Chim. Phys. 61, 1493 (1964).

35. S. Leach, J. Chim. Phys., Special N°, Transitions nonradiatives dans les molécules, ppg. 74-83 (1970).

36. C. Branciard-Larcher, S. Leach and J. Rostas, paper to be presented Fourth Colloquium on High-Resolution Molecular

Spectroscopy, Tours (1975); to be published.

37. S. Leach, Proc. Int. Conf. Molecular Energy Transfer, Ein Bokek (1973), (R. D. Levine and J. Jortner, eds.), Wiley, N.Y., in press (1975).

38. C. A. Barth, W. G. Fastie, C. W. Hord, J. B. Pearce, G. E. Thomas, G. P. Anderson and O. F. Raper, Science 165, 1004 (1969); C. A. Barth, C. W. Hord, J. B. Pearce, K. K. Kelly, G. P. Anderson and A. I. Stewart, J. Geophys. Res. 76, 2213 (1971).

39. M. Horani, S. Leach and J. Rostas, VIème Conférence Internationale sur les phénomènes d'ionisation dans les gaz, S.E.R.M.A., Paris 1, 45.

40. N. Jonathan, A. Morris, M. Okuda, K. J. Ross and D. J. Smith, Faraday Discussions Chem. Soc. 54, 48 (1972); S. Leach, ibid. 54, 68 (1972).

41. P. Natalis, J. Delwiche and J. E. Collin, Faraday Discussions Chem. Soc. 54, 98 (1972); S. Leach, ibid. 54, 139 (1972).

SOME PROPERTIES OF THE EXCITED STATES OF MOLECULAR CRYSTALS*

R. C. Powell

Department of Physics, Oklahoma State University

Stillwater, Oklahoma

The excited states of molecular ions have been discussed in other papers in this conference. Here we would like to focus our attention on what happens to the excited states of molecular ions when they are placed in a crystalline environment. The changes in their properties are derived from two physical effects: the symmetry of the crystal field and the resonant interaction with neighboring ions of the same type. These effects lead to a splitting of the free ion levels and allow the electronic excitation energy to migrate as a molecular exciton. The two molecular ions we will discuss are the WO_4^{2-} ion in calcium tungstate and the VO_4^{3-} ion in yttrium vanadate (1,2).

The properties of these crystals were studied by making absorption, fluorescence, excitation and pulsed fluorescence measurements as a function of temperature from 8K to room temperature. The integrated fluorescence intensities and lifetimes were obtained from the data for different excitation wavelengths.

Four distinct excitation and fluorescence bands with varying intensities were observed for $CaWO_4$. Above about 220°K the integrated fluorescence intensities of all of the $CaWO_4$ fluorescence bands decrease. Below this temperature the intensity remains constant for 2400Å excitation but for longer wavelength excitation the intensity decreases down to about 120°K and then remains constant at lower temperature.

An interesting feature of the $CaWO_4$ fluorescence resulting from 3150Å excitation is the appearance of two sharp lines near 3680Å and 3750Å. At 10°K their peak intensity is much greater than that of the broad band fluorescence but above $100^{o}K$ they rapidly decrease in intensity. The integrated intensity of the high energy line is about five times greater than that of the low energy line at $10^{o}K$ but as temperature is raised, the low energy line increases in intensity with respect to the high energy line.

The fluorescence decay times of $CaWO_4$ for different excitations are quite different.

The excitation spectrum of YVO_4 exhibits five peaks while the fluorescence spectrum consists of only one broad band. The fluorescence decay time decreases from about 326 μs at 11K to about 13 μs at room temperature.

The data presented here can be understood by picturing these crystals as being made of molecular anions loosely bound to a cation. The observed spectra are due to transitions within the tungstate or vanadate molecular ions (3-5). The free ion molecular terms (6,7), their splittings in the crystal fields, and the allowed transitions have been determined. By comparing these considerations with the observed spectra, schematic configuration-coordinate diagrams can be constructed to explain the different excitation and emission bands and their temperature dependences.

In order to account for the appearance of two sharp lines in the fluorescence spectrum of $CaWO_4$ and the temperature dependence of the fluorescences decay times, it is necessary to assume that fluorescence occurs from both normal WO_4^{2-} ions and tungstate ions near to structural defects which have excited states with slightly lower energies. Also, to explain the temperature dependences of the fluorescence lifetimes of both types of crystals, it is necessary to account for the thermally-activated migration of the self-trapped excitation among the molecules of the lattice. To do this we monitored the quenching of the fluorescence intensities and lifetimes of the host crystals as a function the concentration of trivalent rare earth impurity ions and temperature in doped samples (2,8). The resulting parameters characterizing energy transfer in these systems are summarized in Table I. The thermal activation energy for migration is ΔE, while D and ℓ are the diffusion coefficient and diffusion length for the migration at high temperatures. The number of steps in the random walk of the exciton is n and the average time for each step is t_h. The time for the trapping step at an activator and the diffusion time are separately determined from intensity and lifetime measurements as t_{trap} and t_{diff}. This is possible because trapping is the rate limiting step in the transfer process.

For the trapping step, dipole-dipole interaction predicts a

TABLE I

Parameter	$CaWO_4:Sm^{3+}$	$YVO_4:Eu^{3+}$
ΔE (cm^{-1})	36	156
D (cm^2 sec^{-1})	1.2×10^{-7}	3.4×10^{-9}
ℓ (cm)	1.2×10^{-6}	1.3×10^{-6}
t_h (sec)	2.1×10^{-9}	7.5×10^{-8}
n	8,500	3,333
t_{trap} (sec)	2.9×10^{-5}	4.6×10^{-5}
t_{diff} (sec)(1%)	1.4×10^{-7}	4.8×10^{-6}

much smaller rate than the observed rate for $YVO_4:Eu^{3+}$ but approximately the observed rate for $CaWO_4:Sm^{3+}$. For both systems exchange interaction with $L \approx 1\mathring{A}$ and $R_o \approx 4\mathring{A}$ predicts the correct values for the trapping rates. It is not surprising that exchange effects are important for VO_4^{3-} since the charge distribution in the $^1A_1$ excited state lies outside the molecule thus providing the possibility of strong wavefunction overlap with the activator ions. Similar calculations have not been made for the WO_4^{2-} molecular ion but the transition from the ground to excited state has been described as a charge transfer transition with the electron leaving a 2p oxygen orbital and going into a tungsten 5d orbital. Thus the charge distribution may be more localized in this case decreasing the importance of exchange interaction. Although in principle it should be possible to also predict the hopping time from the single-step energy transfer equations considering the interaction between two neighboring molecular ions, the importance of thermal activation energy makes it impossible to determine a meaningful value for the spectral overlap integral.

ACKNOWLEDGEMENTS

The work described here was done in collaboration with Drs. M. J. Treadaway and C. Hsu.

REFERENCES

1. M. J. Treadaway and R. C. Powell, J. Chem. Phys. 61, 4003 (1974).

2. C. Hsu and R. C. Powell, accepted for publication in J. Luminescence.

3. A. M. Gurvich, E. R. Il'mas, T. I. Savikhina, and M. I. Tombak, Z. Prik. Spek. 14, 1027 (1971).

4. F. A. Kroger, Some Aspects of the Luminescence of Solids, Elsevier, New York (1948).

5. G. Blasse, Philips Res. Repts. 24, 131 (1969); ibid. 23, 344 (1968); and G. Blasse and A. Bril., Philips Tech. Rev. 31, 304 (1970).

6. K. H. Butler, in Proceedings of the International Conference on Luminescence, 1966, Akademiai Kiado, Budapest, p.1313 (1968).

7. D. S. Boudreaux and T. S. LaFrance, J. Phys. Chem. Solids 35, 897 (1974).

8. M. J. Treadaway and R. C. Powell, Phys. Rev. B 11, 862 (1975).

*Research supported by the U.S. Army Research Office, Durham, N.C.

TIME-RESOLVED SPECTROSCOPY OF SELF-TRAPPED EXCITONS IN ALKALI-HALIDE CRYSTALS

R. T. Williams

Naval Research Laboratory

Washington, D.C.

Lattice distortion induced by the creation of an exciton in an alkali-halide crystal provides an interesting solid-state analog of vibrational relaxation following Franck-Condon transitions in molecules. Whereas the lowest-energy lattice configuration of an alkali-halide crystal in its electronic ground state is a regular cubic array, the first electronic excited state (exciton) achieves its minimum energy if the lattice is locally distorted by contraction of the internuclear distance between two nearest-neighbor halide ions. Once the vibrational relaxation has occurred, the exciton is effectively removed from resonance with neighboring lattice sites and is therefore rendered immobile, or "self-trapped." To the extent that this pure-crystal excitation can be treated approximately as a molecular impurity complex, and certainly in view of corresponding methods of the spectroscopy of metastable states which are involved in its study, it is useful to consider this system in the context of molecular spectroscopy. A further analog to the photochemistry of molecules is found in the creation of crystal defects (vacancies and interstitials on the halide sublattice) upon nonradiative decay of self-trapped excitons.

Because of the vibrational relaxation mentioned above, the excitonic luminescence in pure alkali halides is Stokes-shifted by as much as 5 eV relative to the unrelaxed exciton level (1). The lowest-energy emission in all alkali halides studied is a phosphorescent transition from the first exciton triplet (T_1) to the crystal ground state (S_0). An excitonic fluorescence band is found roughly 1 eV higher in energy in some alkali halides, and is not seen at all in others. Data to be discussed provides evidence for assigning the fluorescence to the transition $S_2 - S_0$; no fluorescence is observed from the first excited singlet.

We have performed excited-state absorption spectroscopy of transitions from T_1 to higher triplet states following creation of the excitons by a 5 nsec pulse of energetic electrons (2). The spectra can be interpreted in terms of transitions to states involving successively more diffuse hydrogenic electron orbitals, with an identifiable series limit corresponding to the ionized self-trapped exciton (self-trapped hole); and in terms of excitation of the hole between molecular orbitals localized on the relaxed pair of halide ions. The former set of transitions is analogous to the Rydberg series of the unrelaxed exciton, although the exciton binding energy increases by a factor of 2 to 4 upon relaxation. The latter set of transitions of the self-trapped exciton bears close resemblence to spectra observed in steady-state spectroscopy of a self-trapped hole (V_k center) which has been stabilized by trapping the electron elsewhere in the crystal.

The electronic structure of self-trapped excitons in some of the alkali chlorides and alkali fluorides has been calculated by Stoneham and co-workers using two different theoretical approaches. A Hartree-Fock calculation treating a model system comprising a cluster of two alkali and two halide ions surrounded by an array of point charges has given encouraging agreement with the experimental spectra pertaining to the "hole transitions" of the self-trapped exciton (3). Pseudopotential calculations were more successful in treating transitions between the extended, hydrogen-like states of the self-trapped exciton (4).

Intense excitation in the triplet-triplet absorption bands (e.g. by a laser) is observed to quench the T_1 - S_0 phosphorescence and to enhance S_2 - S_0 fluorescence (5). For example, while electron-pulse irradiation of a KI crystal produces a characteristic flash of 300-nm fluorescence, a subsequent pulse of infrared (1.06 μm) light induces a second, substantially more intense, flash of the 300 nm fluorescence while quenching the 370 nm phosphorescence. If the infrared pulse (which excites the T_1 - T_2 transition) is delayed beyond the lifetime of the population in T_1, it has no effect on the fluorescence. Intersystem crossing efficiencies deduced from these experiments are high, in qualitative agreement with the fact that spin-orbit interaction is strong relative to exchange interaction in this system.

The magnetic sublevels of triplet T_1 are split in zero field by the combined effects of spin-orbit interaction and the D_{2h} point symmetry of relaxed excitons in the rock-salt lattice. Using time-resolved spectroscopic techniques, the crossing of sublevels in an applied magnetic field can be observed as a change of the phosphorescence decay time (6). Related experiments on magnetic circular polarization by Marrone and Kabler (7), and on optically-detected electron spin resonance in the metastable exciton triplet state by

Marrone, Patten, and Kabler (8), have provided detailed information on the triplet sublevels.

It has been demonstrated that self-trapped excitons are involved in the photochemical production of vacancy-interstitial pairs on the halide sublattice (9, 10). The time-resolved spectroscopy presently being described has demonstrated that when fast recombination of near-neighbor vacancies and interstitials is allowed for, the non-radiative channel of exciton decay via defect production can be quite highly efficient. In order to further explore the dynamics of lattice defect production by exciton decay, techniques of picosecond spectroscopy have been applied to measure the rise of F-center (halide vacancy trapping an electron) absorption following the generation of free electrons and holes in pure KCl (11). Two-photon absorption of a short pulse of 266 nm light (the fourth harmonic of a pulse from a mode-locked Nd:YAG laser) served to generate electron-hole pairs. The delayed 532 nm second harmonic pulse served to monitor absorption at the F-band peak, while limited spectral measurements of absorption were made using lines obtained by stimulated Raman scattering of 532 nm light in benzene. The characteristic time for formation of F centers in their ground state in KCl at a temperature of 25 K is found to be about 11 psec. Current theories of the defect formation process may be viewed in essence as analogs of dissociation (9) or of translation (12) of a diatomic molecule within the crystal lattice. The rapidity of F-center formation which has been observed is substantial evidence in favor of the latter view, and suggests furthermore that the defect formation process begins in fairly high-lying states of the self-trapped exciton.

REFERENCES

1. M. N. Kabler, Phys. Rev. A136, 1296 (1964); M. N. Kabler and D. A. Patterson, Phys. Rev. Lett. 19, 652 (1967).

2. R. G. Fuller, R. T. Williams, and M. N. Kabler, Phys. Rev. Lett. 25, 446 (1970); R. T. Williams and M. N. Kabler, Phys. Rev. B9, 1897 (1974).

3. A. M. Stoneham, J. Phys. C: Solid St. Phys. 7, 2476 (1974).

4. K. S. Song, A. M. Stoneham, and A. H. Harker, J. Phys. C: Solid St. Phys. 8, 1125 (1975).

5. R. T. Williams and M. N. Kabler, to be published.

6. W. Beall Fowler, M. J. Marrone, and M. N. Kabler, Phys. Rev. B8, 5909 (1973).

7. M. J. Marrone and M. N. Kabler, Phys. Rev. Lett. 27, 1283 (1971).

8. M. J. Marrone, F. W. Patten, and M. N. Kabler, Phys. Rev. Lett. 31, 467 (1973).

9. D. Pooley, Solid St. Commun. 3, 241 (1965).

10. F. J. Keller and F. W. Patten, Solid St. Commun. 7, 1603 (1969).

11. J. N. Bradford, R. T. Williams, and W. L. Faust, Phys. Rev. Lett., in press.

12. Y. Toyozawa, in Vacuum Ultraviolet Radiation Physics (E.E. Koch, R. Haensel, and C. Kunz, eds.), Pergamon Press, New York, p. 317, (1974); M.N. Kabler, in Proceedings of the NATO Advanced Study Institute on Radiation Damage Processes in Materials (C.H.S. Dupuy, ed.), Corsica, France, in press (1973).

MOLECULAR ASPECTS OF PHOTOCHEMICAL DISSOCIATIONS

J. R. Wiesenfeld

Cornell University

Ithaca, New York

The use of time-resolved atomic absorption spectroscopy in probing primary photodissociative processes is discussed. A detailed investigation of the photochemistry of alkyl iodides is reported in order to illustrate the application of this technique. The role of charge transfer in the molecular excited state leading to dissociation is shown to play an important role in determining the yields of electronically excited atoms in the photolysis.

The development of modern spectroscopic techniques has greatly enhanced our understanding of the electronic structure of atoms and molecules. While the spectroscopist usually probes the structure of electronically excited states through the analysis of discrete spectra, the photochemist is often concerned with chemical and physical processes initiated by the absorption of actinic radiation in regions of continuous or heavily predissociated spectra where ordinary analytical methods are of little value. Thus, new techniques must be sought in order to permit consideration of the nature of the potential surfaces which govern the course of photodissociative processes in the gas phase. The outcome of such events, which may at least in part be viewed as the inverse of a collision between electronically excited species, are of direct interest to workers in allied fields of aeronomy, chemical laser design and laser-initiated chemical reactions.

Although a great deal of mechanistic data is now known

about the course of chemical reactions following the initial photolytic event (1), very little detailed information concerning this first step is available. Major experimental contributions in this area have been made over the last decade in the laboratories of R. Bersohn (2) and K. R. Wilson (3) who have developed techniques which permit the direct observation of the symmetry of the dissociative excited state and yield some information concerning the internal energy distributions in the photofragments immediately following dissociation. Other experimental techniques involving the measurement of small-signal gain and time-dependent output in photo-elimination laser systems (4) have also yielded new data on nascent internal energy distributions following photolysis of suitable molecules.

In our laboratory, we have been especially interested in the production of electronically excited atoms and molecules following photodissociation. An example of such a process is the observation (5) of laser output on the $I(5^2P_{1/2}) \rightarrow I(5^2P_{3/2})$ transition at 1315 nm following flash photolysis of small alkyl iodides. This observation clearly demonstrates that the initial branching ratio, $\Phi = \frac{[I(5^2P_{1/2})]_0}{[I(5^2P_{3/2})]_0}$, must be greater than the ratio of the degeneracies of the excited to the ground state, 1/2. The laser output appears to be highly dependent upon the structure of the alkyl group as the photolysis of iso-C_3H_7I does not result in stimulated emission while that of n-C_3H_7I does (6).

The technique of translational spectroscopy has been applied to measurement of Φ for small alkyl iodides, and a value of 3-4 obtained in the case of CH_3I (7). Similar values were reported for C_2H_5I and n-C_3H_7I although these represented estimates only. Results obtained by kinetic spectroscopy (8), suggested that the failure to observe laser output following i-C_3H_7I photolysis was not caused by more efficient deactivation of $I(5^2P_{1/2})$,

$$I(5^2P_{1/2}) + RI \rightarrow I(5^2P_{3/2}) + RI^{\ddagger},$$

by i-C_3H_7I than by n-C_3H_7I, but rather was due to a lower yield of excited atoms in the photolysis of the branched isomer.

At Cornell, Dr. Terence Donohue has employed the technique of time-resolved atomic absorption analysis to investigate the photochemistry of simple iodides (9). In this method, a very dilute sample (typically 1/1000) of iodide in an inert buffer gas is subjected to a 100 J photolysis flash of $\sim$ 20 μs duration. The atoms so produced are then monitored by following the absorption of appropriate resonance radiation produced in a microwave-powered electrodeless discharge lamp. The transient absorption signal corresponding to the temporal profiles of the atomic concentrations is digitized and transferred to a storage memory where a number of experiments (8-64) are averaged in order to enhance the signal-to-noise ratio of the measured signal.

By observing the temporal profiles of $I(5^2P_{1/2})$ decay and $I(5^2P_{3/2})$ build-up and decay, it is clear that the overall kinetic process may be represented by

$$I(5^2P_{1/2}) \xrightarrow{k_1} I(5^2P_{3/2}) \xrightarrow{k_2} I/2I_2$$

where k_1 is composed of contributions by spontaneous emission (via a magnetic dipole transition), diffusion to the vessel walls and collisional quenching while k_2 involves diffusion to and heterogeneous recombination on the walls of the vessel.

Analysis of the observed profiles yields values of the branching ratio for a wide variety of iodides (Table 1) (9). A number of general observations may be made.

Firstly, the fluorinated alkyl iodides have very high (>11) branching ratios for formation of $I(5^2P_{1/2})$. Secondly, the alkyl iodides branched on the α-carbon produce very small quantities of electronically excited atoms following photolysis ($\Phi<0.10$).

These measurements must be reconciled with the observed spectra of the alkyl iodides which display weak continua in the region of photolysis (10). The observed branching ratios appear to be strongly correlated with the ionization potential of the radical. Upon examination of the electronic structure of the iodides in the context of angular momentum conservation rules (9), the role of perturbations by valence structures with the form $[R^+I^-]$ appears to be vital in breaking down correlations between the excited state of $RI(A_1^*)$ which is reached by an electric dipole transition and electronically excited atoms $I(5^2P_{1/2})$. Such perturbations would be expected to be strongest for iodides whose radicals display the lowest ionization potential, in agreement with observation.

Table I

Source	Φ	p^{*a}	I. P. of R (eV)
CH_3I	11.5	0.92 ± 0.02	9.84
C_2H_5I	2.2	0.69 ± 0.05	8.25
n - C_3H_7I	2.0	0.67 ± 0.04	8.15
i - C_3H_7I	<0.1	<0.10	7.52
n - C_4H_9I	4.5	0.82 ± 0.04	8.64
s - C_4H_9I	<0.1	<0.10	7.93
i - C_4H_9I	2.2	0.69 ± 0.04	8.35
t - C_4H_9I	<0.1	<0.10	7.42
CD_3I	80.	0.99	9.83
CF_3I	10.0	0.91 ± 0.03	10.10
C_2F_5I	>50.	>0.98	9.98
n - C_3F_7I	>100.	>0.99	10.06
i - C_3F_7I	8.8	0.90 ± 0.02	10.5
HI	0.10	0.10 ± 0.05	13.6

[a] $p^* = [I^*]_o / ([I^*]_o + [I]_o)$

REFERENCES

1. W. G. Dauben, L. Salem and N. J. Turro, Accts. Chem. Res. 8, 41 (1975).

2. M. Dzvonik, S. Yang and R. Bersohn, J. Chem. Phys. 61, 4408 (1974).

3. G. Hancock and K. R. Wilson, in Applied and Fundamental Laser Physics (M.S. Feld, et al., eds.), Wiley, New York, p. 257 (1973).

4. M. J. Berry and G. C. Pimentel, J. Chem. Phys. 53, 3453 (1970).

5. J. V. V. Kasper and G. C. Pimentel, Appl. Phys. Letters 5, 231 (1964).

6. J. V. V. Kasper, J. H. Parker and G. C. Pimentel, J. Chem. Phys. 43, 1827 (1965).

7. S. J. Riley and K. R. Wilson, Disc. Faraday Soc. 53, 132 (1972).

8. R. J. Donovan, F. G. M. Hathorn and D. Husain, Trans. Faraday Soc. 64, 3192 (1968).

9. T. Donohue and J. R. Wiesenfeld, Chem. Phys. Lett. 33, 176 (1975); T. Donohue and J. R. Wiesenfeld, J. Chem. Phys., in press.

10. K. Kimura and S. Nagakura, Spectrochim. Acta 17, 166 (1961).

ENERGY MIGRATION IN SOLIDS*

R. C. Powell

Dept. of Physics, Oklahoma State University

Stillwater, Oklahoma

Energy migration in solids can occur in two different ways, by a single long step or by many short steps. The former is a long range quantum mechanical "resonance" process caused by the interactions between the charge distribution multipoles of the sensitizer and activator ions or molecules. The latter process is the migration of a coupled electron-hole pair or exciton. If the mean free path of the exciton is on the order of a lattice spacing, a random walk of nearest-neighbor hops can be used to describe the kinematics of the motion. In the limit of many steps in the random walk it is equivalent to use diffusion mathematics. If the mean free path of the exciton is much longer, its coherent motion over many lattice sites must be taken into account and the theoretical description becomes much more complicated. For this discussion we will consider only cases where diffusion mathematics are appropriate. In this situation each step in the random walk can be thought of as a quantum mechanical "resonance" transfer between two sensitizer ions or molecules.

The questions to be answered in an investigation of energy migration are: Is it a single- or multi-step process? What interaction mechanism causes the transfer? What is the range of the energy transfer? To answer these questions, the typical measurements which are made determine the dependence of the energy transfer rate on activator concentration, temperature, and time.

The greatest amount of work has been done by monitoring the quenching of the sensitizer fluorescence intensity or decay time as a function of activator concentration. The concentration dependence of the measured energy transfer rate can be used to distinguish between single-and multi-step transfer and to determine the physical mechanism causing the transfer. Some problems arise in these investigations in trying to accurately measure small activator concentrations and in reproducing exact experimental conditions on a series of different samples.

Temperature dependence measurements have been less useful in characterizing energy transfer processes because of the complicated results that can be obtained. In single-step energy transfer the temperature dependence of the transfer rate can be due to changes in the spectral overlap integral or to phonon absorption or emission probabilities if the transfer is phonon assisted. In exciton diffusion theory the temperature dependence of the transfer rate is determined by the mechanism limiting the mean free path of the exciton. This will be different for scattering from defects, acoustical phonons and optical phonons and for trapping.

Time resolved spectroscopy measurements yield the time dependence of the energy transfer rate which can be used to distinguish between single- and multi-step processes and to ascertain the mechanism of energy transfer.

Two important points must be made concerning multi-step energy transfer processes(1).The first is that it is really made up of two different physical processes: migration among the sensitizers and trapping at an activator site. It is important to distinguish between the contributions made to experimental results due to migration characteristics and those due to trapping characteristics. When activator ions distort the surrounding lattice, activator-induced host traps can be formed and trapping characteristics can become very important and very complicated. In general, bulk quenching measurements yield only one primary experimental parameter and the only way characterize the migration part of the transfer is to make some basic theoretical assumption concerning trapping characteristics. In certain specific cases trapping may become the rate limiting step in the trasnfer process. In this situation the results of lifetime quenching measurements are indicative of the trapping step rate while intensity quenching results yield diffusion rates. Host-sensitized energy transfer in tungstate and vanadate crystals doped with rare earth ions is an example of this situation(2). If time-resolved spectroscopy measurements can be done with high enough resolution to observe the time dependence of the energy transfer rate, two primary experimental parameters can be obtained and thus the characteristics of migration and trapping can be separately ascertained. Host-sensitized energy transfer in doped aromatic hydrocarbon crystals is a good example of this situation.

The second important point to be made concerns the effects of radiative energy migration on the experimental data(1). This depends on the exciton and activator distributions. Bulk quenching experiments generally involve a random distribution of both excitons and activators and in this case the rate of energy transfer is proportional to the rate at which the exciton samples previously unsampled sites. Radiative migration simply superimposes a random walk of long, slow step on the radiationless random walk of short, fast steps. The only significant result is the lengthening of the fluorescence lifetime of the exciton which can be measured. On the other hand, surface quenching experiments involve non-uniform exciton and activator distributions, and the rate of energy transfer depends on the displacement of the exciton over a macroscopic distance in the sample. In this case it is possible for one long radiative step to dominate many short radiationless steps. Accounting for the effects of radiative migration in this case can be more complicated and very important.

As an example of an investigation of energy migration, let us consider energy transfer between trivalent samarium ions in different crystal field sites in calcium tungstate crystals (3).The different sites arise from different types of charge compensation, and ions in each type of site can be selectively excited by a tunable dye laser. Two different cases of energy transfer can be observed. The first of these involves transfer between ions in sites whose transitions exhibit some spectral overlap. For this case efficient transfer occurs even at low temperature. The time dependence of the energy transfer rate does not vary as $t^{0.5}$ as predicted by electric dipole-dipole interaction, but rather goes as $t^{0.3}$ which is consistent with electric quadrupole-quadrupole interaction.

The second type of transfer occurs between ions in sites whose transitions do not have any spectral overlap. A constant intensity ratio is observed at low temperatures which indicates a lack of energy transfer, whereas at high temperatures the observed time dependence is consistent with a multi-step diffusion process with back transfer from activator to sensitizer included. The temperature dependence of the fluorescence intensity ratio for this case is also consistent with a multistep process and is opposite to the predictions of phonon-assisted energy transfer. This temperature dependence is found to be consistent with the temperature dependence of the linewidth of the sensitizer transition which changes from a Gaussian shape at low temperatures to a shape between that of a Gaussian and Lorentzian line at high temperatures.

The effects of pressure on this intensity ratio were obtained. Pressure along the a_1 axis squeezes certain orientations of samarium-charge compensator pairs and leaves others unaffected. This skews the line shape to produce a tail on the low energy side. When polarized light is used to excite the paris which are squeezed the transfer is quenched. Otherwise it is unaffected.

These results can be understood in terms of the localization of energy in a system with an inhomogeneously broadened transition, as predicted by Anderson(4). In this case the inhomogeneous broadening arises from the slight differences in the microscopic crystal fields at the site of each samarium ion. The energy level of each individual ion is homogeneously broadened due to phonon processes. At low temperatures the homogeneous broadening is small, and an excited ion may not be in resonance with near neighbor ions. Thus no transfer occurs and the energy is localized. At high temperatures phonon processes broaden the lines enough for an excited ion to be in resonance with a neighboring ion and energy migration occurs.

ACKNOWLEDGEMENTS

The thoughts on trapping and reabsorption were developed in collaboration with Dr. Z. G. Soos and the work on energy migration among Sm^{3+} ions was done in collaboration with Dr. C. Hsu.

REFERENCES

1. R. C. Powell and Z. G. Soos, accepted for publication in J. Luminescence.

2. R. C. Powell, p. 377, this volume.

3. C. Hsu and R. C. Powell, (to be published).

4. P. W. Anderson, Phys. Rev. 109, 1492 (1958).

*Research supported by the U.S. Army Research Office, Durham, N.C.

PHOTOIONIZATION RESONANCE SPECTRA

E. W. Schlag

Institut für Physikalische Chemie

Technische Universität München

We describe here a new method for observing resonant states in an ionization continuum of a polyatomic molecule.

As a molecule is excited by light of ever shorter wavelength, it passes the ionization potential, at which point the molecule generally forms a positive ion and an electron. If this process is carried out between two collection plates to which a field is applied, a current will flow. Hence, generally, one can study this photoionization current as a function of irradiating wavelength. The long-wavelength onset of this current is generally the ionization potential of the molecule. This wavelength can be accurately determined and Watanabe et al. (1) have employed this as an accurate and general method for determining ionization potentials.

The question may be raised if structure in the photoionization current vs. wavelength spectrum is to be expected. We could expect transitions to higher-lying states of the molecular ion but unfortunately such transitions do not have sharp structures, the reason being that energy conservation is not only satisfied at the onset of the transition, but also at all higher energies. Hence the transition is not a δ-function, but rather a step function. This is so since energy can always be conserved at higher energies due to the relative kinetic motion of the two particles produced.

The various transitions will then overlap for all vibrational overtones and combinations of the ions. The result is that the ion spectrum consists simply of a current gradually increasing from the ionization potential. Any structure observed in such spectra is generally due to autoionizing transitions resulting from highly excited states of the neutral, lying above the ionization energy of the molecule. Hence we have the situation that although the energy scale is precisely known from the setting of the vacuum ultraviolet monochromator, no corresponding spectrum can be observed due to the many overlying states.

The question we wish to raise is the revival of this experiment, by asking the question whether it is not possible to identify the resonant states in this photoion current continuum by constructing perhaps a special detector. The principle of such a detector would be to respond only to the resonant transitions in the ion, and hence such a detector should produce a signal if and only if the Bohr frequency condition is obeyed. Hence only if the setting of the monochromator equals the onset of a new level in the ion should a signal be observable. If the energy exceeds this onset, no signal should be observed. Such a detector should then be able to convert the step function excitation back to the δ-functions of normal spectroscopy. If this is possible one should be able to observe many resonant states in the continuum observed in the photoion current spectrum.

We conceived and constructed such a detector by pulling the photoelectrons out through a homogeneous drawout field and injecting them into a long drift tube. The only way electrons can make it through the long drift tube is if they possess no perpendicular velocity component. However, one would expect all photoelectrons ejected with some finite velocity to have such a component and hence not be transmitted by the drift tube. Also autoionizing transitions will in general have a mismatch with the ion states, and hence be rejected by the drift tube. Should the autoionizing state be degenerate with an ion state, a signal will be expected, but hence again an ion state must be present at that energy as a precondition for observing a signal through the drift tube. Hence the only photoelectrons that can be transmitted by the drift tube will be those having no kinetic energy. Such photoelectrons will be produced only at the onset of new transitions in the ion. Hence this detector should be able to pull sharp states out of the ionizing continuum. These states can be energy-indexed exactly since the peaks can directly be read off on the wavelength scale of the monochromator. Hence this detector converts photoion current spectra back to sharp line spectra as in conventional spectroscopy. It should be emphasized that this zero-kinetic-energy detection functions without any calibration of an energy scale, since the transmission can only be attained for resonant transitions. Furthermore the apertures are large, hence avoiding slits of fine apertures for energy selection, and therefore allowing for an optimal flux of photoelectrons. In short, only threshold electrons can all be made to proceed in a straight line through the drift tube. All hot electrons will have steradiency limitations due to the poor optical aperture of the drift tube. It is analogous to a poor light-gathering instrument, still transmitting directed light, but rejecting diffuse light. Hence we have termed this instrument a steradiency detector.

Thus, the threshold method has the following characteristics:

1. Threshold electrons are intrinsically transmitted; no calibration of the electron optics is required and no alteration of the electron optics is made during the course of an experiment.
2. The only energy calibration required is that of the photon beam, i.e. calibration of the monochromator by means of well-known atomic and/or molecular lines.
3. The threshold electron analyzer has a transmission function approaching unity for threshold electrons which are produced over an extended volume.
4. The theoretical resolution of the analyser is comparable with that obtained from optical spectroscopy. This, combined with the fact that it is essentially impossible by means of the latter to determine molecular parameters of gaseous molecular ions except via Rydberg series extrapolations, makes the threshold technique particularly powerful.

Our initial experiments (2,3) were applied to nitric oxide, benzene, and methyl chloride. These showed that the vibrational fine structure of the molecular ions is now readily observable at threshold. The threshold method may also be used in conjunction with a mass spectrometer to act as a state selector for a mass spectrum. Furthermore this can be applied to state-selected ion-molecule reactions. Various such applications employing this method of resonant state detection have recently been carried out in various laboratories (4-7).

One of the classic problems of molecular ions is the question of the states of the hydrogen molecule ion. Some hints of these states can be gleaned from high-resolution photoelectron spectroscopy (8). These states are virtually unobservable in photocurrent spectra due to the overpowering autoionization structure in such a spectrum.

A most dramatic demonstration of the singular advantage of the new threshold principle is the recent, successful, direct observation of the H_2^+ rotational fine structure for the first seven vibrational states of the ion (9). In these experiments, thirty-one different vibrational-rotational states of the ion have been directly observed and molecular parameters obtained. Some thirteen of these thirty-one states have been calculated via extensive Rydberg-series extrapolations, involving complex corrections for perturbations (10). This unequivocally demonstrates the ability of this new principle to extract pure, sharp ion states imbedded in the ion continuum and thus avails one of sharp line spectroscopy, even for ion-states.

REFERENCES

1. K. Watanabe, F.F. Marmo and E.C.Y. Inn, Phys. Rev. 91, 1155 (1953).
2. W.B. Peatman, T.B. Borne and E.W. Schlag, Chem. Phys. Lett. 3, 492 (1969).
3. T. Baer, W.B. Peatman and E.W. Schlag, Chem. Phys. Lett. 4, 243 (1969).
4. R. Spohr, P.M. Guyon, W.A. Chupka and J. Berkowitz, Rev. Sci. Instrum. 42, 1872 (1971).
5. R. Stockbauer, J. Chem. Phys. 58, 3800 (1973).
6. R. Stockbauer and M.G. Inghram, J. Chem. Phys. 62, 4862 (1975).
7. a) A.S. Werner, B.P. Tsai and T. Baer, J. Chem. Phys. 60, 3650 (1974); b) A.S. Werner and T. Baer, ibid. 62, 2900 (1975); c) B.P. Tsai, T. Baer, A.S. Werner and S. F. Lin, J. Phys. Chem. 79, 570 (1975).
8. L. Asbrink, Chem. Phys. Lett. 7, 549 (1970).
9. W.B. Peatman, to be published.
10. G. Herzberg and Ch. Jungen, J. Molec. Spect. 41, 425 (1972).

ELECTRON SPECTROSCOPY

A. J. Yencha

Department of Chemistry, State University of New York
at Albany
Albany, New York

Electron spectroscopy refers to the measurement of the kinetic energies of electrons that are ejected from molecular species by various processes. Two of the most important of these processes are photoelectron spectroscopy and Penning ionization electron spectroscopy. Both involve the formation of cations in various states of electronic excitation, and through such studies the structures and energetics of molecular ions can be explored. The difference between the two processes lies in the mode of ionization. The former is due to photon impact while the latter involves high energy metastable atom impact. It seems appropriate at this time to compare the results of these two related fields.

Photoelectron spectroscopy has been shown to be an invaluable technique in the elucidation of both ground and excited state properties of molecular ions. For example, from an analysis of the vibrational structure observed in photoelectron spectra it is often quite possible to assign to the various ionization potentials the type of electron being ionized. This is, of course, extremely important in attempting to properly construct molecular orbital diagrams of both neutral and ionic molecular species.

Consider, for example, the photoelectron spectrum of H_2O. The ground state of this molecule is of $^1A_1$ symmetry with gas phase active vibrations of ν_1 = 3652 cm^{-1} (symmetric stretching mode) and ν_2 = 1595 cm^{-1} (bending mode). The observed vibrational frequencies of the various states of H_2O^+ are given in Table I (1). The ground state of H_2O^+ is characterized by two, short vibrational series, but ν_2 is considerably weaker than ν_1 and is barely detectable. The fact that the vibrational energies change so little in this ionic state,

TABLE I. Vibrational frequencies of the various states of H_2O^+.

	Ionic state	Vibrational frequencies (cm^{-1}) ν_1	ν_2
Ground state of H_2O^+	$^2B_1$	3200 ± 50	1380 ± 50
1st excited state of H_2O^+	$^2A_1$	---	975 ± 50
2nd excited state of H_2O^+	$^2B_2$	2990 ± 100	1610 ± 100

plus the fact that the first member of each series is by far the strongest, indicates that the origin of the ionized electron was from an orbital of weakly bonding or nonbonding character, probably from a $1b_1$ molecular orbital. The first excited state of H_2O^+ exhibits a long progression of ν_2 vibrations with a maximum in the Franck-Condon envelope occurring at 0,9,0. This indicates that the ionized electron had its origin in a strongly bonding orbital - one probably involving H-H bonding character. Based on this fact and the fact the isoelectronic radical NH_2 in its first excited state (having the same electron configuration) has a bond angle of 144°, it is expected that the most probable bond angle in this state should be approximately 180°. The second excited state of H_2O^+ is characterized by broad, complex vibrational structure with only a few of each of the vibrational mode members clearly discernible. To the high energy side of the Franck-Condon envelope extensive broadening is observed suggesting a dissociative characteristic. This conclusion is supported by the fact that the OH^+ ion (2,3) and the O^+ ion (4) are observed mass spectrometrically in this same energy range. The assignment of the electron being ionized to the $1b_2$ molecular orbital seems quite reasonable.

In a manner similar to this a large number of photoelectron spectra have been analyzed, and our knowledge of both neutral and ionic molecular orbital structures has increased manifold in recent years. It is now of interest to see what we can learn about the collisional processes of Penning ionization through a comparison of photoelectron and Penning ionization spectra.

In Penning ionization electron spectroscopy a beam of atomic metastables, usually helium or neon, is generated in a source and allowed to flow into a separate ionization cavity where they encounter a rather low, static pressure of neutral molecules. Electrons that are produced through ionization are then sampled through a slit perpendicular to the flow of metastables. The slit is elongated in the direction of the metastable beam. Analysis of the kinetic energy of the electrons is accomplished in the usual way using an electrostatic energy analyzer and particle-counting techniques.

The Penning ionization process can be described by the following equation:

$$X^* + AB \rightarrow X + AB^+ + e^- , \qquad (1)$$

where X^* is the atomic metastable, AB is the molecule to be ionized, X is the ground state of the atom, AB^+ is the ionized molecule, and e^- is the ejected electron. The electron kinetic energy relationship is as follows:

$$KE(e^-) = E(X^*) - IP - \Delta E_{vib} - \Delta E_{rot} + \Delta E_{shift} , \qquad (2)$$

where $KE(e^-)$ is the measured kinetic energy of the ionized electron in electron volts, $E(X^*)$ is the energy of the metastable atom, IP is the energy of the adiabatic ionization potential (1st, 2nd, 3rd, etc.) of the gaseous molecule, ΔE_{vib} and ΔE_{rot} are the energies of vibrational and rotational excitation, respectively, of the molecular ion, and ΔE_{shift} is an energy associated with particle-particle interactions during the ionization process. It is this last term that distinguishes the Penning process from the photoelectron process.

There are four basic sources for this ΔE_{shift} in Penning ionization. They are as follows:

1. Conversion of a part of the potential energy of reactants into kinetic energy of products. This makes ΔE negative.
2. Conversion of the collision energy of reactants into the energy of electrons released. This makes ΔE positive by about 10 meV.
3. Associative ionization: $X^* + AB \rightarrow XAB^+ + e^-$. The ΔE value is positive in this case and may attain values up to the dissociation energy of the XAB^+ complex ion.
4. Collision complex formation in which the geometrical configuration of the AB molecule is altered. Small negative ΔE values can result.

It is convenient to visualize the Penning process in terms of a potential energy curve model (5) which is shown in Fig. 1. The upper curve $V^*(R)$ is the potential of interaction that may occur as a result of the collisional process between the metastable projectile and the molecular target. $V^+(R)$ describes the potential energy of interaction between the ground state atom and the molecular ion. It should be noted that only at large values of the inter-particle distance, R, where the potential energy curves are parallel, will the ΔE_{shift} values be zero, and hence the Penning results will be

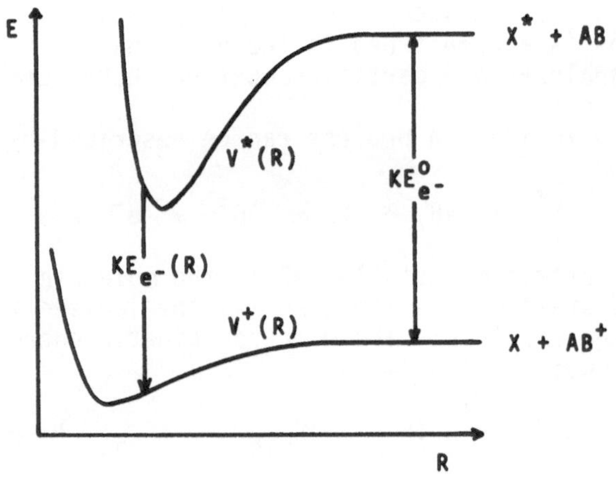

$$KE_{e^-}(R) = V^*(R) - V^+(R)$$

$$KE_{e^-}(R\to\infty) \to KE^{0}_{e^-}$$

Fig. 1. Potential curve model for Penning ionization showing both incoming and outgoing channels.

identical to those of photoelectron spectroscopy. The only exceptions to this statement would be if accidentally the upper and lower curves were identically parallel at small values of R or if neither the metastable-molecule or atom-ion curves were attractive at all values of R. Both of these exceptions, however, seem highly unlikely. Therefore, we might expect in most cases to observe particle-particle interaction effects in Penning ionization electron spectra.

By way of an example, consider the Penning ionization spectrum of cyanogen obtained using helium metastables as shown in Fig. 2. As can be seen, the spectrum appears to be quite broad and somewhat complicated. The complexity of the spectrum arises because there are three different sources of ionizing energy, namely 584 Å He I radiation, 2^1S He metastable, and 2^3S He metastable. Fortunately, the relative proportion of He I radiation to metastables is not large and in general does not interfere in the analysis. Furthermore, it provides a very convenient internal calibration of the electron kinetic energy scale. From a careful inspection of the spectrum, four different photoionization peaks can be identified.

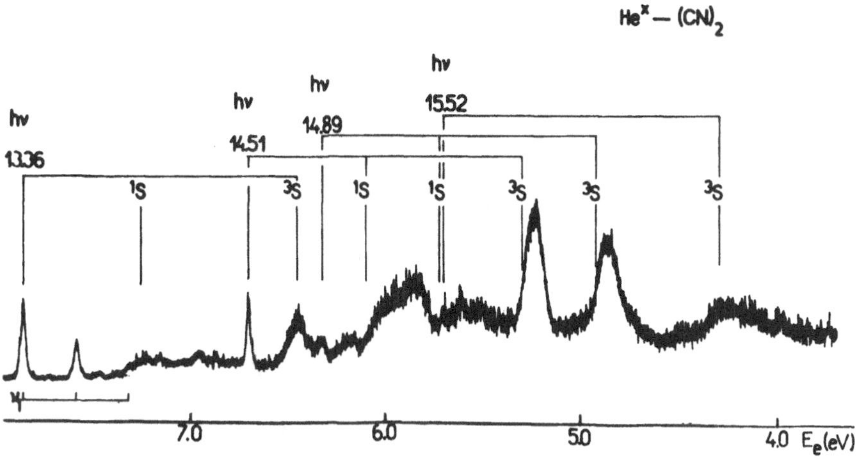

Fig. 2. Penning ionization electron spectrum of cyanogen obtained using helium metastables.

With the help of these reference points the corresponding 2^1S and 2^3S positions can be identified as shown in Fig. 2. This enables a rather complete assignment of all major bands in the spectrum including some vibrational structure. It is of importance to note that the positions of the 2^3S peaks for the second and third ionization potentials (around 5.0 eV) are significantly shifted to lower energies as compared to the predicted values, whereas ionization by the same metastables in the first and fourth ionization potentials appear not to be shifted at all. The former ionization potentials are assigned to the sigma nonbonding electrons on the nitrogen atom while the latter are identified with pi-bonding electrons from the CN bond.

Two further points need to be made regarding this spectrum. The first is that in all cases the intensity of the 2^3S produced ionization potentials is greater than that of the corresponding 2^1S ionization potentials. Secondly, in a general way, all peak intensities are comparable.

It is now of interest to see what type of spectrum is obtained for cyanogen using neon metastables. Such a spectrum is shown in Fig. 3. As can be seen, the same four ionization potentials are observed as was the case with helium metastables and are numbered 1-4. It is of particular interest to note the marked increase in intensity of peaks 2 and 3, which correspond to nonbonding electron ionization. This is apparently due to the increased polarizability of Ne^* over that of He^* in their interaction with relatively high charge density molecular orbitals. This effect is born out in

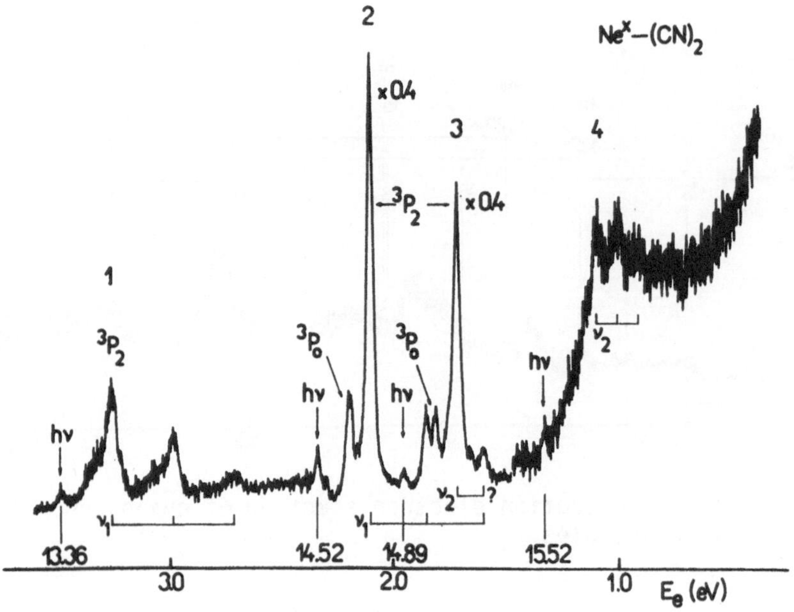

Fig. 3. Penning ionization electron spectrum of cyanogen obtained using neon metastables.

studies of other cyano compounds (6), and provides a relatively easy method for assigning the nonbonding ionization potentials. It is also important to note that neon metastables produce much sharper spectra than do helium metastables, and that the ΔE_{shift} values are considerably smaller. Both of these effects suggest that ionization occurs at much greater values of the inter-particle distance.

In conclusion, it appears that Penning ionization like photoionization proceeds by a direct, vertical excitation process, and that in both cases the target molecule being ionized acts as a rigid body. The observance of significant shifts in the ionization potentials in Penning ionization electron spectroscopy as compared to photoelectron spectroscopy confirms the notion of considerable particle-particle interaction either before, during, or after the actual ionizing event. Thus, we see that significant information is now becoming available on molecular systems through combined photoelectron spectroscopy and Penning ionization electron spectroscopy, and this is providing the basis of an understanding of the fundamental collisional processes of Penning ionization.

REFERENCES

1. D. W. Turner, C. Baker, A. D. Baker, and C. R. Brundle, Molecular Photoelectron Spectroscopy, Wiley-Interscience, London (1970).
2. M. Cottin, J. Chem. Phys. 56, 1024(1959).
3. V. H. Dibeler, J. A. Walker, and H. M. Rosenstock, J. Res. Nat. Bur. Std. A 70, 459(1966).
4. K. E. McCulloh and V. H. Dibeler, Abs. of Papers, 23rd Annual Conference on Mass Spectrometry and Allied Topics, Houston, Texas, May, 1975, p. 41, paper D-3.
5. H. Hotop, Radiation Res. 59, 379(1974).
6. V. Čermák and A. J. Yencha, in press, J. Electron Spectr. Rel. Phen.

SHORT SEMINARS (ABSTRACTS)

THE 2 ^2P STATE OF THE Li ATOM

P. Van Leuven

University of Antwerp

Antwerp, Belgium

This state serves as an example for the study of spatial deformation in atomic orbits. We described how one takes into account the deformation of the core by the valence electrons. Numerical calculations show that the effect is small but that it has an important influence on certain physical quantities like hyperfine structure constants. Some insight is gained in the significance of angular momentum projection.

THE "GENERATOR COORDINATE" METHOD AND MOLECULAR VIBRATIONS

P. Van Leuven

University of Antwerp

Antwerp, Belgium

We presented a method for the theoretical description of coupled electronic and nuclear motion. The method conserves the intuitive picture of collective coordinates but goes beyond the adiabatic approximation. It was illustrated on the H_2^+ molecular ion.

MAGNETIC DICHROISM SPECTROSCOPY

E. Krausz

Inorganic Chemistry Laboratory, University of Oxford

Oxford, England

The technique of magnetic circular dichroism (m.c.d.) has been well established for many years (since ~1965) and has proved useful in the assignment of electronic energy levels as well as the amplification of vibronic, magnetic and spin-orbit effects. Recent experimental advances have enabled second-order magnetically induced linear dichroism (m.l.d.) to be measured which provide complementary information in many areas. For systems that emit, the magnetically induced circular polarization of emission (m.c.p.e.) may be remarkably easily measured in favorable systems. Furthermore optical modulation of microwave (e.p.r.) may greatly increase the overall sensitivity of magnetic optical spectroscopy.

THE CHEMILUMINESCING PRODUCTS OF THE DISILANE-FLUORINE, DISILANE-CHLORINE AND DISILANE-OZONE REACTIONS

G. W. Stewart

Department of Chemistry, West Virginia University

Morgantown, West Virginia

and

J. Gole and D. Linsay

Department of Chemistry, Massachusetts Institute of Technology
Cambridge, Massachusetts

Measurements of formation and deactivation rates for electronically excited species that are potential laser media are being made. Particular attention is given to product state distributions and branching ratios in the elementary formation reactions. The promising nature of the silicon-ozone and germanium-ozone reactions, which produce relatively long-lived $a^3\Sigma^+$ and $b^3\Pi$ excited state species, has heightened the importance of the disilane system. In the disilane-ozone reaction we have found the fluorescence is characterized by emission from either the $a^3\Sigma^+$ and $b^3\Pi$ excited states of SiO or an excited state of the previously unobserved HSiO molecule analogous to the HCO $\tilde{C}$-$\tilde{X}$ hydrocarbon flame bands. The disilane-fluorine system is characterized by SiF* $A^2\Sigma$ - $X^2\Pi$ emission from v' = 0 through 4 and from the SiH* $A^2\Delta$ - $X^1\Pi$ emission, whereas, the disilane-chlorine system is characterized by the $^3A''$ - $^1A'$ transition in HSiCl. Studies have been carried out over a pressure range 50 to 1500m torr. Studies on the digermane-ozone and digermane-fluorine systems are also being conducted.

COLLISIONAL TRANSFER OF EXCITATION IN HIGH-PRESSURE RARE GASES AND MIXTURES*

P. E. Thiess

Nuclear Engineering Program, University of Illinois

Urbana, Illinois

Emission spectroscopy of high-pressure pure rare gas and rare gas mixture plasmas shows that energy transfer takes place to create enhanced fluorescence. In pure rare gases the transfer is between nearly resonant states (e.g. $n^1P \rightarrow n^{3,1}F$ in He) and by association of short-lived atomic states formed from higher atomic states than the metastable excited molecules (e.g. $2P \rightarrow 3\Sigma_u^+$). In mixtures energy transfer via collisions-of-the-second-kind occurs from both long-lived molecular as well as atomic states (e.g. $2^3\Sigma_u^+$ He $\rightarrow N_2^+$ B) and in some cases from short-lived states (e.g. Ne $2p_1 \rightarrow N_2^+$ B). Most previous work done at low pressure (1) (< 20 torr) has only shown the atomic metastable transfer and not the molecular metastable transfer which predominates at high pressures. Enhanced emission in molecular ions of atmospheric species (N_2, O_2, H_2, CO, NO, N_2O, NO_2, NH_3, etc.) was observed in helium. In neon only H_2 and N_2 show enhanced fluorescence. Atomic metastable transfer was found in high pressure He-Ne, Ar-N_2, Ar-Kr, Ar-Xe but it is often swamped by two-step excitation from the metastable. The immediate application appears to be to high-power, short-pulse lasers and plasma display devices.

(1) C. B. Collins and W. W. Robertson, J. Chem. Phys. 40, 701 (1964).

*Supported in part by USAEC Contract AT(11-1)2007 and equipment grants from NSF, DNET-USAEA, Kettering Foundation and the University of Illinois.

A SHOCK TUBE-LASER SCHLIEREN MEASUREMENT OF THE DISSOCIATION OF MOLECULAR CHLORINE

R. J. Santoro* and G. J. Diebold

Department of Physics, Boston College

Chestnut Hill, Massachusetts

Shock tubes are a commonly used technique for the production of high temperatures (500-10,000°K) in gases under adiabatic conditions. Measurements of vibrational relaxation and dissociation of molecules using a variety of diagnostic techniques have been reported. In this brief seminar the use of a shock tube-laser schlieren combination to measure the dissociation of molecular chlorine over the temperature range of 1700-3000°K was presented. The measured rate constant for collisions involving Ar, He, and Cl_2 was given. A brief discussion of the application of laser schlieren techniques to vibrational relaxation was also given.

*Present address: Department of Aerospace and Mechanical Sciences, Princeton University, Princeton, New Jersey.

CORE EXCITONS IN SYNCHROTRON RADIATION ABSORPTION SPECTRA

A. Bianconi

Istituto di Fisica, Università di Camerino

Camerino (MC), Italy

Synchrotron radiation is a fantastic light source for spectroscopy because of its high degree of polarisation and its continuum spectrum from infrared to x-rays and its narrow pulses ($\sim 10^{-10}$ sec). By the use of the high energy accelerators as photon sources interest in soft x-ray absorption spectroscopy (SXA) has developed rapidly. SXA has shown a lot of new effects like the Fano-Cooper effect, Mahan-Nozieres infrared singularity at the absorption edge of metals due to the interaction of the electron Fermi gas with the hole, and core excitons in insulators and semiconductors. As Prof. Sugano (1) has shown core or soft x-ray excitons are excitons with their holes lying in a deep crystal level. In order to ascertain the existence of the excitons and to determine the energy range above the absorption edge where the excitonic structures are, it is necessary to associate the SXA measurements to the x-ray photoemission spectroscopy (XPS) (2).

From XPS it is possible to measure the energy distance from the top of the valence band to the core level E_c. By adding to E_c the energy gap of the insulator E_g, it is possible to assign the threshold of transitions to the continuum of the conduction band. When this method is used in the absorption spectrum of aluminum oxide (an ionic large gap insulator $E_g \approx 9.5$ eV) the peaks at the edge of the $L_{2,3}$ spectrum (3) can be assigned to core excitons. The first two excitons arise from the configuration $\tilde{t}_{1u}a_{1g}$(1). These excitons are observed also in the amorphous aluminum oxide. Their intensity ratio is largely affected by the crystalline order; on the contrary the binding energy (∿3 eV) and the line width (∿0.6 eV) are nearly independent of the crystalline order.

(1) S. Sugano, "Core Excitation and Electron Correlation in Crystals," this volume.

(2) S. Pantelides, F. C. Brown, Phys. Rev. Lett. 33, 298 (1974).

(3) A. Balzarotti, A. Bianconi, E. Burattini, M. Grandolfo, R. Habel, M. Piacentini, Phys. Status Solidi (b) 63, 77 (1974).

REACTIONS OF SINGLET EXCITED DYES IN SENSITIZED PHOTO-OXIDATIONS

I. Kraljic

Laboratoire de Physico-Chime des Rayonnements,

Université Paris-Sud

Orsay, France

Two major pathways in photo-oxidations (photodynamic action) sensitized by dyes (S) are the free radical mechanism (the triplet dye reacting with a substrate A) and the singlet oxygen ($^1O_2$) mechanism ($^1O_2$ reacting with A):

$$S_o + h\nu \rightarrow {}^1S^* \rightarrow {}^3S \begin{cases} \xrightarrow{+A} S^- + A^+ \xrightarrow{^3O_2} \text{Products } (AO_2, S_o) \\ \xrightarrow{+^3O_2} S_o + {}^1O_2 \xrightarrow{+A} AO_2 \, . \end{cases}$$

Some results of such reactions and the method for identification of primary reactions were presented. However, the scavenging (interception) of short-lived singlet excited dye ($^1S^*$) by some substrates is also possible. Thus, we find that some sensitizing dyes (eosine, phenosafranine, etc.) are bleached by the allylthiourea (ATU) via the reactions with $^1S^*$. This bleaching reaction can be followed at low ATU concentrations at which the fluorescence quenching cannot be measured accurately. The reaction $^1S^*$ + ATU begins at lower ATU concentrations than the reaction ^3S + ATU. The explanation of this phenomenon was given. The possibility of applying sensitized photo-oxidation to solar energy conversion and storage was also briefly discussed.

A MODEL SYSTEM FOR PHOTOBIOLOGY: THE AZOALDOLASE

J. L. Houben

Consiglio Nazionale delle Ricerche

Pisa, Italy

Most of the photobiological systems can be described as a dye covalently bound to a protein. To study a simple model system presenting these characteristics, a diazonium salt has been bound to aldolase, an allosteric enzyme, without loss of enzymatic activity. The azoaldolase thus obtained has been characterized partly through comparison with the spectroscopic properties of azoaminoacids and azopolypeptides. Shining visible or near UV light on azoaldolase causes photoisomerization of the azodyes and affects the enzymatic activity. To determine the mechanism of action of light on the enzymatic activity, two types of studies are developed: (a) role of substituents on the azodyes; (b) conformational changes in the protein induced by the dye excitation. This last point, or at least its possibility, is searched for in polypeptides to which azodyes are bound as side or lateral chains. Photoisomerization and thermal *cis* to *trans* isomerization are studied in simple azobenzene derivatives.

RELAXED EXCITED STATE OF F-CENTERS

G. Baldacchini

Laboratori Nazionali di Frascati

Frascati, Rome, Italy

After a general discussion on the nature of the R.E.S. some results on the magnetic circular dichroic effects were given. The importance of these small but significant effects was outlined in view of a complete understanding of the R.E.S.

LIST OF CONTRIBUTORS

G. Baldacchini, Laboratori Nazionali di Frascati, Rome, Italy
A. Bianconi, Università di Camerino, Camerino (MC), Italy
S. Claesson, Uppsala University, Uppsala, Sweden
B. Dellinger, Florida State University, Tallahassee, Florida, U.S.A.
B. Di Bartolo, Boston College, Chestnut Hill, Massachusetts, U.S.A.
G. J. Diebold, Boston College, Chestnut Hill, Massachusetts, U.S.A.
J. Gole, M.I.T., Cambridge, Massachusetts, U.S.A.
J. L. Houben, Consiglio Nazionale delle Ricerche, Pisa, Italy
M. Kasha, Florida State University, Tallahassee, Florida, U.S.A.
I. Kraljic, Université Paris-Sud, Orsay, France
E. Krausz, University of Oxford, Oxford, England
S. Leach, Université Paris-Sud, Orsay, France
D. Linsay, M.I.T., Cambridge, Massachusetts, U.S.A.
D. S. McClure, Princeton University, Princeton, New Jersey, U.S.A.
R.G.W. Norrish, Cambridge University, Cambridge, England
R. C. Powell, Oklahoma State University, Stillwater, Oklahoma, U.S.A.
D. A. Ramsay, National Research Council of Canada, Ottawa, Ont., Canada
R. J. Santoro, Princeton University, Princeton, New Jersey, U.S.A.
E. W. Schlag, Technische Universität, München, West Germany
G. W. Stewart, West Virginia University, Morgantown, W. Va., U.S.A.
S. Sugano, University of Tokyo, Roppongi, Minato-Ku, Tokyo, Japan
P. E. Thiess, University of Illinois, Urbana, Illinois, U.S.A.
B. A. Thrush, University of Cambridge, Cambridge, England
P. Van Leuven, University of Antwerp, Antwerp, Belgium
J. Wiesenfeld, Cornell University, Ithaca, New York, U.S.A.
R. T. Williams, Naval Research Laboratory, Washington, D.C., U.S.A.
A. J. Yencha, State University of New York, Albany, New York, U.S.A.

SUBJECT INDEX